Mastering System Identification in 100 Exercises

Mastering System Identification in 100 Exercises

Johan Schoukens

Rik Pintelon

Yves Rolain

IEEE PRESS

A JOHN WILEY & SONS, INC., PUBLICATION

Published by John Wiley & Sons, Inc., Hoboken, New Jersey.
Published simultaneously in Canada.

For general information on our other products and services please contact our Customer Care Department within the United States at (800) 762-2974, outside the United States at (317) 572-3993 or fax (317) 572-4002.

Wiley publishes in a variety of print and electronic formats and by print-on-demand. Some material included with standard print versions of this book may not be included in e-books or in print-on-demand. If this book refers to media such as a CD or DVD that is not included in the version you purchased, you may download this material at http://booksupport.wiley.com. For more information about Wiley products, visit www.wiley.com.

Library of Congress Cataloging-in-Publication Data:

Schoukens, J. (Johan)
  Mastering system identification in 100 exercises / Johan Schoukens, Rik Pintelon, Yves Rolain.
    p. cm.
  ISBN 978-0-470-93698-6 (pbk.)
  1. System identification—Problems, exercises, etc. 2. Linear systems—Mathematical models—Quality control.
I. Pintelon, R. (Rik) II. Rolain, Yves. III. Title. IV. Title: Mastering system identification in one hundred exercises.
  QA402.S3556 2012
  519.5—dc23   2011038051

10 9 8 7 6 5 4 3 2 1

To Annick, Sanne, Maarten, and Ine

Johan

To my grandchildren Lamine, Aminata, Alia-Rik, Malick, Khadidiatou, Babacar, Ibrahima, and those to be born

Rik

To Marian, Sofie, Iris, and Boempa

Yves

Contents

2   Generation and Analysis of Excitation Signals 29

4   Identification of Linear Dynamic Systems 91

5   Best Linear Approximation of Nonlinear Systems 137

7   Identification of Parametric Models in the Presence of Nonlinear Distortions   239

Preface

Since the ascent of man, mankind has wanted to understand, predict, and control the environment. As scientists and engineers we want to do the same, but in a more systematic and quantified approach. To that end, the scientific community makes use of mathematical models intensively because this allows for the simulation of the world on the computer. These mathematical models are often obtained from first principles, making use of detailed knowledge about the physical laws that describe systems. The major advantage of such an approach is that it provides detailed physical models that give much insight into the problems studied, however, at the cost of a long and difficult modeling process. In such models there also remain, often, unknown parameters to be tuned such that a good match between the model and reality is obtained. At the other end of the possible modeling strategies we find the data-driven approach, where all information is retrieved from experimental data. These models are called black box models, and it is usually less expensive and less time-consuming to get them. However, since all experimental data are disturbed by noise disturbances, it is important to use well-designed methods to remove the impact of the noise on the final result as much as possible. The mere reliance on intuition can lead to erroneous results, even when a lot of data are used. As an example we note that linear regression methods underestimate the slope in the presence of noise on the regressor (the input of the modeled system).

System identification theory was developed to address the need for good methods to estimate mathematical models from noisy data. Many excellent books are written, guiding the user to the best solution for his problem. However, for many engineers and scientists, the modeling task is only an intermediate step towards the solution to their problems, and they do not have enough time to learn the theory in full detail. With this book, the authors want to circumvent this problem. We guide the user to good solutions to built a mathematical model for a linear dynamic system by making a series of well-selected MATLAB® exercises that highlight many of the important steps in the identification process, point out possible pitfalls to the reader, and illustrate the powerful tools that are available.

The exercises are kept as simple as possible to lower the technical barrier as much a possible. On the other hand, we selected the problems such that we can still guarantee that the lessons learned from the examples are generally valid. For each set of problems we start with a brief introduction to the specific aspects that will be highlighted. Next, we define for each exercise a series of intermediate steps to be followed by the user, and finally the results are graphically presented together with a discussion that emphasizes what should be observed in the results. In some sections we add general conclusions that can be learned from the examples.

The book covers the whole identification process, from data to model. First we introduce the basic ideas and methods of system identification in Chapter 1. Here least squares, weighted least squares, and maximum likelihood estimation are introduced. The impact of noise disturbances on the regressor variables is studied, and the reader learns how to select the model complexity. In Chapter 2 we instruct the user to deal with random and periodic signals in the time- and the frequency domain. In Chapter 3 we show how the impulse and frequency response function (FRF) can be measured for linear systems, using random and periodic excitation signals. We also show how to get a nonparametric estimate of the noise variance as a function of the frequency. This allows for uncertainty bounds to be generated together with the measured FRF. In Chapter 4, the reader will be able to learn how to estimate a parametric model to estimate plant and noise dynamics. Because many real-life systems suffer from nonlinear distortions, we should also be able to characterize the level and the nature of these errors, so that the user can decide at the very beginning of the complex modeling process if a linear modeling approach will do or a more involved nonlinear model is needed. The latter is more difficult to built and will be more expensive. In Chapters 5 and 6 we provide simple and more advanced methods to address these questions. The presence, the actual level, and the nature of nonlinear distortions are measured simultaneously with the FRF of the best linear approximation of the system for the selected class of excitation signals. Eventually, we show in Chapter 7 how parametric models can be obtained under these conditions.

In many chapters, the authors took a frequency domain approach to the problem. This is partly due to the authors' background and experience, but also due to the fact that the state-of-the-art nonparametric preprocessing methods are done in the frequency domain. For the identification of parametric linear dynamic models (Chapters 4 and 7), we took a more balanced vision on the time and frequency domain methods, illustrating the similarities and differences of both approaches.

The exercises in this book can be used as illustrations in a bachelor degree course on Signal Processing (Chapter 2 and 3). Chapter 1 can also be used as an introduction to a basic course on the general concepts of system identification, while Chapter 4 is dedicated to the identification of linear dynamic systems. Chapters 6 and 7 can be used as a first introduction to understand the behavior of nonlinear systems and their impact on the linear system identification framework, with an emphasis on the practical user aspects.

The MATLAB® solutions of the exercises are available on the book support site, booksupport.wiley.com. With the help of these files the reader can verify her/his solutions and debug her/his own files. For some of the exercises, we used the commercially available MATLAB® toolboxes, on the one hand, the "System Identification Toolbox" developed by Lennart Ljung and distributed by the Mathworks, and on the other hand, the "Frequency Domain System Identification Toolbox" developed by the authors of this book together with Istvan Kollar and distributed by Gamax.

Department of Electrical Engineering
Vrije Universiteit Brussel
BELGIUM
Johan Schoukens
Rik Pintelon
Yves Rolain

Acknowledgments

This book is based on the experience in system identification that has been accumulated over the last 30 years in the identification group at the Department of Fundamental Electricity and Instrumentation (ELEC) of the Vrije Universiteit Brussel (VUB). We were able to draw results and insights from more than 35 Ph.D. theses related to system identification that were presented during that period at the department ELEC. Also, over the years, many visitors from all over the world spent short or long visits and contributed significantly to our work and to our knowledge, making us learn how to tackle the identification problem from different perspectives. These many different inputs were indispensable to the fruition of this book on system identification exercises.

The exercises in the book were solved by many students who followed the system identification courses at the VUB. Also the researchers in the department ELEC who started for a Ph.D. worked through earlier versions of this book. In the last 3 years, the exercises were used intensively at the annual doctoral school on "Advanced modelling and simulation techniques of (non)linear dynamic systems, starting from experimental data." The inputs of all these visitors and participants helped us a lot to improve the earlier versions of this book.

We are very indebted to the former heads of the department ELEC, Jean Renneboog and Alain Barel, to the R&D Department of the VUB, to the FWO-Vlaanderen (the Fund for Scientific Research in Flanders), and to the Flemish and the Federal Government (Methusala, GOA, and IAP research programs). Without their sustained support, this work would never have seen the light of day.

Last but not least, we want to thank Professors Mikaya Lumori, Keith Godfrey, and Gerd Vandersteen for their enthusiastic support to bring this work to fruition. We are also grateful to Professors Ljung and Kollar for their input and advice on the exercises that illustrate the use of the time and frequency domain system identification toolboxes.

Johan Schoukens
Rik Pintelon
Yves Rolain

Abbreviations

AIC	Akaike Information Criterion
AR	AutoRegressive
ARMAX	AutoRegressive Moving Average with eXternal input
ARX	AutoRegressive with eXternal input
BJ	Box-Jenkins (model structure)
BLA	Beste Linear Approximation
DIBS	Discrete Interval Binary Sequence
DITS	Discrete Interval Tinary Sequence
DFT	Discrete Fourier Transform
EIV	Errors-In-Variables
FIR	Finite Impulse Response
FFT	Fast Fourier Transform
FPE	Final Prediction Error
FRF	Frequency Response Function
FRM	Frequency Response Matrix
GUI	Graphical User Interface
GWH	Generalized Wiener–Hammerstein
iid	Independent Identically Distributed
IV	Instrumental Variables
LAV	Least Absolute Values
LPM	Local Polynomial Method
LS	Least Squares
MLE	Maximum Likelihood Estimate
MLBS	Maximum Length Binary Sequence
MIMO	Multiple Input Multiple Output
NL	NonLinear
NFIR	Nonlinear Finite Impulse Response
OE	Output Error (model structure)
PISPO	Periodic In Same Period Out

pdf	Probability Density Function
PE	Prediction Error
PISPO	Periodic In Same Period Out
PRBS	PseudoRandom Binary Sequence
rms	Root Mean Square Value
SISO	Single-Input Single-Output
WLS	Weighted Least Squares
ZOH	Zero-Order Hold

1

Identification

What you will learn: This chapter introduces the reader to the basic methods that are used in system identification. It shows what the user can expect from the identification framework to solve her/his modelling problems. For that purpose the following topics are addressed:

- An estimator is a random variable that can be characterized by its mean value and covariance matrix (see Exercises 1.a, 7).

- The stochastic characteristics of an estimator depend on the number of raw data, the experiment design, the choice of the cost function, ... (see Exercises 1.b, 8, 9).

- The estimates are asymptotically normally distributed when a growing amount of raw data is processed (see Exercise 2).

- Noise disturbances on the regressor (e.g., the input data) can create a systematic error on the estimate, and the choice of the regressor variable has an impact on the final result. Specific methods are needed to deal with that problem (see Exercises 3, 4, 12, 13).

- The choice of the cost function influences the properties of the estimator (see Exercises 5.a, 5.b, 8, 9).

- The "optimal" choice of the cost function depends on the disturbing noise probability density function (see Exercise 9).

- Least squares problems can be explicitly solved if the model is linear-in-the-parameters (see Exercise 6).

- The numerical properties of the algorithms are strongly affected by the choice of the model parameters (see Exercise 6).

- The model complexity (number of unknown model parameters) should be carefully selected. Systematic tools are available to help the user to make this choice (see Exercise 10, 11).

1.1 INTRODUCTION

The aim of system identification is to extract a mathematical model $M(\theta)$ from a set of measurements Z. Measurement data are disturbed by measurement errors and process noise, described as disturbing noise n_z on the data:

$$Z = Z_0 + n_z. \tag{1-1}$$

Since the selected model class M does not, in general, include the true system S_0, model errors appear:

$$S_0 \in M_0 \text{ and } M_0 = M + M_\varepsilon, \tag{1-2}$$

with M_ε the model errors. The goal of the identification process is to select M and to tune the model parameters θ such that the "distance" between the model and the data becomes as small as possible. This distance is measured by the cost function that is minimized. The selection of these three items (data, model, cost function) sets the whole picture; all the rest are technicalities that do not affect the quality of the estimates. Of course this is an oversimplification. The numerical methods used to minimize the cost function, numerical conditioning problems, model parameterizations, and so on are all examples of very important choices that should be properly addressed in order to get reliable parameter estimates. Failing to make a proper selection can even drive the whole identification process to useless results. A good understanding of each of these steps is necessary to find out where a specific identification run is failing: Is it due to numerical problems, convergence problems, identifiability problems, or a poor design of the experiment?

In this chapter we will study the following issues:

- What is the impact of noise on the estimates (stochastic and systematic errors)?
- What are the important characteristics of the estimates?
- How to select the cost function?
- How does the choice of the cost function affect the results?
- How to select the complexity of the model? What is the impact on the estimates?

1.2 ILLUSTRATION OF SOME IMPORTANT ASPECTS OF SYSTEM IDENTIFICATION

In this book almost all the estimators that will be studied and used are based on the minimization of a cost function. It is a measure for the quality of the model and it is calculated starting from the errors: the differences between the actual measurements and their modeled values. We will use mostly a least squares cost function that is the sum of the squared errors.

In this section we present a simple example to illustrate some important aspects of system identification. Specifically, the impact of the noise on the final estimates is illustrated. It will be shown that zero mean measurement noise can result in systematic errors on the estimates (the mean of the parameter errors is not equal to zero!). Also the uncertainty is studied. Depending on the choice of the cost function, a larger or smaller noise sensitivity will be observed. All these aspects are studied using a very simple example: the measurement of the value of a resistance starting from a series of voltage and current measurements.

1.2.1 Least squares estimation: A basic approach to system identification

Exercise 1.a (Least squares estimation of the value of a resistor) Goal: Estimate the resistance value R_0 starting from a series of repeated current and voltage measurements:

$$u_0(t) = R_0 i_0(t), \quad t = 1, 2, \ldots, N \qquad (1\text{-}3)$$

with u_0, i_0 the exact values of the voltage and the current.

Generate an experiment with $N = 10, 100, 1000$, and $10{,}000$ measurements. The current i_0 is uniformly distributed in $[-i_{max}, i_{max}]$ with $i_{max} = 0.01$ A (use the MATLAB® routine rand(N,1)), $R_0 = 1000$. The current is measured without errors; the voltage is disturbed by independent, zero mean, normally distributed noise n_u with $N(0, \sigma_u^2 = 1)$.

$$\begin{aligned} i(t) &= i_0(t) \\ u(t) &= u_0(t) + n_u(t) \end{aligned}, \quad t = 1, 2, \ldots, N \qquad (1\text{-}4)$$

To measure the distance between the model and the data, we select in this exercise a least squares cost function: $V(R) = \dfrac{1}{N}\sum_{t=1}^{N}(u(t) - Ri(t))^2$. Notice that many other possible choices can be made.

The least squares estimate $\hat{R}$ is defined as the minimizer of the cost function $V(R)$:

$$\hat{R} = \arg\min_R V(R) \qquad (1\text{-}5)$$

■ Show that the minimizer of (1-5) is given by

$$\hat{R} = \sum_{t=1}^{N} u(t)i(t) \Big/ \sum_{t=1}^{N} i(t)^2. \qquad (1\text{-}6)$$

■ Generate 100 data sets with a length $N = 10, 100, 1000$, and $10{,}000$, and calculate the estimated value $\hat{R}$ for each N.

■ Plot the 100 estimates, together with the exact value for each N, and compare the results.

Observations (see Figure 1-1) From the figure it is seen that the estimates are scattered around the exact value. The scattering decreases for an increasing number N. It can be shown that under very general conditions, the standard deviation of the estimates decreases as $1/\sqrt{N}$. This is further elaborated in the next exercise.

Exercise 1.b (Analysis of the standard deviation) In this exercise, it is verified how the standard deviation varies as a function of N. Consider the resistance

$$u_0(t) = R_0 i_0(t), \; t = 1, 2, \ldots, N. \qquad (1\text{-}7)$$

with a constant current $i_0 = 0.01$ A, and $R_0 = 1000\,\Omega$. Generate 1000 experiments with $N = 10, 100, 1000$, and $10{,}000$ measurements. The current is measured without errors, the voltage is disturbed by independent, zero mean Gaussian distributed noise $n_u \sim N(0, \sigma_u^2 = 1)$ (use the MATLAB® routine randn(N,1)):

$$\begin{aligned} i(t) &= i_0 \\ u(t) &= u_0(t) + n_u(t) \end{aligned}, \quad t = 1, 2, \ldots, N \qquad (1\text{-}8)$$

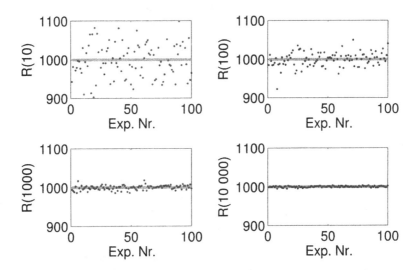

Figure 1-1 Estimated resistance values $\hat{R}(N)$ for N = 10, 100, 1000, and 10,000 for 100
repeated experiments. Gray line: exact value; dots: estimated value.

■ Calculate for the four values of N the standard deviation of $\hat{R}$ using the MATLAB®
routine std(x). Make a loglogplot of the standard deviation versus N.

■ Compare it with the theoretical value of the standard deviation that is given in this
simplified case (constant current) by

$$\sigma_R = \frac{1}{\sqrt{N}} \frac{\sigma_u}{i_0}. \tag{1-9}$$

Observations (see Figure 1-2) From the figure it is seen that the standard deviation decreases as $1/\sqrt{N}$. Collecting more data makes it possible to reduce the uncertainty. To get a
reduction with a factor 10 in uncertainty, an increase of the measurement time with a factor

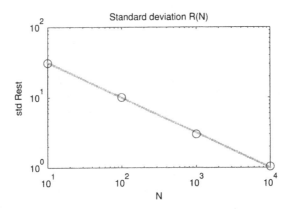

Figure 1-2 Experimental (black circles) and theoretical (gray dots) standard deviation on $\hat{R}$ as
a function of N. The error drops with $\sqrt{10}$ if the number of data N grows with a
factor 10.

100 is needed. This shows that the measurement time grows quadratically with the required noise reduction, and hence it still pays off to spend enough time on a careful setup of the experiment in order to reduce the level of the disturbing noise σ_u on the raw data.

Remark: For the general situation with a varying current, the expression for the standard deviation σ_R for a given current sequence $i_0(t)$ is

$$\sigma_R = \frac{\sigma_u}{\sqrt{\sum_{t=1}^{N} i_0^2(t)}} \tag{1-10}$$

Exercise 2 (Study of the asymptotic distribution of an estimate) The goal of this exercise is to show that the distribution of an estimate is asymptotic for $N \to \infty$ normally distributed, and this is independent of the distribution of the disturbing noise (within some regularity conditions, like finite variance, and a restricted "correlation" length of the noise).

Consider the previous exercise for $N = 1, 2, 4, 8$, and 10^5 repetitions. Use a constant current $i_0 = 0.01$ A, measured without errors. For the voltage we consider two situations. In the first experiment, the voltage is disturbed by independent, zero mean Gaussian distributed noise $N(0, \sigma_u^2 = 0.2^2)$. In the second experiment the voltage noise is uniformly distributed in $[-\sqrt{3}\sigma_u, \sqrt{3}\sigma_u]$.

- Verify that the standard deviation of the uniformly distributed noise source also equals σ_u.

- Calculate the least squares solution [see equation (1-6)] for $N = 1, 2, 4, 8$ and repeat this 10^5 times for both noise distributions. Plot the estimated pdf for the eight different situations. The pdf can be estimated by making a proper normalization of the histogram of the estimates (use the MATLAB® routine `hist(x)`). The fraction of data in each bin should be divided by the width of the bin.

- Calculate the mean value and the standard deviation over all realizations (repetitions) for each situation, and compare the results.

□

Observations (see Figure 1-3) From the figure it is seen that the distribution of the estimates depends on the distribution of the noise. For example, for $N = 1$, the pdf for the Gaussian disturbed noise case is significantly different from that corresponding to the uniformly disturbed experiment. These differences disappear for a growing number of data per experiment (N increases), and for $N = 8$ it is impossible to identify a different shape visually. The uniform distribution converges to the Gaussian distribution for growing values of N. This is a general valid result.

In this case, the mean value and the variance is the same for both disturbing noise distributions, and this for each value of N. This is again a general result for models that are linear in the measurements (e.g., $y_0 = au_0$ is linear in u_0, while $y_0 = au_0^3$ is nonlinear in the measurements). The covariance matrix of the estimates depends only on the second-order properties of the disturbing noise. This conclusion cannot be generalized to models that are nonlinear in the measurements. In the latter case, the estimates will still be Gaussian distributed, but the mean value and variance will also depend on the distribution of the disturbing noise.

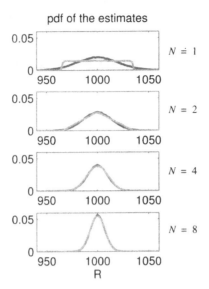

pdf of the estimates

Figure 1-3 Evolution of the pdf of $\hat{R}$ as a function of N, for $N = 1, 2, 4, 8$. Black: Gaussian disturbing noise; Gray: uniform disturbing noise.

1.2.2 Systematic errors in least squares estimation

In the previous section it was shown that disturbing noise on the voltage resulted in noisy estimates of the resistor value, the estimated value of the resistor varies from one experiment to the other. We characterized this behavior by studying the standard deviation of the estimator. The mean value of these disturbances was zero: the estimator converged to the exact value for a growing number of experiments. The goal of this exercise is to show that this behavior of an estimator cannot be taken for granted. Compared with the Exercises 1.a–2, we add in the next two exercises also disturbing noise on the current. The impact of the current noise will be completely different from that of the voltage noise, besides the variations from one experiment to the other, a systematic error will arise. This is called a bias.

Exercise 3 (Impact of noise on the regressor (input) measurements)   Consider Exercise 2 for $N = 100$, and 10^5 repetitions. The current i_0 is uniformly distributed between $[-10, 10]$ mA. It is measured this time with white disturbing noise added to it: $i(t) = i_0 + n_i(t)$, with a normal distribution $N(0, \sigma_i^2)$. The voltage measurement is also disturbed with normally distributed noise: $N(0, \sigma_u^2 = 1)$.

■ Repeat the simulations of the previous exercise once without and once with noise on the current. Vary the current noise standard deviation in 3 successive simulations: $\sigma_i = 0, 0.5, 1$ mA.

■ Calculate the least squares solution [see eq. (1-6)] for $N = 100$ and repeat this 10^5 times for all situations and plot the pdf for each of them.

■ Calculate the mean value and the standard deviation over all realizations (repetitions) for each situation, and compare the results.

□

Observations (see Figure 1-4) From the figure it is seen that the distribution of the estimates depends strongly on the presence of the noise on the current measurement. Not only is the standard deviation affected, but also a visible bias grows with the variance of the current noise. This result is closely linked to the fact that the current is used as regressor or independent variable that makes the voltage a dependent variable: We used a model where the current is the input, and the voltage is the output. Whenever the measurement of the input variable is disturbed by noise, bias problems will appear unless special designed methods are used. These will be studied in Section 1.6.

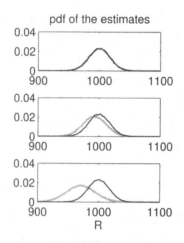

Figure 1-4   Evolution of the pdf of $\hat{R}$ as a function of the noise level at the current. Black: Only noise on the voltage $\sigma_u = 1$ V. Gray: Noise on the voltage $\sigma_u = 1$ V and the current. $\sigma_i = 0, 0.5, 1$ mA (top, middle, bottom).

Exercise 4 (Importance of the choice of the independent variable or input)   In Exercise 3 it became clear that noise on the input or independent variable creates a bias. The importance of this choice is explicitly illustrated by repeating Exercise 2, where the disturbing noise is only added to the voltage: $i(t) = i_0(t)$, $u(t) = u_0 + n_u(t)$ with $n_u(t) \sim N(0, \sigma_u^2 = 1)$. In this exercise the same data are processed two times:

■ Process the data using the current as independent variable, corresponding to the function $u(t) = Ri(t)$ and an estimate of R:

$$\hat{R} = \sum_{t=1}^{N} u(t)i(t) \Big/ \sum_{t=1}^{N} i(t)^2. \tag{1-11}$$

■ Process the data using the voltage as independent variable, corresponding to $i(t) = Gu(t)$, with G the conductance:

$$\hat{G} = \sum_{t=1}^{N} u(t)i(k) \Big/ \sum_{t=1}^{N} u(t)^2 \text{ and } \hat{R} = 1/\hat{G}. \tag{1-12}$$

■ Repeat each experiment 10^5 times for $N = 100$, then calculate and plot the pdf of the estimated resistance for both cases.

□

Discussion (see Figure 1-5) Whenever the measurement of the variable that appears squared in the denominator of (1-11) or (1-12) is disturbed by noise, a bias will become visible. This shows that the signal with the highest SNR should be used as independent variable or input in order to reduce the systematic errors. The bias will be proportional to the inverse SNR (noise power/signal power).

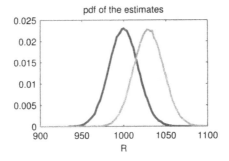

Figure 1-5 Study of the impact of the selection of the independent variable for the estimation of the resistance. Only the voltage is disturbed with noise. The pdf of the estimated resistance is shown for the independent variable being the current (black) or the voltage (gray).

1.2.3 Weighted least squares: optimal combination of measurements of different quality

The goal of this section is to combine measurements with different quality. A first possibility would be to throw away the poorest data, but even these poor data contain information. It is better to make an optimal combination of all measurements taking into account their individual quality. This will result in better estimates with a lower standard deviation. The price to be paid for this improvement is the need for additional knowledge about the behavior of the disturbing noise. While the least squares (LS) estimator does not require information at all about the disturbing noise distribution, we have to know the standard deviation (or in general, the covariance matrix) of the disturbing noise in order to be able to use the improved weighted least squares (WLS) estimator, illustrated in this exercise.

Exercise 5.a (combining measurements with a varying SNR: Weighted least squares estimation)   Estimate the resistance value starting from a series of repeated current and voltage measurements:

$$u_0(t) = R_0 i_0(t) , \quad t = 1, 2, ..., N \tag{1-13}$$

with u_0, i_0 the exact values of the voltage and the current. Two different voltmeters are used, resulting in two data sets, the first one with a low noise level, the second one with a high noise level.

- Generate an experiment with N measurements, i_0 uniformly distributed in $[-0.01, 0.01]$ A, $R_0 = 1000 \, \Omega$. The current is measured without errors, the voltage measured by the 2 voltmeters is disturbed by independent, zero mean, normally distributed noise n_u with $N(0, \sigma_u^2 = 1)$ for the first, good voltmeter and $N(0, \sigma_u^2 = 16)$ for the second, bad one.

$$
\begin{aligned}
i(t) &= i_0(t) \\
u(t) &= u_0(t) + n_u(t)
\end{aligned}, \quad t = 1, 2, \dots, N \tag{1-14}
$$

■ Calculate the weighted least squares solution as the minimizer of

$$
V_{\text{WLS}} = \frac{1}{N} \sum_{t=1}^{2N} \frac{(u(t) - Ri(t))^2}{w(t)} \tag{1-15}
$$

using (1-16), given below:

$$
\hat{R} = \frac{\sum_{t=1}^{2N} \dfrac{u(t)i(t)}{w(t)}}{\sum_{t=1}^{2N} \dfrac{i(t)^2}{w(t)}}, \tag{1-16}
$$

with $w(t)$ the weighting of the tth measurement: $w(t) = \sigma_{u1}^2$ for the measurements of the first voltmeter and $w(t) = \sigma_{u2}^2$ for the measurements of the second one.

■ Repeat this exercise 10^5 times for $N = 100$. Estimate the resistance also with the least squares method of Exercise 1.a. Make an histogram of both results.

Observations (see Figure 1-6) From the figure it is seen that the estimates are scattered

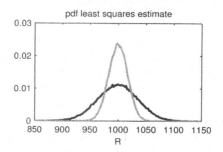

Figure 1-6 Estimated resistance values for $N = 100$, combining measurements of a good and a bad voltmeter. Black: pdf of the least squares; gray: pdf of the weighted least squares estimates.

around the exact value. However, the standard deviation of the weighted least squares is smaller than that of the least squares estimate. It can be shown that the inverse of the covariance matrix of the measurements is the optimal weighting for least squares methods.

Exercise 5.b (Weighted least squares estimation: A study of the variance) In this exercise we verify by simulations the theoretical expressions that can be used to calculate the variance of a least squares estimator and a weighted least squares estimator. It is assumed that there is only noise on the voltage. The exact, measured current is used as regressor (input). The theoretical variance of the linear least squares estimator (no weighting applied) for the resistance estimate is given by

$$
\hat{\sigma}_{\text{LS}}^2 = \frac{\sum_{t=1}^{2N} \sigma_u^2(t) i^2(t)}{\left(\sum_{t=1}^{2N} i(t)^2 \right)^2}, \tag{1-17}
$$

and the variance of the weighted least squares estimator using the variance on the output (the voltage) as weighting is

$$\hat{\sigma}^2_{\mathrm{WLS}} = \frac{1}{\displaystyle\sum_{t=1}^{2N} \frac{i(t)^2}{\sigma_u^2(t)}} . \tag{1-18}$$

■ Consider Exercise 5.a, calculate the theoretical value for the standard deviation, and compare this with the results obtained from the simulations.

Observations A typical result of this exercise is:

theoretical standard deviation LS, 35.6; experimental standard deviation, 35.8

theoretical standard deviation WLS, 16.8; experimental standard deviation, 16.8

Remark: The expressions (1-17) and (1-18) for the theoretical values of the variance are valid for a given input sequence. If the averaged behavior over all (random) inputs is needed, an additional expectation with respect to the input current should be calculated.

1.2.4 Models that are linear-in-the-parameters

The least squares estimates of the resistor that have been studied thus far were based on the minimization of the weighted cost function

$$V(R) = \frac{1}{N}\sum_{t=1}^{N} \frac{(u(t) - Ri(t))^2}{w(t)}, \tag{1-19}$$

with u, i the measured voltage (output) and current (input), respectively.

In general, the difference between a measured output $y(t)$ and a modeled output $\hat{y}(t) = g(t, u_0, \theta)$ is minimized for a given input signal $u_0(t)$. All model parameters are grouped in $\theta \in \mathrm{R}^{n_\theta}$. This can be formulated under a matrix notation for models that are linear-in-the-parameters. Define the signal vectors $\hat{y}, u_0, g \in \mathrm{R}^N$, for example:

$$y^T = \{y(1), ..., y(N)\} \tag{1-20}$$

and a positive weighting matrix $W \in \mathrm{R}^{N \times N}$. Then the weighted least squares cost function becomes

$$V_{\mathrm{WLS}} = (y - g(u_0, \theta))W^{-1}(y - g(u_0, \theta))^T. \tag{1-21}$$

For a diagonal matrix $W_{tt} = w(t)$, $W_{ij} = 0$ elsewhere, and equation (1-21) reduces to

$$V_{\mathrm{WLS}} = \frac{1}{N}\sum_{t=1}^{N} \frac{(y(t) - g(t, u_0, \theta))^2}{w(t)} . \tag{1-22}$$

The estimate $\hat{\theta}$ is found as the minimizer of this cost function:

$$\hat{\theta} = \arg\min_{\theta} V_{\mathrm{WLS}}(\theta) . \tag{1-23}$$

In general it is impossible to solve this minimization problem analytically. However, if the model is linear-in-the-parameters, then it is possible to formulate the solution explicitly, and it is also possible to calculate it in a stable numerical way with one instruction in MATLAB®. A model is called linear-in-the-parameters if the output is a linear combination of the model parameters:

$$y = K(u_0)\theta \text{ with } K \in \mathbb{R}^{Nxn_\theta}. \tag{1-24}$$

Note that K can be a nonlinear function of the input. The explicit solution of the linear (weighted) least squares problem becomes

$$\hat{\theta}_{\text{WLS}} = (K^T W K)^{-1} K^T W y \text{ and } \hat{\theta}_{\text{LS}} = (K^T K)^{-1} K^T y. \tag{1-25}$$

Solutions that are numerically stable for expression (1-25) exclude the explicit calculation of the product $K^T W^{-1} K$ or $K^T K$, thus improving the numerical conditioning. This can be done with the MATLAB® solution given by

$$\hat{\theta}_{\text{WLS}} = (W^{1/2}K)\backslash(W^{1/2}y) \text{ with } W = W^{1/2}W^{1/2}$$

$$\hat{\theta}_{\text{LS}} = K\backslash y. \tag{1-26}$$

Exercise 6 (Least squares estimation of models that are linear in the parameters) Consider the model y0 = `tan(u0*0.9*pi/2)`, evaluated for the inputs u0 = `linspace(0,1,N)`. Use the model

$$\hat{y}(t) = \sum_{i=0}^{n} \theta_i u_0^i(t) \tag{1-27}$$

to describe these data. Note that this is a nonlinear model that is linear-in-the-parameters θ_i.

- Generate a data set $y = y_0$. Put $N = 100$, and vary $n = 1$ to 20.
- Calculate the least squares solution ($W = I^{NxN}$) for the different values of n, using the stable MATLAB® solution (1-26) and the direct implementation (1-25).
- Compare the solutions, and calculate the condition number of K and $K^T K$. This can be done with MATLAB® instruction `cond( )`
- Compare the modeled output with the exact output and calculate the rms value of the error.

□

Observations (see Figure 1-7) From this figure, it can be seen that the condition number of the numerical unstable method (1-25) grows two times faster on a logarithmic scale than that of the stable method (1-26). The number of digits required to make the equations is given by the exponent of the condition number. From order 10 or larger, more than 15 digits are needed which is more than the calculation precision of MATLAB®. As a result, the obtained models are no longer reliable, even if there was no disturbing noise in the experiment. This shows that during system identification procedures, it is always necessary to verify the numerical conditions of the calculations. The condition number of the stable numerical im-

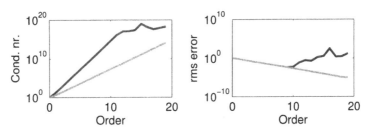

Figure 1-7 Identification of a polynomial model that is linear-in-the-parameters using a
method that is numerical stable $\hat{\theta}_{WLS} = K \backslash y$ (gray lines) or numerical unstable
$\hat{\theta}_{WLS} = (K^T K)^{-1} K y$ (black lines). Left: Condition number as a function of the
model order. Right: The rms error as a function of the model order.

plementation grows slower, making it possible to solve higher order polynomial approximations.

Remark: If very high order polynomial approximations are needed, other more robust
polynomial representations can be applied using orthogonal polynomials. The nature of these
polynomials will depend upon the applied input signal.

1.2.5 Interpretation of the covariance matrix & Impact on experiment design

In this section, a one- and a two-parameter model will be considered. It is shown that: (1) The
variance of a set of parameters is not enough to make conclusions on the model uncertainty;
the full covariance matrix is needed. (2) The covariance matrix (and the correlation between
the parameters) is strongly influenced by the design of the experiment.

Exercise 7 (Characterizing a 2-dimensional parameter estimate) Generate a set
of measurements:

$$y(t) = au_0(t) + n_t. \tag{1-28}$$

In the first experiment, $u_0(t)$ is generated by `linspace(-3,3,N)`, distributing N points
equally between -3 and 3. In the second experiment, $u_0(t)$ is generated by
`linspace(2,5,N)`.

- Choose $a = 0.1$, $N = 1000$, and $n_t \sim N(0, \sigma_n^2)$ with $\sigma_n^2 = 1$.
- Use as a model $y = au_0 + b$, and estimate the parameters (a, b) using the method
  of Exercise 6.
- Repeat this experiment 10^5 times.
- Estimate the LS-parameters for both experiments, calculate the covariance matrix,
  and plot $\hat{a}(i)$ as a function of $\hat{b}(i)$.
- Plot also the estimated lines for the first 50 experiments.

Observations (Figure 1-8) In Figure 1-8 top, the parameters are plotted against each
other. For the second experiment ($u \sim$ uniform in [2,5]), the parameters are strongly correlated, as can be seen from the linear relation between the estimated values $\hat{a}(i)$ and $\hat{b}(i)$.
This is not so for the first experiment ($u = [-3, 3]$), the black cloud has its main axis paral-
lel to the horizontal and vertical axis which is the typical behavior of an uncorrelated variable. This can also be seen in the covariance matrices:

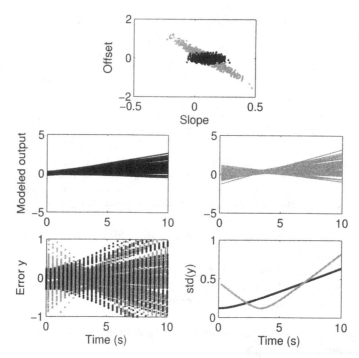

Figure 1-8 Black: Experiment in time interval $[-3, 3]$. Gray: Experiment in time interval $[2, 5]$. Top: Scatter plot (slope, offset). Middle: Modeled output. Bottom: Error on modeled output (left) and its standard deviation (right).

$$C_{\text{exp1}} = \begin{bmatrix} 3.2\times10^{-3} & 0.85\times10^{-4} \\ 0.85\times10^{-4} & 10.5\times10^{-3} \end{bmatrix}, \text{ and } C_{\text{exp2}} = \begin{bmatrix} 1.31\times10^{-2} & -4.6\times10^{-2} \\ -4.6\times10^{-2} & 16.9\times10^{-2} \end{bmatrix}, \tag{1-29}$$

or even better from the correlation matrices

$$R_{\text{exp1}} = \begin{bmatrix} 1 & 0.02 \\ 0.02 & 1 \end{bmatrix}, \text{ and } R_{\text{exp2}} = \begin{bmatrix} 1 & -0.97 \\ -0.97 & 1 \end{bmatrix}. \tag{1-30}$$

The correlation in the first matrix is almost zero, while for the second experiment it is almost one, indicating that a strong linear relation between the offset and slope estimate exists. This means that both variables vary considerably (a large standard deviation), but they vary together (a large correlation) so that the effect on the modelled output is small in the input range that the experiment was made (see Figure 1-8, middle and bottom). In that range, the variations of $\hat{a}$ are mostly canceled by those of $\hat{b}$. Outside this range, the standard deviation of the modeled output will be larger compared to that obtained with the first experiment because there the offset-slope compensation is no longer valid. This shows that the covariances play an important role in the model uncertainty.

1.2.6 What have your learned in Section 1.2? Further reading

In this section we studied the properties of linear (weighted) least squares estimators. Because these models are linear-in-the-parameters, it is possible to calculate the weighted least squares solution explicitly.

It is important to use numerical stable algorithms to calculate this explicit solution because otherwise the numerical noise can jeopardize the theoretical properties. The bias and covariance matrix of the estimates can be explicitly calculated provided that the covariance matrix of the disturbing noise is known.

The parameter uncertainty is directly influenced by the choice of the input signals. Experiment design methods provide systematic tools to get the best results within user specified constraints, for example for a given input power the determinant of the covariance matrix of the parameter estimates should be minimized.

The selection of the weighting matrix in the weighted least squares method influences the covariance matrix of the parameters. The smallest covariance matrix is obtained by choosing the weighting matrix as the inverse of the disturbing noise covariance matrix.

The parameter estimates are asymptotically Gaussian distributed under very weak conditions of the disturbing noise.

The books of Sorenson (1980) and van den Bos (2007) provide a general introduction to system identification, spending a lot of attention to weighted least squares estimation. Also the first chapter of the book of Pintelon and Schoukens (2001) introduces the reader to the general ideas of identification theory. More information on the numerical issues can be found in Golub and Van Loan (1996). Experiment design is covered in the books of Federov (1972), Goodwin and Payne (1977), and Zarrop (1979). Recently a new interest in this topic emerged, for example in the work of Hjalmarsson (2009), Gevers *et al.* (2009), and Bombois *et al.* (2006). The new design methods aim for an integrated design that optimizes the model for its final purpose, like, for example, the design of a controller.

1.3 MAXIMUM LIKELIHOOD ESTIMATION FOR GAUSSIAN AND LAPLACE DISTRIBUTED NOISE

In Sections 1.2 and 1.3, Gaussian distributed noise was added as disturbances to the measurements. It is shown in theory that least squares estimators, where the cost function is a quadratic function of the errors, perform optimal under these conditions. The smallest uncertainty on the estimators is found if a proper weighting is selected. This picture changes completely if the disturbing noise does not have a Gaussian distribution. In the identification theory it is shown that for each noise distribution, there corresponds an optimal choice of the cost function. A systematic approach to find these estimators is through the maximum likelihood theory, which is not within scope of this book, but some of its results will be illustrated on the resistance example. The disturbances will be selected once to have a normal distribution, and once to have a Laplace distribution. The optimal cost functions corresponding to these distributions are a least squares and a least absolute value cost function.

Exercise 8 (Dependence of the optimal cost function on the distribution of the disturbing noise)   Consider a set of repeated measurements:

$$u_0(t) = R_0 i_0(t), \quad t = 1, 2, ..., N \tag{1-31}$$

with u_0, i_0 the exact values of the voltage and the current. Two different voltmeters are used, resulting in two data sets, the first one is disturbed by Gaussian (normal) distributed noise, the second one is disturbed with Laplace noise.

Generate an experiment with $N = 100$ measurements, with i_0 uniformly distributed in $[0, i_{max} = 0.01 \text{ A}]$, and $R_0 = 1000 \ \Omega$. The current is measured without errors. The voltage measured with the first voltmeter is disturbed by independent, zero mean, normally distributed noise $n_u \sim N(0, \sigma_u^2 = 1)$, the second voltmeter is disturbed by Laplace distributed noise with zero mean, and $\sigma_u^2 = 1$.

$$\begin{aligned} i(t) &= i_0(t) \\ u(t) &= u_0(t) + n_u(t) \end{aligned}, \quad t = 1, 2, ..., N.$$

(1-32)

For the Gaussian noise the maximum likelihood solution reduces to a least squares (LS) estimate as in (1-6); for the Laplace distribution the maximum likelihood estimator is found as the minimizer of

$$V_{LAV}(R) = \frac{1}{N}\sum_{t=1}^{N} |u(t) - Ri(t)| \text{ and } \hat{R}_{LAV} = \arg \min_R V_{LAV}(R),$$

(1-33)

called the least absolute values (LAV) estimate.

- Repeat this exercise 10,000 times for $N = 100$.
- Apply both estimators also to the other data set.
- Calculate the mean value, the standard deviation, and plot for each case the histogram.

Help 1: Laplace distributed noise with zero mean and standard deviation *stdu* can be generated from uniformly distributed noise $[0, 1]$ using the following MATLAB® implementation:

```
x = rand(NData,1);       % generate uniform distributed noise
nLap = zeros(size(x));   % vector used to store Laplace noise
nLap(x<=0.5) = log(2*x(x<=0.5))/sqrt(2)*stdU;
nLap(x>0.5)  = - log(2*(1-x(x>0.5)))/sqrt(2)*stdU;
```

Help 2: to minimize $V_{LAV}(R)$, a simple scan can be made over R belonging to [800:0.1:1200]

Observations (see Figure 1-9) From Figure 1-9, it is seen that the estimates are scattered around the exact value. For the Gaussian case, the LS squares estimate is less scattered

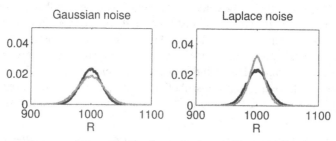

Figure 1-9 PDF of the Gaussian $\hat{R}_{LS}$ and Laplace $\hat{R}_{LAV}$ Maximum Likelihood estimators, applied to a Gaussian (left) and Laplace (right) noise disturbance. Black line: $\hat{R}_{LS}$; gray line: $\hat{R}_{LAV}$.

than the LAV estimate. For the Laplace case the situation is reversed. The estimated mean values $\hat{\mu}$ and standard deviations $\hat{\sigma}$ are given in Table 1-1. This shows that the maximum

TABLE 1-1 Mean and standard deviation of the Gaussian and Laplace maximum likelihood estimators, applied to a Gaussian and Laplace noise disturbance

	$\hat{\mu}_{LS}$	$\hat{\sigma}_{LS}$	$\hat{\mu}_{LAV}$	$\hat{\sigma}_{LAV}$
Gaussian noise	1000.040	17.5	999.94	22.0
Laplace noise	1000.002	17.3	999.97	13.7

likelihood estimator is optimal for the distribution that it is designed for. If the noise distribution is not known a priori, but the user can guarantee that the variance of the noise is finite, then it can be shown that the least squares estimate is optimal in the sense that it minimizes the worst possible situation among all noise distributions with a finite variance.

Further reading: The books of Sorenson (1980) and van den Bos (2007) give an introduction to maximum likelihood estimation, including illustrations on non-Gaussian distributions. The properties of the maximum likelihood estimator (consistency, efficiency, asymptotic normal distribution) are studied in detail.

1.4 IDENTIFICATION FOR SKEW DISTRIBUTIONS WITH OUTLIERS

In Section 1.3, it was shown that the optimal choice of the cost function depends on the distribution of the disturbing noise. The maximum likelihood theory offers a theoretical framework for the generation of the optimal cost function. In practice a simple rule of thumb can help to select a good cost function. Verify if the disturbing noise has large outliers: large errors appear to be more likely than expected from a Gaussian noise distribution.

In Exercise 9, the LS and the LAV estimates are applied to a χ^2 distribution with 1 degree of freedom: this is nothing other than a squared Gaussian distributed variable. Compared to the corresponding Gaussian distribution, the extremely large values appear too frequently (due to the squared value). Neither of both estimates (LS, LAV) is the MLE for this situation. But from the rule of thumb we expect that the LAV will perform better than the LS estimator. It will turn out that a necessary condition to get good results is to apply a proper calibration procedure for each method, otherwise a bias will become unavoidable.

Exercise 9 (Identification in the presence of outliers) Consider a set of repeated measurements:

$$u_0(t) = R_0 i_0(t), \quad t = 1, 2, ..., N \tag{1-34}$$

with u_0, i_0 the exact values of the voltage and the current. The voltage measurement is disturbed by noise, generated from a χ^2-distribution with 1 degree of freedom (= squared Gaussian noise).
Generate an experiment with N measurements, i_0 uniformly distributed in $[0, i_{max} = 0.01$ A] (use the MATLAB® routine `rand`), $R_0 = 1000\ \Omega$. The current is measured without errors. The measured voltage $u(t)$ is disturbed by χ^2 distribution distributed noise n_u with

$n_u = n^2$, with n generated as $N(0, \sigma_u^2 = 1)$.

Note that the mean value of $(\mathbb{E}\{n_u\})$ is 1, and median(n_u) is 0.455.

$$\begin{aligned} i(t) &= i_0(t) \\ u(t) &= u_0(t) + n_u(t) \end{aligned}, \quad t = 1, 2, ..., N \tag{1-35}$$

■ In order to reduce the systematic errors, calibrate the data first. To do so, the mean value or the median of the noise should be extracted from the measurements. First, perform a measurement with zero current, so that $u(t) = n_u(t)$. The calibration is done using this record.

■ Repeat the exercise 10,000 times and estimate, each time, the LS and the LAV estimate for both data sets.

■ Estimate the pdf of the estimates, and calculate their mean value and standard deviation.

Observations (see Figure 1-10) From the figure it is seen that the estimates are no

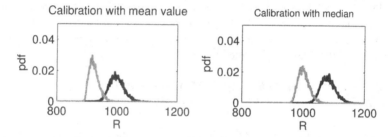

Figure 1-10 PDF of the Gaussian $\hat{R}_{LS}$ and Laplace $\hat{R}_{LAV}$ applied to χ^2 disturbed data. Left: Calibration with the mean value. Right: Calibration with the median value. Black line: $\hat{R}_{LS}$. Gray line: $\hat{R}_{LAV}$.

longer scattered around the exact value $R = 1000 \, \Omega$. Only the combinations (LS estimate, mean value calibration) and the (LAV estimate, median value calibration) work well. The other combinations show a significant bias.

The mean and standard deviations are given in Table 1-2. Observe that the standard deviation of the LAV estimate is smaller than that of the LS estimate. LAV estimates are less sensitive to outliers! Note that the mean value of the LAV estimator, combined with the median calibration has still a small systematic error of 1.85, which is larger than the uncertainty on the mean value: 18.62/sqrt(10,000) = 0.18. If instead of using the mean, the median value is selected to average the 10,000 estimates, the bias disappears completely.

TABLE 1-2 Mean and standard deviation of the Gaussian and Laplace maximum likelihood estimators, applied to a Gaussian and Laplace noise disturbance

	$\hat{\mu}_{LS}$	$\hat{\sigma}_{LS}$	$\hat{\mu}_{LAV}$	$\hat{\sigma}_{LAV}$
Calibr.: mean value	999.84	24.30	924.29	16.26
Calibr.: median	1081.86	24.43	1001.85	18.62

·Conclusion: The LS estimate should be combined with a calibration based on the mean, and the mean should be used to average the results. It is sensitive to outliers.

The LAV estimate should be combined with a calibration based on the median, the median should be used to average the results, and it is less sensitive to the presence of outliers.

1.5 SELECTION OF THE MODEL COMPLEXITY

1.5.1 Influence of the number of parameters on the uncertainty of the estimates

In this exercise it will be shown that, once the model includes all important contributions, the uncertainty grows if the number of model parameters is still increased.

Exercise 10 (Influence of the number of parameters on the model uncertainty)
In order to measure the flow of tap water, the height $y_0(t) = a_0 t$ of the water level in a measuring jug is recorded as a function of time t. However, the starting point of the measurements is uncertain. Hence two models are compared:

$$y(t) = at \quad \text{and} \quad y(t) = at + b. \tag{1-36}$$

The first model estimates only the flow, assuming that the experiment started at time zero, while the second one also estimates the start of the experiment.

Generate a set of measurements:

$$y(t) = a_0 t + n(t), \quad \text{with} \quad t = [0:N]/N. \tag{1-37}$$

- Choose $a_0 = 1$, $N = 1000$, and $n_t \sim N(0, \sigma_n^2)$ with $\sigma_n^2 = 1$.
- Repeat this experiment 10^5 times.
- Estimate the LS parameters of both models, and compare $\hat{a}$ for the one parameter model $y(t) = at$ and two-parameter model $y(t) = at + b$, by estimating the pdf of $\hat{a}$.
- Calculate the mean value and the standard deviation of the slope.
- Plot also the error $y_0(t) - \hat{y}(t)$ for the first 50 experiments, for $t \in [0, 2]$, with $\hat{y}(t)$ the modeled output.

□

TABLE 1-3 Mean and standard deviation of $\hat{a}$ in the one- and two-parameter model

	One-Parameter Model	Two-Parameter Model
mean	0.996	0.987
std. dev.	0.057	0.113

Observations The results are shown below in Table 1-3 and Figure 1-11. From the table it is seen that the uncertainty of the one-parameter estimate is significantly smaller than that of the two-parameter model. The mean values of both estimates are unbiased; the error equals the exact value within the uncertainty after averaging 100 experiments. Also in Figure 1-11 the same observations can be made. Note that due to the prior knowledge of the one-parameter model (at time zero the height is zero), a significantly smaller uncertainty on $\hat{a}$ is found for small values of t, and the one-parameter model is less scattered than the two-parameter model. If prior knowledge is incorrect, systematic errors would be made on the flow estimate; if it is correct, better estimates are found. An analysis of the residuals can guide the user to find out which of both cases s/he is faced with.

1.5.2 Model selection

The goal of this section is to show how to select an optimal model for a given data set. Too simple a model will fail to capture all important aspects of the output, and this will result in errors that are too large in most cases. Too complex models use too many parameters. As was illustrated in the previous section, such models also result in a poor behavior of the modeled output because the model becomes too sensitive to the noise. Hence, we need a tool that helps us to select the optimal complexity that balances the model errors against the sensitivity to the noise disturbances. It is clear that this choice will depend on the quality of the data. All these aspects are illustrated in the next exercise where we propose the Akaike information criterion (AIC) as a tool for model selection.

Consider a single input single output linear dynamic system, excited with an input $u_0(t)$ and output $y_0(t) = g_0(t)*u(t)$. The system has an impulse response $g_0(t)$ that is infinitely long (infinite impulse response or IIR system). For a stable system, $g_0(t)$ decays exponentially to zero, so that the IIR system can be approximated by a system with a finite length impulse response $g(t)$, $t = 0, 1, ..., I$ (finite impulse response or FIR system). For $t > I$, the remaining contribution can be considered to be negligible. The choice of I will depend not only on $g(t)$, but also on the signal-to-noise-ratio (SNR) of the measurements as well as on the length of the available data record.

$$\hat{y}(t) = \hat{g}(t)*u_0(t) = \sum_{k=0}^{I} \hat{g}(k)u_0(t-k), \quad \text{with } u_0(k) = 0 \text{ for } k < 0. \tag{1-38}$$

In (1-38) it is assumed that the system is initially at rest. If this is not the case, transient errors will appear, but these disappear in this model for $t > I$ (why?).

The model parameters θ are, in this case, the values of the impulse response. θ is estimated from the measured data $u_0(t), y(t)$, $t = 0, 1, ..., N$, with $y(t)$ the output measurement that is disturbed with i.i.d. (independent and identically distributed) noise $v(t)$ with zero mean and variance σ_v^2:

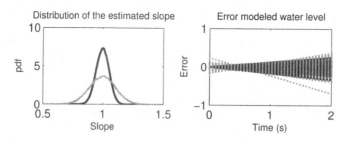

Figure 1-11 Impact of the number of parameters on the uncertainty of the slope estimate and the variability of the model. Black: One-parameter model $y = au$. Gray: Two-parameter model $y = au + b$. Left: The pdf of the estimated slope. Right: The error on the modeled output as a function of time.

$$y(t) = y_0(t) + v(t).$$ (1-39)

The estimates $\hat{\theta}$ are estimated by minimizing the least squares cost function:

$$V_{est}(\theta, Z^N) = \frac{1}{N}\sum_{t=0}^{N} |y(t) - \hat{y}(t)|^2, \text{ with } \hat{y}(t) = \hat{g}(t)*u_0(t) = u_0(t)*\hat{g}(t).$$ (1-40)

Note that this model is linear-in-the-parameters, and solution (1-26) can be used.

In order to find the "best" model, a balance is made between the model errors and the noise errors using a modified cost function that accounts for the complexity of the model. Here we propose to use of, amongst others, the AIC criterion:

$$V_{AIC} = V_N(\theta)\left(1 + 2\frac{\dim\theta}{N}\right).$$ (1-41)

Exercise 11 (Model selection using the AIC criterion)   Consider the discrete time system $g_0(t)$ given by its transfer function:

$$G_0(z) = \sum_{k=0}^{n_b} b_k z^{-k} / \sum_{k=0}^{n_a} a_k z^{-k},$$ (1-42)

Generate the filter coefficients a_k, b_k using the MATLAB® instruction

```
[b,a] = cheby1(3,0.5,[2*0.15   2*0.3])
```                                                                                                        (1-43)

This is a band pass system with a ripple of 0.5 dB in the pass band. Generate two data sets D_{est} and D_{val}, the former with length N_e being used to identify the model, the latter with length N_v to validate the estimated model. Note that the initial conditions for both sets are zero! Use the MATLAB® instruction

```
y0 = filter(b,a,u0),  y = y0+ny
```                                                                                                        (1-44)

with u_0 zero mean normally distributed noise with $\sigma_{u_0} = 1$, and $n_y \sim N(0, \sigma^2 = 0.5^2)$ for a first experiment, and $n_y \sim N(0, \sigma^2 = 0.05^2)$ for a second experiment. Put $N_{est} = 1000$, and $N_{val} = 10,000$ in both experiments.

■ Use the linear least squares procedure (1-26) to estimate the model parameters of an approximating FIR model, for varying orders from 0 to 100.

■ Calculate for each of the models the simulated output $\hat{y} = \text{filter}(\hat{g}, 1, u_0)$, and calculate the cost function (1-40) on D_{est} and on D_{val}.

■ Calculate V_{AIC}.

■ Calculate $V_0 = \frac{1}{N_{val}}\sum_{t=0}^{N_{val}} |y_0(t) - \hat{y}(t)|^2$ on the undisturbed output of the validation set.

■ Plot V_{est}, V_{AIC}, V_{val} as a function of the model order. Normalize the value of the cost function by $\sigma_{n_y}^2$ to make an easier comparison of the behavior for different noise levels.

■ Plot $\sqrt{V_0/\sigma_{n_y}^2}$ as a function of the model order.

□

Observations The results are shown in Figure 1-12, and the following observations can be made:

(i) Increasing the model order results in a monotonically decreasing cost function V_{est}. This result was to be expected because a simpler model is always included by the more complex model, and the linear LS always retrieves the absolute minimum of the cost function, so that no local minima of the cost function as a function of the model order exist. Hence, increasing the complexity of the model should reduce the value of the cost function (it is a monotonic not increasing function of the model complexity).

(ii) On the validation data we observe first a decrease and then an increase of V_{val}. In the beginning, the additional model complexity is mainly used to reduce the model errors, a steep descent of the cost function is observed. From a given order on, the reduction of the model errors is smaller than the increased noise sensitivity due to the larger number of parameters, resulting in a deterioration of the capability of the model to simulate the validation output. As a result the validation cost function V_{val} starts to increase.

(iii) V_{AIC} gives a good indication, starting from the estimation data only, where V_{val} will be minimum. This reduces the need for long validation records, and it allows us to use as much data as possible for the estimation step.

(iv) The optimal model order increases for a decreasing disturbing noise variance. Since the plant is an IIR system with an infinite long impulse response, it is clear that in the absence of disturbing noise $\sigma_n = 0$, the optimal order would become infinite. In practice this value is never reached due to the presence of calculation errors that act also as a disturbance.

(v) A fair idea about the quality of the models is given by V_0. The normalized rms value $\sqrt{V_0/\sigma_v^2}$ is plotted on the right side of Figure 1-12. This figure shows that a wrong selection of the model can result in much larger simulation errors. The good news is that the selection of the best model order is not so critical, the minimum is quite flat and all model orders in the

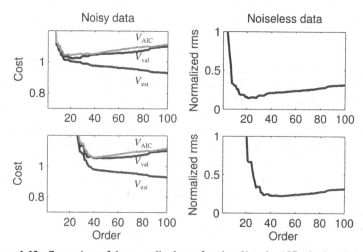

Figure 1-12 : Comparison of the normalized cost function V_{est}, the AIC criterion V_{AIC}, and the validation V_{val} for $\sigma_n = 0.5$ (top) and $\sigma_n = 0.05$ (bottom).

neighborhood of the minimum result in good estimates. Note that in real-life experiments, V_0 is not available.

Remark: In practice the validation set is chosen always (much) smaller than the test data. The data should be primaly used to estimate a good model. In this exercise we selected an extremely large data set to eliminate the noise variation in the validation to better visualize the quality of the AIC model selection. The AIC method makes it possible to select the model on the test data without using a validation set. In practice it is however still advisable to test the final model on a validation test to verify if the model can explain also fresh data that were not used to build it.

1.5.3 What have we learned in Section 1.5? Further reading

In this section we learned that the choice of the model complexity is an important issue. Too simple models lead to bias errors, while models with too many parameters suffer from an increased variability. Model selection tools balance the bias and variance errors. A first possibility is to verify the identified model on a data set that was not used during the parameter estimation step, this is called a validation set. Instead of saving a part of the available data for this test it is also possible to predict the behavior of an identified model on a new data set. We have learned that the Akaike information criterion AIC is able to predict the value of the cost function on the validation set. This allows the user to use all the data in the estimation step, leading to a smaller variance. In the literature a lot of results are published on this topic. Besides the original paper of Akaike (1974), we refer the reader to the classical textbooks on system identification, — for example, Ljung (1999), Söderström and Stoica (1989), and Johansson (1993) — to learn more about this topic.

1.6 NOISE ON INPUT AND OUTPUT MEASUREMENTS: THE IV METHOD AND THE EIV METHOD

In Section 1.2.2 it was shown that the presence of disturbing noise on the input measurements creates a systematic error. In this set of exercises, more advanced identification methods are illustrated that can deal with this situation. Two methods are studied: The first is called the instrumental variables method (IV), the second is the errors-in-variables (EIV) method. The major advantage of the IV method is its simplicity. No additional information is required from the user. The disadvantage is that this method does not always perform well. Both situations are illustrated in the exercises. The EIV performs well in many cases, but in general additional information from the user is required. The covariance matrix of the input–output noise should be known. All methods are illustrated again on the resistance example with measured current and voltage $i(t)$, $u(t)$, $t = 1, 2, ..., N$. Both measurements are disturbed by mutually uncorrelated Gaussian noise, $n_i(t)$, $n_u(t)$:

$$\begin{aligned} i(t) &= i_0(t) + n_i(t), \\ u(t) &= u_0(t) + n_u(t). \end{aligned} \tag{1-45}$$

The least squares estimate is given by

$$\hat{R}_{LS} = \frac{\sum_{t=1}^{N} u(t)i(t)}{\sum_{t=1}^{N} i(t)^2}, \tag{1-46}$$

the instrumental variables estimator (IV) is

$$\hat{R}_{IV} = \frac{\sum_{t=1}^{N} u(t)i(t+s)}{\sum_{t=1}^{N} i(t)i(t+s)}, \tag{1-47}$$

with s a user selectable shift parameter. Note that the IV estimator equals the LS estimator for $s = 0$.

The EIV estimator is given by

$$\hat{R}_{EIV} = \frac{\dfrac{\sum u(t)^2}{\sigma_u^2} - \dfrac{\sum i(t)^2}{\sigma_i^2} + \sqrt{\left(\dfrac{\sum u(t)^2}{\sigma_u^2} - \dfrac{\sum i(t)^2}{\sigma_i^2}\right)^2 + 4\dfrac{\left(\sum u(t)i(t)\right)^2}{\sigma_u^2 \sigma_i^2}}}{2\dfrac{\sum u(t)i(t)}{\sigma_u^2}}, \tag{1-48}$$

with σ_u^2, σ_i^2 the variances of the voltage and current noise respectively, the covariance is assumed to be zero in this expression: $\sigma_{ui}^2 = 0$.

Exercise 12 (Noise on input and output: The instrumental variables method applied on the resistor estimate)   Generate the current $i_0(k)$ from a Gaussian white noise source e_1 filtered by a first order Butterworth filter with cutoff frequency f_{Gen}

$$i_0 = \text{filter}(b_{Gen}, a_{Gen}, e_1), \tag{1-49}$$

with $[b_{Gen}, a_{Gen}] = \texttt{butter}(1, 2*f_{Gen})$. Next this filtered sequence is scaled to get a signal i_0 with standard deviation σ_{i_0}.

Generate the measured current and voltage (1-45), where $n_u(k)$ is white Gaussian noise: $N(0, \sigma_{n_u}^2)$. The current noise $n_i(k)$ is obtained from a Gaussian white noise source filtered by a second-order Butterworth filter with cut-off frequency f_{Noise}:

$$n_i = \text{filter}(b_{Noise}, a_{Noise}, e_2), \tag{1-50}$$

with $[b_{Noise}, a_{Noise}] = \texttt{butter}(2, 2*f_{Noise})$, and e_2 white Gaussian noise. Its variance is scaled to $\sigma_{n_i}^2$.

■ Experiment 1: Generate three sets of 1000 experiments with $N = 5000$ measurements each on a resistor $R_0 = 1000\ \Omega$, and the following parameter settings:

$$f_{Gen} = 0.05, \quad f_{Noise} = [0.4995, 0.475, 0.3],$$
$$\sigma_{i_0} = 0.1, \quad \sigma_{n_i} = 0.1, \quad \sigma_{n_u} = 1. \tag{1-51}$$

- Process these measurements with the LS estimator, as well as with the IV-estimator with the shift parameter $s = 1$.

- Experiment 2: Generate 1000 experiments with $N = 5000$ measurements each, and the following parameter settings:

$$f_{\text{Gen}} = 0.05, \quad f_{\text{Noise}} = 0.3, \quad \sigma_{i_0} = 0.1, \quad \sigma_{n_i} = 0.1, \quad \sigma_{n_u} = 1. \tag{1-52}$$

- Process these measurements with the LS estimator, and with the IV estimator with the shift parameter $s = 1, 2, 5$.

Plot for both experiments:

- the pdf of $\hat{R}_{\text{LS}}$ and $\hat{R}_{\text{IV}}$,

- the autocorrelation function of i_0 and n_i (*hint*: use the MATLAB® instruction `xcorr`),

- the FRF of the generator and the noise filter.

□

Observations   The results are shown below in Figure 1-13 and Figure 1-14. In Figure 1-13, the results are for a fixed generator filter and a varying noise filter. The shift parameter for the IV is kept constant at 1. From this figure it is clearly seen that the LS results are

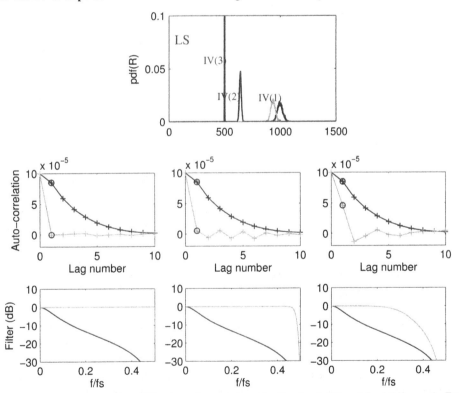

Figure 1-13 Study of the LS and IV estimate for a varying noise filter bandwidth and fixed shift $s = 1$. Top: The LS (black line) and IV estimate (black or gray line). IV(1), IV(2), and IV(3) correspond to the first, second, and third filter. Middle: The auto correlation of i_0 (black) and n_i (gray) for the different noise filters. Bottom: The filter characteristics of i_0 (black) and the noise n_i (gray).

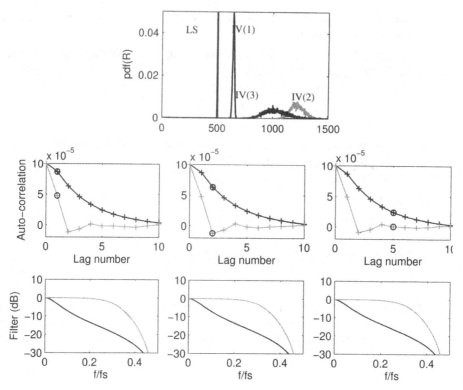

Figure 1-14 Study of the LS and IV estimate for a fixed noise filter bandwidth and a varying shift $s = 1, 2, 5$. Top: The LS (black) and IV (black and gray) estimate. IV(1), IV(2), and IV(3) correspond to a shift of 1,2, and 5 tabs. Middle: The auto correlation of i_0 (black) and n_i (gray). Bottom: The filter characteristics of i_0 (black) and the noise n_i (gray).

strongly biased. This is due to the noise on the input, the relative bias is in the order of $\sigma_{n_i}^2 / \sigma_{i_0}^2$. For the IV results, the situation is more complicated. For the white noise situation, no bias is visible. However, once the output noise is filtered, a bias becomes visible. The relative bias is proportional to the ratio of the autocorrelation functions of the noise and the current $R_{n_i n_i}(s) / R_{i_0 i_0}(s)$.

The same observations can also be made in Figure 1-14. In this figure, the shift parameter is changed while the filters are kept constant. It can be seen that the bias becomes smaller with increasing shift s, because $R_{n_i n_i}(s) / R_{i_0 i_0}(s)$ is getting smaller. At the same time the dispersion is growing, mainly because $R_{i_0 i_0}(s)$ is getting smaller. Observe also that the sign of the bias depends on the sign of $R_{n_i n_i}(s)$. The IV method works well if the bandwidth of the generator signal is much smaller than that of the noise disturbances.

Exercise 13 (Noise on input and output: the errors-in-variables method)   In   this exercise the EIV method is used as an alternatives for IV method to reduce/eliminate the bias of the least squares estimate. This time no constraint is put on the power spectra (bandwidth) of the excitation and the disturbing noise, but instead the variance of the input and output disturbing noise should be given in advance (prior information). This is illustrated again on the resistance example with measured current and voltage $i(t)$, $u(t)$, $t = 1, 2, ..., N$.

The least squares estimate is given in (1-46), the EIV-estimator is given in (1-48), where the sum runs over $t = 1, ..., N$. It is shown to be the minimizer of the following cost function:

$$V_{\mathrm{EIV}} = \frac{1}{N}\sum_{t=1}^{N}\left\{\frac{(u(t)-u_0(t))^2}{\sigma_{n_u}^2} + \frac{(i(t)-i_0(t))^2}{\sigma_{n_i}^2}\right\}, \tag{1-53}$$

with respect to u_0, i_0, R_0 under the constraint that $u_0(t) = R_0 i_0(t)$.

■ Setup: Generate the current $i_0(t)$ from a white zero mean Gaussian noise source $N(0, \sigma_{i_0}^2)$.

Generate the measured current and voltage as

$$\begin{aligned} i(t) &= i_0(t) + n_i(t), \\ u(t) &= u_0(t) + n_u(t); \end{aligned} \tag{1-54}$$

$n_u(t)$ and $n_i(t)$ are white Gaussian noise sources with zero mean and variance $\sigma_{n_u}^2$ and $\sigma_{n_i}^2$, respectively.

■ Generate a set of 1000 experiments with $N = 5000$ measurements each, and the following parameter settings:

$$R_0 = 1000, \quad \sigma_{i_0} = 0.01, \quad \sigma_{n_i} = 0.001, \quad \sigma_{n_u} = 1. \tag{1-55}$$

Calculate the LS and EIV estimate. Plot the histogram with $\hat{R}_{\mathrm{LS}}$ and $\hat{R}_{\mathrm{EIV}}$.

Observations   The results are shown below in Figure 1-15, From this figure it is clearly

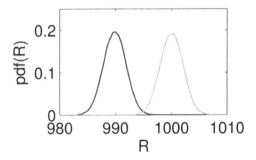

Figure 1-15 Comparison of the pdf of the LS (black) and the EIV estimate (gray), calculated with prior known variances.

seen that the LS estimates are strongly biased (mean value is 990.15). This is due to the noise on the input, the relative bias is in the order of $\sigma_{n_i}^2 / \sigma_{i_0}^2$. No systematic error can be observed in the EIV results (mean value is 999.96). The IV estimate would fail completely in this situation (why?).

Discussion   We learned that noise on the input (regressor) results in a bias on the estimates that depends on the inverse SNR of the input (regressor). Both methods, the IV and the EIV method, have pros and cons. While the IV method requires a lot of insight of the user (are the tight experimental conditions met?), the EIV method needs knowledge of the covariance matrix of the input and output noise. In the literature, alternative methods are proposed to deal with this problem. A first possibility is to use repeated experiments. From the variations from one experiment to the other, it is possible to estimate the noise covariance matrix from the data. This is later illustrated in this book; see, for example, Exercise 47 (Schoukens *et al.*, 1997). Alternative methods estimate at the same time a parametric plant and noise model (Söderström, 2007). This allows to relax the experimental conditions (no repeated experiments are required), at a cost of more difficult optimization problem to be solved (estimate the plant- and noise model parameters).

2

Generation and Analysis of Excitation Signals

What you will learn: This chapter makes the reader familiar with the fast Fourier transform (FFT) algorithm that is intensively used to make a spectral analysis of measured signals. Next we study how to generate good excitation signals to solve nonparametric and parametric system identification problems. The following topics are addressed:

- Representing a continuous time signal by a discrete signal without loss of information (see Exercise 14).

- Setting the parameters of the FFT and understanding the impact on the spectral resolution, the leakage errors, and the relation between frequency and FFT lines (see Exercises 15, 16, 17, 18, 22.a, 22.b).

- Generating periodic signals using the IFFT (inverse FFT) (see Exercises 19.a, 19.b, and 20).

- Generation and tuning the properties of broadband and multi-harmonic periodic excitation signals (see Exercises 20, 21, 23, 24, 25, 26, 27).

- Generation, tuning, and processing random excitations (see Exercises 28, 29, 30, 31).

- Amplitude distribution of filtered random signals (see Exercise 32).

- Calculating the derivative of a measured periodic signal (see Exercise 33).

2.1 INTRODUCTION

Each system identification process starts with experiments that provide information about the system to be modeled. Within a given time as much information as possible should be collected. The amount of information that can be retrieved within the operational constraints (e.g., maximum excitation level, minimal power consumption, ...) depends strongly on the selection of the excitation signal. For that reason we study in this chapter a number of potential excitation signals like sines, multisines and binary sequences (periodic excitation signals) on the one hand, and random noise excitations on the other hand. The measurements are often analyzed in the time and in the frequency domain, using the FFT as an algorithm to move from one domain to the other. For that reason we learn the reader also how to use this algorithm, and how to understand its results in order to avoid the pitfalls that are linked to its use. We start this chapter with a number of exercises on the use of the FFT; next we learn how to generate and analyze periodic excitation signals. Finally we study how random noise excitation signals can be generated and processed.

2.2 THE DISCRETE FOURIER TRANSFORM (DFT)

The Fourier integral transforms a continuous, analog signal $u(t)$ from the time to the frequency domain:

$$U(j\omega) = \frac{1}{2\pi}\int_{-\infty}^{\infty} u(t)e^{-j\omega t}dt, \quad \text{where } \omega = 2\pi f. \tag{2-1}$$

Although such a transform does not create new information, it might reveal some of the properties of the signals. In practice this transform is calculated on a digital computer starting from a signal measured in a discrete, finite set of equidistant samples. Hence, the continuous time Fourier transform (2-1) is replaced by the discrete Fourier transform (DFT):

$$U(l) = S\sum_{k=0}^{N-1} u(kT_s)e^{-j2\pi\frac{kl}{N}}, \tag{2-2}$$

where l is the frequency line number.

In this expression, S is an arbitrary scaling factor that is often equated to:

- $1/\sqrt{N}$ for random noise sequences and signals where the number of frequency components grow proportional to N, e.g. a random noise sequence `u = randn(N,1)`.

- $1/N$ for sequences where the number of frequency components does not depend upon N, e.g. a sine wave: `u = sin(2πt)`, $t = 0, 1, ..., N-1$.

Replacing the continuous Fourier transform (2-1) by the discrete Fourier transform involves the following actions:

- Discretization in time: The continuous signal $u(t)$ is replaced by the discrete set of samples $u(kT_s)$. The sample frequency is $f_s = 1/T_s$.

- Windowing: The infinite integration interval $]-\infty, \infty[$ is replaced by a finite window $[0, NT_s[$. Note that the end point NT_s does not belong to the interval.

- Discretization in frequency: The integral is calculated at a discrete set of frequencies. $U(l)$ is the DFT at frequency $l/NT_s = lf_s/N$. Observe that the frequency resolution is set by the window length NT_s of the previous step.

The DFT can be calculated rapidly using the MATLAB® instruction: `U = fft(u)`, and the inverse transform is given by `u = ifft(U)`. Note that the scaling factor S of the MATLAB® `fft` routine is $S = 1$. The `fft` and `ifft` are calculated fastest if the length of u and U is $N = 2^n$. However, the MATLAB® implementations allow also for signal lengths that are different from this optimal value, while the calculation time is still very small.

The major aspects of the three basic steps discussed before (discretization in time, windowing, discretization in frequency) are illustrated in the next two exercises. In the rest of this chapter and this book, the `fft` and `ifft` routines will be intensively used.

The frequency axis of the DFT is often expressed in "line numbers". The zero frequency ($f = 0$) corresponds to line number zero, the lth line number corresponds to the frequency lf_s/N.

Exercise 14 (Discretization in time: Choice of the sampling frequency: ALIAS)
Consider two sines with frequency $f_1 = 1\,\text{Hz}$ and $f_2 = 9\,\text{Hz}$, for example: $u_{c1}(t) = \sin(2\pi f_1 t)$ and $u_{c2}(t) = -\sin(2\pi f_2 t)$. Sample both signals in the time interval $[0, 1[$ with a sampling frequency $f_s = 10\,\text{Hz}$: for example, $u_{d1}(t) = u_{c1}(t/f_s) = u_{c1}(tT_s)$, with $t = 0, 1, ..., N-1$. Make a time domain figure with the continuous time and discrete time signals on one plot. Make a frequency domain plot with the spectrum of the continuous time signals, and the result of the DFT for both signals. Compare the DFT of both signals. □

Discussion In the time domain, it can be observed (Figure 2-1, middle left side) that the samples of both signals completely coincide, it is no longer possible to distinguish them on the basis of the discrete time representation. The fast signal results in a slowly varying discrete time sequence. This effect will appear each time when the signal frequency is larger than half the sample frequency. It is the alias effect. In order to avoid alias, the maximum frequency f_{max} of the sampled signals should be smaller than half the sample frequency: $f_{max} < f_s/2$.

In the frequency domain, the spectra of the original continuous time signals can be compared to that of the DFT. Observe that the amplitude spectrum of the DFT is symmetric around $N/2$ corresponding to the frequency $f_s/2$, and the phases are changing sign, viz.,

$$U(l) = \overline{U}(N-l). \tag{2-3}$$

It can also be seen that the fast frequency (9 Hz) is mapped to a low frequency of $f_s - 9\,\text{Hz} = 1\,\text{Hz}$. This is the alias effect in the frequency domain. Sampling maps the original frequencies $-f$ and f of a sine with frequency f to a set of frequencies $f_k = \pm f + kf_s$.

The spectrum between 0 and $f_s/2$ corresponding to the fft lines between 0 and $N/2$ gives all information of the DFT of a real signal: the remaining DFT lines $[N/2 + 1, N-1[$ are the complex conjugate of those between $[0, N/2[$, $U(N-l) = \overline{U}(l)$.

Discussion What you learned in this exercise

- When a continuous time signal is sampled, the sample frequency should be larger than 2 times the maximum frequency, otherwise alias errors occur.

- In order to avoid alias, also for signals with a bandwidth that is larger than $f_s/2$, it is strongly advised to filter the signals first with a low-pass filter, called an anti-alias filter, that surpresses the high-frequency content.

- The spectrum of a DFT analysis is periodic, with period N which is the number of samples in the time domain window. The spectrum is mirrored around half the sample frequency: $U(N-l) = \overline{U}(l)$.

- More information on the DFT and FFT can be found in Brigham (1974).

Exercise 15 (Windowing: Study of the leakage effect and the frequency resolution) Consider a sine wave: $u(t) = \sin(2\pi ft)$ with $f = 2\,\text{Hz}$. Sample it with a sample frequency $f_s = 10$ Hz in N points, with N being [10 11 12 13 14 15]. Calculate for each of these values of N the DFT $U = \text{fft}(u)$, and draw the amplitude spectrum $|U|$ with the x-axis scaled in Hz. Add to this plot also the amplitude spectrum of the original sine wave. □

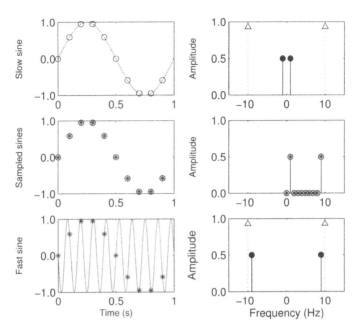

Figure 2-1 Illustration of the alias effect due to the sampling of a continuous time signal in the time domain (left) and frequency domain (right). Top: Slow sine and its Fourier transform. Bottom: Fast sine and its Fourier transform. Middle: The sampled sine and its DFT spectrum.

Observations (see Figure 2-2) Depending upon the value of N the lower half of the DFT spectrum either coincides with the original sine spectrum or does not. The spectrum is perfectly retrieved for $N = 10$ and 15. For the other values the DFT spectrum differs from that of the sine, although it is still concentrated around the original spectral line at 2 Hz. This effect is called leakage; it is present if for a periodic signal the window length is not an integer multiple of the period length.

Observe also that the spectral resolution of the DFT varies inversely proportional to the window length. The longer the window, the higher the spectral resolution.

It is important to remark that the sample on the right extreme value of the window does not belong to the window because it is an open interval on the right side, beyond that sample.

Discussion What you learned in this exercises The lessons to be taken from this exercise are:

■ In order to avoid leakage in a DFT analysis, an integer number of periods of a periodic signal should be measured.

■ The frequency resolution of a DFT analysis is influenced by the length of the window in the time domain; longer measurements result in a finer frequency resolution.

■ The first sample of the next period does not belong to the time domain window, it is an open interval: $[0, NT_s[$.

Further reading: The book of Brigham (1974) gives a very didactical introduction to the theory and use of the FFT. It discusses the Fourier transform, the DFT, the FFT, and basic FFT applications. Also the book by Bendat and Piersol (1980) provides a lot of useful information on the use of FFT methods in daily practice.

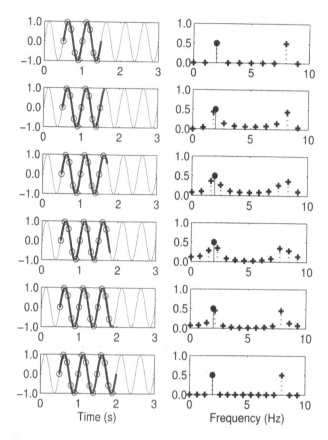

Figure 2-2 Illustration of the windowing-leakage effect of the DFT on a sine wave. On the left side, different window lengths are used $N = [10, 11, ..., 15]$ samples corresponding to a window length of $[1.1, 1.2, ..., 1.5]$ seconds. On the right side, the amplitude of the DFT spectrum is shown (+), together with the positive frequency component of the sine (filled circle).

2.3 GENERATION AND ANALYSIS OF MULTISINES AND OTHER PERIODIC SIGNALS

2.3.1 Generation and analysis of sine wave signals

In this series of exercises, it will be shown using a sine wave signal:

- How to avoid leakage.

- How to read the frequency axis when line numbers, instead of a scaling in Hz are used.

- What is the impact of the FFT scaling factor?

- How to calculate a sine wave using the inverse fast Fourier algorithm.

Exercise 16 (Generate a sine wave, noninteger number of periods measured)   Generate a signal

$$u(t) = A \sin(\omega t T_s + \varphi), \quad t = 0, 1, ..., N-1 \tag{2-4}$$

with $\omega = 2\pi f$ and $f = 80\,\text{Hz}$, $A = 1$, $\varphi = \pi/2$, $f_s = 1/T_s = 1000$, and $N = 16$. Calculate $U = \text{fft}(u)$. Put the FFT scale factor equal to $1/N$ (Why?). Make four plots:

- Plot 1: $u(t)$ as a function of time.
- Plot 2: The amplitude spectrum (dB) as a function of the frequency.
- Plot 3: The amplitude spectrum (linear scale) as a function of the FFT line number (DC is line 0).
- Plot 4: The amplitude spectrum (linear scale) as a function of the frequency.

□

Observations (see Figure 2-3) It is seen that since a non integer number of periods is measured, leakage errors appears. The amplitude axis can be shown on a linear or a logarithmic (db) scale: $A_{\text{dB}} = 20\log_{10}A$. The frequency axis can be scaled in DFT-line numbers or in

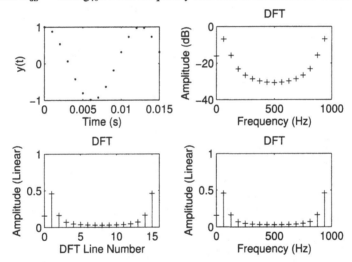

Figure 2-3   Results of exercise 16: Generate a sine wave, no integer number of periods measured. Number of samples $N = 16$.
Top left: Discrete time signal.
Top right: DFT-amplitude spectrum (dB), frequency axis in Hz.
Bottom left: DFT-amplitude spectrum (linear), frequency axis: line numbers.
Bottom right: DFT-amplitude spectrum (linear), frequency axis in Hz.

Hz. In the first case, the position will depend on the number of measured periods, as is illustrated in the next exercise. In the latter case, the position is fixed (see Exercise 18).

Exercise 17 (Generate a sine wave, integer number of periods measured)   Generate a signal

$$u(t) = A \sin(\omega t T_s + \varphi), \quad t = 0, 1, ..., N-1 \tag{2-5}$$

with $f_s = 1/T_s = 1000\,\text{Hz}$, $A = 1$, $\varphi = \pi/2$, and $N = 16$. Choose ω such that an integer number of periods is measured, for example 1 period. Calculate U = fft(u). Put the FFT scale factor equal to $1/N$. Make four plots:

■ Plot 1: $u(t)$ as a function of time.
■ Plot 2: The amplitude spectrum (dB) as a function of the frequency.
■ Plot 3: The amplitude spectrum (linear scale) as a function of the FFT line number (DC is line 0).
■ Plot 4: The amplitude spectrum (linear scale) as a function of the frequency.

□

Observations (see Figure 2-4) From the spectral analysis it turns out that no leakage is present, which should be the case since an integer number of periods is measured. It can be seen that the unexcited frequencies have an amplitude which is about 300 dB below the amplitude of the excited frequencies. This corresponds to a calculation precision of 15 digits (1 digit is 20 dB).

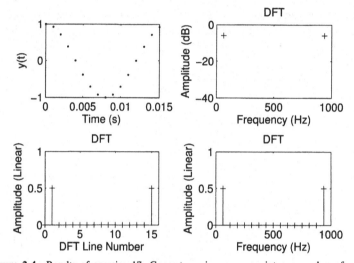

Figure 2-4   Results of exercise 17: Generate a sine wave, an integer number of periods measured. Number of samples $N = 16$.
Top left: discrete time signal.
Top right: DFT-amplitude spectrum (dB), frequency axis in Hz.
Bottom Left: DFT-amplitude spectrum (linear), frequency axis: line numbers.
Bottomr Right: DFT-amplitude spectrum (linear), frequency axis: line numbers.

Exercise 18 (Generate a sine wave, doubled measurement time)   Generate a signal

$$u(t) = A\sin(\omega t T_s + \varphi), \quad t = 0, 1, ..., N-1. \tag{2-6}$$

Choose ω, φ, A, f_s equal to the values of Exercise 17, but double the number of data points $N = 32$.
Calculate U = fft(u). Make four plots:

■ Plot 1: $u(t)$ as a function of time.
■ Plot 2: The amplitude spectrum (dB) as a function of the frequency.

- Plot 3: The amplitude spectrum (linear scale) as a function of the FFT line number (DC is line 0).

- Plot 4: The amplitude spectrum (linear scale) as a function of the line number.

What are the excited spectral lines (the FFT line numbers)? What are the excited frequencies? Explain the different behavior: the line numbers shifted, the frequencies remained the same. What is the relation between line number, frequency, the sample frequency f_s, spectral resolution of the FFT, and the data length?   □

Observations (see Figure 2-5) From the spectral analysis it turns out that no leakage is present, as expected since an integer number of periods is measured. The frequency of the excited spectral lines did not change by increasing the measurement time. The line number changed: the excited frequency line is now 2 because two periods are now measured. It can also be seen that the frequency resolution of the plots is doubled to $f_s/32$. Due to the FFT scaling factor of $1/N$, the amplitude of the spectrum did not change when N is varied.

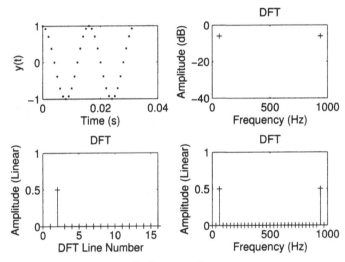

Figure 2-5   Results of exercise 18: Generate a sine wave, an
integer number of periods measured. Number of samples $N = 32$..
Top left: discrete time signal.
Top right: DFT-amplitude spectrum (dB), frequency axis in Hz.
Bottom left: DFT-amplitude spectrum (linear), frequency axis: line numbers.
Bottom right: DFT-amplitude spectrum (linear), frequency axis: line numbers.

Discussion   What you learned in this exercise

- The position of a harmonic component is independent of the window length if the frequency axis is scaled in Hz (mHz, kHz, ...). If the frequency axis is labeled with the DFT-line number, then the position changes with a varying window length.

- At the unexcited frequencies, the level of the (calculation) noise is visible.

- The amplitude of the spectrum does not vary with the window length if the proper FFT scaling factor is used. For a signal where the number of harmonic components does not depend upon the window length, the proper choice for the scaling factor is $S = 1/N$.

Exercise 19.a (Generate a sine wave using the MATLAB® IFFT instruction)
Generate a signal:

$$u(k) = A\sin(\omega kT_s + \varphi), \quad k = 0, 1, ..., N-1 \tag{2-7}$$

with $\varphi = 0$ and 1 period in a record of $N = 16$ data points ($u(k) = u(k+N)$). Use the inverse fast Fourier transform (IFFT) to make the calculations.
Hints: Define the DFT spectrum $U(k)$, $k = 0, 1, 2, ..., N-1$ and calculate u = ifft(U). Apply a scale factor $N/2$ to compensate for the internal scaling of the FFT. Notice that the FFT coefficient of a sinusoid with zero phase is exp(-j*pi/2) for the positive frequency and exp(j*pi/2) for the negative frequency.

Exercise 19.b (Generate a sine wave using the MATLAB® IFFT instruction, defining only the first half of the spectrum)   Make the same exercise, but define this time only $U(k)$, $k = 0, 1, 2, ..., N/2-1$, and put $U(k) = 0$ for $k = N/2, ..., N-1$.
Hint: Make use of the fact that $U(k) = \overline{U}(N-k)$, and $xy + \bar{x}\bar{y} = 2\text{real}(xy)$.
Plot the result of Exercise 19.a and 19.b on top of each other.
Analyze the number of operations to calculate a (sum or) sines directly in the time domain, or using the FFT routine. How does this answer depend upon the number of generated harmonics in the signal?   □

Observations (see Figure 2-6). Both results coincide completely after proper scaling of the amplitude by $1/N$. This is again due to the scaling of the MATLAB® IFFT command.

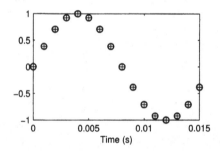

Figure 2-6   Results of Exercise 19.a and 19.b: Generate a sine wave using IFFT. o: direct calculation; + calculation via u=2real(ifft(U)).

Discussion What you learned in this exercise To calculate with a minimal numerical effort a time domain signal starting from its spectrum, the MATLAB® instruction u = N*(real(ifft(U)) can be used. Only the first half of the spectrum should be defined, the second half is put to zero.

For a single sine generation, this method is not very efficient from computational point of view. However, for a sum of harmonically related sines with many components, a huge reduction of the calculation time is obtained. Because the ifft is an orthogonal transformation from the frequency to the time domain, a maximal numerical stability is maintained.

2.3.2 Generation of multisine signals

A multisine is a sum of harmonically related sines. It is an example of a broadband or multiharmonic signal. Instead of measuring at one frequency at a time, the system is excited simultaneously at a whole set of frequencies. This allows often to reduce the measurement time because the transients that appear at the output at each change of the excitation frequency disappear. Initial transient effects will only be present at the start of the experiment. In the following sections the reader will be able to learn how to design and analyze multisine measurements. The user can set the amplitude and the phase of each frequency component of a multisine individually:

$$u(t) = \sum_{k=1}^{F} A_r \cos(2\pi f_0 k t T_s + \varphi_r), \qquad (2\text{-}8)$$

where $T_0 = 1/f_0$ is the period of the multisine and T_s is the sample period of the generator.

Exercise 20 (Generation of a multisine with flat amplitude spectrum)   Generate a multisine, using u = 2*real(ifft(U)), with a period of $N = 1000$ samples, exciting the frequency band $[0, 0.1f_s]$. Choose $f_s = 1000$ Hz. Note that from these settings the number of frequencies F in the multisine is set. First put all amplitudes $A_r = 1$, and then normalize the rms value of the multisine to 1, e.g. u = u/std(u). Make three choices for the phase:

- Signal 1 (impulse multisine): $\varphi_r(k) = -\tau\omega_k$ (a linear phase), with $\omega_k = 2\pi\dfrac{k}{N}f_s$ and $\tau = 0.3$ s.
- Signal 2 (random phase multisine): φ_r uniformly random distributed in $[0, 2\pi[$ (random phase multisine).
- Signal 3 (Schroeder multisine): $\varphi_k = -k(k-1)\pi/F$ (Schroeder phase).

Make two plots: Plot 1 is the impulse and random phase multisine on top of each other; Plot 2 is the Schroeder and random phase multisine on top of each other.   □

Observations (see Figure 2-7) A multisine with a linear phase $\varphi(\omega_k) = -\omega_k\tau$ behaves as an impulse centered in τ. By switching to a random phase a "noisy" behavior is obtained. (What is the asymptotic distribution of $u(t)$ as $F \to \infty$?) The peak value of the random phase multisine is also significantly reduced, compared to the zero phase multisine. The ratio of the peak value u_{peak} over the root mean square value u_{rms} of a signal is called the crest factor:

$$Cr(u) = \frac{u_{\text{peak}}}{u_{\text{rms}}}. \qquad (2\text{-}9)$$

Making the Schroeder phase choice, a sweeping behavior is obtained: In the beginning of the period the signal varies slowly, while at the end of the period fast variations occur. The peak value of the time domain signal is reduced even more, the crest factor of the Schroeder multisine is typically 1.65 for flat low-band power spectra (all frequencies in the band $[0, f_{\text{max}}]$ excited), while for an impulse multisine it is $\sqrt{2N}$. Notice that all these signals have exactly the same amplitude spectrum and rms value. Reducing the peak value for a given rms value allows us to improve the signal-to-noise ratio of an experiment.

Note that in this example $N = 1000 \neq 2^n$. Although the calculations are less fast than with $N = 1024$, the calculation time is still very small.

2.3.3 The swept sine signal

A popular alternative for a multisine signal is the swept sine. This is a periodic signal that continuously varies its frequency between two limits. In contrast to a multisine all its extreme amplitude values are equal ($\pm A$), but the freedom of the user to shape the amplitude spectrum is much more restricted. The signal is given by

$$u(t) = A \sin((at + b)t) \qquad (2\text{-}10)$$

with $a = \pi(k_2 - k_1)f_0^2$, $b = 2\pi k_1 f_0$, $f_0 = 1/T_0$, $k_2 > k_1 \in N$, with $t = T_0[0:N]/N$. Remark: Only one period can be calculated with (2-10). Concatenating this signal allows us to generate multiple periods.

Exercise 21 (The swept sine signal) Generate a swept sine signal with a period of 1 second, having most of its power between 50 and 200 Hz. Choose the sample frequency equal to 1 kHz.

Observations see Figure 2-8 $u(t)$ is a periodic signal with period T_0 and most of its power is in the frequency band $k_1 f_0$ to $k_2 f_0$. The amplitude spectrum shows a ripple of a few dB.

Discussion What you learned in this exercise. In this exercise we showed how to generate periodic signals using computationally efficient methods.

■ Using the `ifft` it is possible to generate multisine signals. The user has a full control over the amplitude spectrum of the signal.

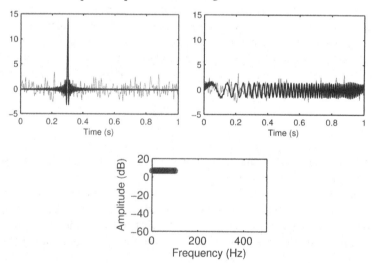

Figure 2-7 Multisines with the same amplitude spectrum and different phases.
Top left: Linear phase (black) and random phase (gray) multisine.
Top right: Schroeder phase (black) and random phase (gray) multisine (notice the difference in scale!).
Bottom: Amplitude of the DFT spectrum of the 3 signals.

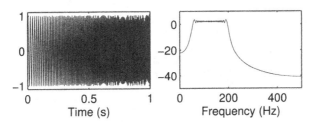

Figure 2-8 Generation of a swept sine signal with a period of 1 second and its power concentrated between 50 and 200 Hz. Left: The time domain signal; Right: The amplitude spectrum (dB).

- The Schroeder multisine results in a sweeping signal that has all its power within the frequency band that is specified by the user.

- The swept sine method is an alternative method to generate a sweeping signal. Compared to the Schroeder multisine, not all power is in the frequency band of interest. The crest factor for a swept sine (typically 1.45) is slightly smaller than that of a Schroeder multisine (typically 1.65).

- In Exercise 24 it will be shown how multisines with a lower crest factor can be generated.

- More information on the design of advanced excitation signals can be found in Pintelon and Schoukens (2001).

2.3.4 Spectral analysis of multisine signals

Exercise 22.a (Spectral analysis of a multisine signal, leakage present) Generate a random phase multisine $u(t)$ (see Exercise 20), using `u = ifft(U)` normalized such that $u_{RMS} = 1$. with a period of $N = 1024$ samples, exciting the frequency band $[0.15f_s, 0.35f_s]$. Put all amplitudes $A_r = 1$ in the frequency band of interest. Make a long signal $u_{long}(t)$ with $t = 1, ..., N_{long}$ by concatenating $M + 0.25$ periods of this signal. Put $M = 4$. Calculate the FFT using the window $w(t)$:

$$z = w.*u, \ \text{and} \ Z = \text{fft}(z) \qquad (2-11)$$

Use two windows:

- Rectangular window: $w(t) = 1$ for all t.
- Hanning (also called Hann) window: $w(t) = 0.5\left(1 - \cos\dfrac{2\pi t}{N_{long}}\right)$, $t = 0, 1, ..., N_{long} - 1$.

Plot both spectra and compare the results.
Hint: use the MATLAB® instruction `HANNING(Nlong,'periodic')` to generate the Hanning window $w(t)$.   □

Exercise 22.b (Spectral analysis of a multisine signal, no leakage present) Repeat Exercise 22.a, but with exactly $M = 4$ periods so that no leakage appears. Plot again the spectra.

Observations (see Figure 2-9 and 2-10) Observe the leakage in the spectra of the first figure. Using the Hanning window, the main lobe (points closely around the spectra lines) is

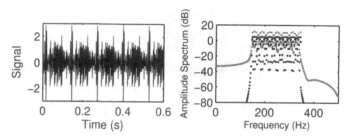

Figure 2-9 FFT of a multisine: no integer number of periods measured. Left: the measured time domain signal. Right: Amplitude of the DFT spectrum. Gray dots: Rectangular window. Black dots: Hanning window.

widened to three points, while the side lobes become much smaller than those obtained with a rectangular window.

In Figure 2-10, the leakage disappeared completely because an integer number of periods is measured. In this case the rectangular window gives the best result, while the Hanning window disturbs the spectral lines.

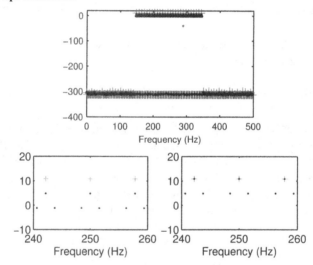

Figure 2-10 DFT spectrum of a multisine measured over an integer number of periods. (+) rectangular window; (·) Hanning window. Top: Global spectrum; Bottom left: Zoom, Hanning window without scale factor; Bottom right: Zoom, Hanning window with a scale factor 2.

Discussion What you learned in this exercise: leakage >< no leakage

■ If possible, measure an integer number of periods of the multisine. Use the rectangular window. This gives an undistorted view of the spectrum. A Hanning window applied to an integer number of periods results in worse results than the rectangular window.

■ No integer number of periods is measured: the first advice is to avoid this situation. If this is impossible, a Hanning window can be used to reduce (or better reshape) the error such that it is less disturbing in most cases.

■ A concise review of a large variety of windows and their properties is given in Harris (1978).

2.3.5 What have we learned in Section 2.3? Further reading

In this section we have illustrated that periodic signals can be very efficiently generated using the FFT algorithm. As a user it is important to be able to link the FFT spectral lines to the frequencies of the actualy generated and measured signals. Multisines are the sum of harmonic related sines. The properties of a multisine are completely set by the choice of the amplitudes and phases of the individual sines. The advantages that can be gained by a proper choice will be illustrated in the next section and also in the next chapter. A special case among the multisines is a swept sine signal. In order to avoid leakage errors during the FFT analysis of these signals, it is necessary to measure an integer number of periods. For more references on this topic, we refer the reader to Section 2.4.3.

2.4 GENERATION OF OPTIMIZED PERIODIC SIGNALS

2.4.1 Optimized multisines

In Exercise 20 a multisine is generated. It turned out that the shape of the signal not only depends on the user defined amplitude spectrum but also on the phase. Proper selection of the phases in (2-8) allows to reduce the peak value or crest factor of the signal for a given amplitude spectrum. In the next exercise two possibilities to minimize $Cr(u)$ are studied: the first is based on the selection of the signal with the smallest crest factor out of a large number of multisines with random generated phases, the second uses a nonlinear optimization approach.

Exercise 23 (Generation of a multisine with a reduced crest factor using random phase generation) Generate a random phase multisine $u(t)$ (see Exercise 20), using u = ifft(U), with a period of $N = 256$ samples, exciting the frequency band $[0, 0.1f_s]$. Set all amplitudes $A_r = 1$. Choose φ_r uniformly random distributed in $[0, 2\pi[$ (random phase multisine). Repeat this 10,000 times and select the realization with the lowest crest factor. Plot the optimized signal and its histogram. Make a plot of the evolution of the smallest crest factor as a function of the number of random phase realizations. Compare the best obtained crest factor to that of a multisine with Schroeder phase (see Exercise 20).
Repeat this exercise for $N = 4096$. How many excited frequencies will be in the frequency band of interest?   □

Observations (see Figures 2-11 and 2-12) Note that the best crest factor slowly decreases when the number of trials is increased. For multisines with more excited frequencies the decrease becomes very slow. Random phase selection is not a good method to generate signals with a small crest factor.

Exercise 24 (Generation of a multisine with a minimal crest factor using a crest factor minimization algorithm) Generate a low crest factor multisine using the following instructions of the FDIDENT toolbox:

```
ExcitationLines = 1+floor([1:0.1*N]');
                    % the excited FFT lines
Amp = ones(size(ExcitationLines));
                    % set the amplitudes
SignalData = fiddata([],Amp,ExcitationLines-1)
                    % creates a MATLAB® object
```

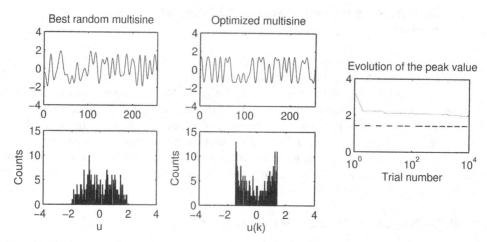

Figure 2-11 Crest factor minimization of a multisine with 25 excited frequencies. Left top: Best result selected from 10,000 random phase realizations. Middle top: Result of a nonlinear search procedure (Exercise 24). Left middle bottom: Histogram of the multisines. Right: Evolution of the crest factor as a function of the number of trials. Gray: random phase generation. Broken black line: the crest factor of the optimized phase multisine.

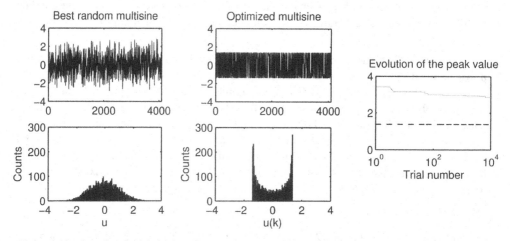

Figure 2-12 Crest factor minimization of a multisine with 400 excited frequencies. Left top: Best result selected from 10 000 random phase realizations. Middle, top: Result of a nonlinear search procedure (Exercise 24). Left, Middle bottom: Histogram of the multisines. Right: Evolution of the crest factor as a function of the number of trials. Gray: Random phase generation. Broken black line: The crest factor of the optimized phase multisine.

```
UMinCrest = crestmin(SignalData);
                        % minimize the crest factor
z = msinprep(UMinCrest,Ndata1);
                        % calculates the multisine
uMinCrest = z.input;
                        % put the signal in a vector
```

Plot the signals on the plot of Exercise 23. Also plot the histograms of multisines for both exercises.   □

Observations (see Figures 2-11 and 2-12) The numerically optimized signal has a very low crest factor (typically 1.4). From its histogram it is seen that this signal has almost a binary behavior while the samples of the best random multisine still have an almost Gaussian distribution. Random phase multisines (without optimization) are asymptotically Gaussian distributed. The phase optimization reduces the size of the tails of the distribution.

Discussion Generating multisines with a small crest factor

- The amplitude distribution of a random phase multisines is asymptotically normally distributed.

- Minimization of the crest factor by selecting the smallest peak value out of a large number of realizations does not work well. The probability to find a very low crest factor drops very fast with the number of excited frequencies.

- Numerical optimization methods are available to generate multisines with a minimal crest factor (typical value 1.4).

- The amplitude distribution of an optimized multisine tends to almost a binary distribution; it is close to a sine distribution.

- More details on the crest factor minimization algorithm `crestmin` is given in Guillaume *et al.* (1991). Random phase multisines are extensively discussed in Pintelon and Schoukens (2001).

2.4.2 Maximum length binary sequences

The amplitude distribution of an optimized multisine almost tends to a binary distribution, with the extreme amplitudes having the highest density. Binary excitations are signals that take only two values: 1 or -1. The power spectrum of these signals is set by the switching sequence between these two values. Special signals in this class are the maximum length binary sequences (MLBS). These are periodic signals with a crest factor equal to 1, that are designed to generate a flat amplitude spectrum. They can be generated using a shift register as shown in Figure 2-13. The spectrum of the discrete time sequence is flat, while the spectrum of the

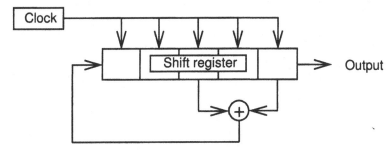

Figure 2-13   Generation of a maximum length binary signal with a shift register (can be initialized with an arbitrary nonzero code).

corresponding continuous time ZOH-reconstructed signal (see Figure 2-14) has a spectrum whose components decrease inversely proportional to the frequency. The amplitude $A(k)$ of the Fourier coefficient U_k of a ZOH-MLBS is given by

$$A(0) = \frac{a}{N} \quad \text{and} \quad A(k) = a\frac{\sqrt{N+1}}{N}\text{sinc}(k\pi/N) \quad \text{for} \quad k = 1, 2, ..., N-1, \tag{2-12}$$

with a the amplitude of the binary signal. The length of an MLBS is $N = 2^n - 1$ with n the register length. The feedback positions in Figure 2-13 depend on the register length.

Exercise 25 (Generation of a maximum length binary sequence)

■ Generate an MLBS with a register length $n = 5$. Plot the discrete time sequence and its ZOH continuous time reconstruction. Select a clock frequency of 100 Hz.

■ Plot the DFT spectrum of the discrete time sequence (with a scaling factor $1/\sqrt{N}$).

■ Approximate the spectrum of the continuous time ZOH-MLBS by upsampling the original sequence 16 times.

Hints
To generate the required signals the following instructions can be used

```
u = mlbs(nRegister);
            % generates a MLBS with registerlength = nRegister
            %        the clock frequency is normalized to 1 Hz

uOverSample = kron(u(:),ones(OverSample,1));
            % over samples u with a factor OverSample        □
```

Observations (see Figures 2-14) The discrete time sequence has a flat amplitude spec-

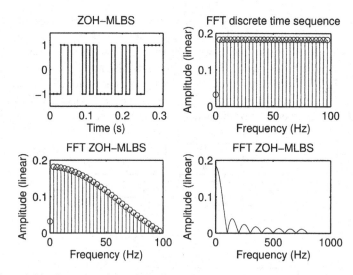

Figure 2-14   Generation of a maximum length binary sequence with a register length of 5. Top left: The discrete time MLBS sequence (dots) together with a ZOH reconstruction. Top right: The DFT amplitude spectrum of the discrete time sequence. Bottom left: Approximation of the spectrum of the continuous time sequence in $f = [0, 100]$ Hz. Bottom right: Approximation of the spectrum of the continuous time sequence in $f = [0, 800]$ Hz.

trum. The ZOH reconstruction introduces zeros at multiples of the clock frequency used to generate the MLBS. In practice, the frequency band of interest is usually restricted to the first lobe between zero and the clock frequency.

Exercise 26 (Tuning the parameters of a maximum length binary sequence)
Select the clock frequency and the register length to generate an MLBS such that at least 100 spectral components are present in the frequency band [0,1000] Hz while the amplitude of the lowest frequency component in that band is not more than 6 dB below that of the largest amplitude.   □

Discussion What you learned in these exercises

■ An MLBS injects a maximum power for a given amplitude, its crest factor equals one.

■ The amplitude spectrum of the discrete time sequence is flat.

■ The amplitude spectrum of the ZOH sequence drops as a sinc-function with its first zero crossing at the clock frequency.

■ Not all power is in the frequency band of interest. A MLBS has a part of its power at much higher frequencies than the clock frequency.

■ Within the frequency band of interest, the amplitude spectrum rolls off slowly.

■ An extensive overview of binary excitation signals and related excitations are discussed in Godfrey (1993). An overview of available software to generate these signals is made in Godfrey *et al.* (2005).

2.4.3 What have we learned in Section 2.4? Further reading

In this section we learned to design multi-frequent periodic excitations with optimized characteristics. In many applications we want to generate a signal with a user defined power spectrum, while at the same time the peak value of the signal should remain small. This can be done using either multisine signals with well-selected phases, or by using binary excitations. Both problems are intensively documented in the literature. In Schroeder (1970) a method is proposed to generate multisines that are very similar to swept sine excitations, the major difference being that all the signal power is confined to the frequency band of interest which is not the case for a swept sine excitation. Guillaume *et al.* (1991) propose a very efficient numerical search algorithm to create multisines with an arbitrary power spectrum and an extremely low crest factor. In Schoukens and Dobrowiecki (1998), a multisine is imposed that not only generates a user imposed power spectrum, the user can also define the amplitude distribution of the signal (e.g., a uniform distribution). Also the design of binary excitations is intensively studied. The papers by Tan and Godfrey (2009), Godfrey *et al.* (2005), and the book Godfrey (1993) give an extensive overview with many reference therein. These authors developed also a toolbox to generate these signals. In Chapter 6, we will also discuss the use of random phase multisines to identify linear approximations of nonlinear systems.

2.5 GENERATING SIGNALS USING THE FREQUENCY DOMAIN IDENTIFICATION TOOLBOX (FDIDENT)

The FDIDENT toolbox (Kollár, 1994) offers a GUI (graphical user interface) to design advanced periodic excitations, allowing the user to optimize the crest factor or the spectral content of the excitation. The user should select the nature of the signal:

- Multisine: See Section 2.3.2

- Pseudo random binary sequence (PRBS): The MLBS of the previous section is a special case of these signals. The restriction on the length ($N = 2^n - 1$) can be relaxed which leads to the PRBS signals.

- Discrete interval binary (DIBS) and ternary (DITS) signals: These are signals that take either two (-1,1) or three (-1,0,1) values, respectively. The signal can only switch on an equidistant discrete time grid. The switching sequences are optimized in order to get as close as possible to the user defined amplitude spectrum.

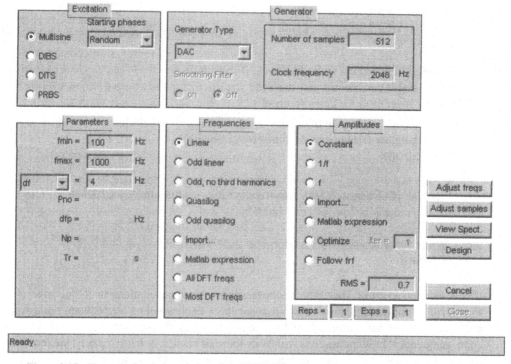

Figure 2-15   The excitation design window of the FDIDENT toolbox.

Exercise 27 (Generation of excitation signals using the FDIDENT toolbox)   Use the graphical user interface (see Figure 2-15) to generate a periodic excitation signal that excites the system with a flat amplitude in the frequency band [100 Hz, 500 Hz] with a frequency resolution of 5 Hz or higher. Normalize all signals to have the same rms value in the frequency band of interest:

$$\sum_{k \in \text{freq. band}} A^2(k). \tag{2-13}$$

Compare the different excitation signals:

- Analyze the ratio of the in-band power to the out-of-band power.
- Calculate the crest factor of the signals.
- Look for the DFT lines with the smallest amplitude in the frequency band of interest.

2.6 GENERATION OF RANDOM SIGNALS

An alternative to the special designed periodic excitation signals is the use of random excitations. Their use is very popular in practical applications because it seems much easier to generate and use random noise excitations than periodic signals. As we learned in the previous sections, we have to choose the amplitude spectrum and the spectral resolution in order to design a periodic excitation while these questions are seemingly not present when dealing with random excitations. However, this impression is misleading because these choices are implicitly made when the user does not address these aspects explicitly. The frequency resolution of a random noise excitation is set by the length of the experiment, and the amplitude spectrum is determined by the noise shaping filters that are used in the generator and actuator. For that reason, we strongly advice the reader to use periodic excitations whenever it is possible because their periodic nature gives access to many additional advantages during the processing in time- and in frequency domain methods. If the reader decides to use random excitations she/he should also understand very well the choices to be made during the generation and processing of these signals. The next series of exercises illustrate how to do that.

2.6.1 Generation, analyzing, and shaping random excitations

Exercise 28 (Repeated realizations of a white random noise excitation with fixed length) Generate a white zero mean random noise sequence, <u>uniformly</u> or normally distributed, with a length of $N = 128$ samples and an rms value $\sqrt{E\{u^2\}}$ equal to 1. Plot the DFT spectrum (scale with $1/\sqrt{N}$). Repeat this four times and compare the amplitude spectra.   □

Observations Note that the spectrum varies from one realization to the other (see Figure 2-16). Although a white noise sequence has a flat power spectrum (0 dB in this example), the spectrum of an individual realization is not flat. Large spikes and dibs occur. At some frequencies the amplitude is 20 dB or more below the average value. This leads to a poor SNR at those frequencies.

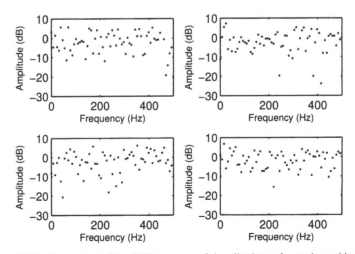

Figure 2-16 Comparison of the DFT spectrum of 4 realizations of a random white noise sequence each with a length of $N = 128$.

Exercise 29 (Repeated realizations of a white random noise excitation with increasing length)   Repeat the previous exercise for an increasing length of the record: $N = 128, 256, 512, 1024$. Plot for each of these realizations the amplitude of the DFT spectrum (scale DFT with $1/\sqrt{N}$).   □

Observations   Observe that a longer record does not result in a smoother spectrum (see Figure 2-17). Irrespective of the record lengths, apparent dips and spikes still appear in the amplitude spectra. From the plots it can be seen that by increasing the record length, the frequency resolution is increasing (see also Exercise 18).

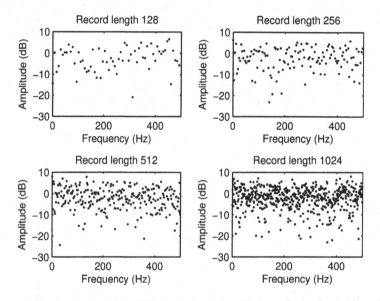

Figure 2-17   Comparison of the DFT spectrum of 4 realizations of a random white noise sequence with increasing length.

Exercise 30 (Smoothing the amplitude spectrum of a random excitation)   Generate n successive realizations $u^{[r]}(k)$, $r = 1, ..., n$ of the white random noise sequence. Average over the different DFT spectra (scale with $1/\sqrt{N}$):

$$\hat{S}_{UU}(\omega_k) = \frac{1}{M}\sum_{r=1}^{M} \bar{U}^{[r]}(k)U^{[r]}(k) = \frac{1}{n}\sum_{r=1}^{n} |U(k)|^2. \qquad (2\text{-}14)$$

Plot the results for $M = 1, 16, 64$. Also add the histogram of $|U(k)|$, $k = 1, ..., N/2$.   □

Observations   Note that averaging over different realizations smooths the spectrum (see Figure 2-18). The peaks and drops disappear and the histogram becomes concentrated around its mean value. What is the underlying distribution of $|U(k)|$?
From the central limit theorem, it follows that under loose conditions, the real and imaginary part of $U(k)$ are zero mean normally distributed. The power spectrum $\hat{S}_{UU}(\omega_k)$ consists of the scaled sum of $2M$ squared normally distributed variables, resulting in a χ^2-distribution with $2M$ degrees of freedom. Such a distribution has a mean value of M and a standard deviation of $\sqrt{2M}$. This shows immediately that the standard deviation of the power spectrum drops as $1/\sqrt{M}$ due to the scaling with $1/M$ in (2-14).

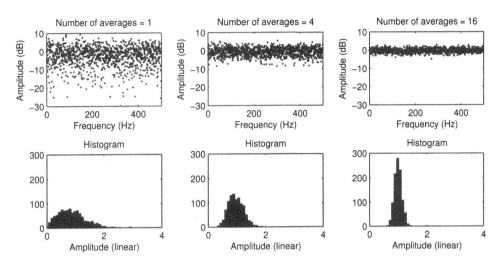

Figure 2-18   Comparison of the DFT spectrum for an increasing number of averages.

Conclusion - What you learned in these exercises Power spectrum of random noise excitations:

- The DFT-amplitude spectrum of a random noise sequence is a random sequence with spikes and dips (of 20 dB or more).

- Making the sequence longer does not reduce the spiky behavior.

- Averaging over multiple realizations allows to average the power spectrum $\hat{S}_{UU}(\omega_k)$.

- The measured power spectrum is χ^2-distributed. For a large number of averages it tends to a normal distribution with mean value $S_{UU}(\omega_k)$ and variance $2S_{UU}(\omega_k)/M$.

- More information on the use and processing of random excitation signals can be found in Bendat and Piersol (1980).

Exercise 31 (Generation of random noise excitations with a user-imposed power spectrum)   Generate a filtered noise sequence with a length $N = 2048$ using the following instructions:

- Noise filter
  ```
  [b,a] = butter(6,CutOff*2), CutOff = 0.1
  ```

- Generate M realizations of a Gaussian white noise sequence with length N each:
  ```
  u = randn(N,M);
  ```
  Choose $M = 1, 16, 64$.

- Generate the filtered signals:
  ```
  u = filter(b,a,u);
  ```

- Plot the amplitude of the DFT spectrum (scale with $1/\sqrt{N}$) and compare it to the amplitude of the filter transfer function.

- Repeat the exercise using a Hanning window (see Exercise 22.a).

Observations The behavior of the amplitude spectrum is similar to that of the previous exercise. But this time the power spectrum is shaped by the filter characteristic (see Figure 2-19). Note that for the rectangular window, the averaged power spectrum is not following the filter characteristic any more from -40 dB on. This is due to the leakage effect. Replacing the

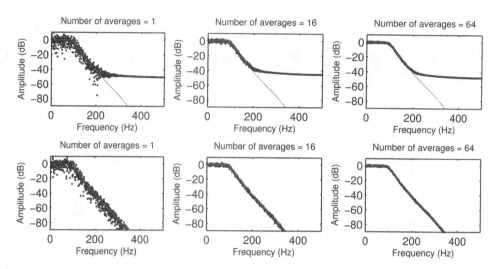

Figure 2-19 Comparison of the DFT spectrum of filtered noise sequences for a different number of averages. Top figures: Rectangular window. Bottom figures: Hanning window. Black dots: The actual realization of the spectrum. Gray line: The filter characteristic.

rectangular window by the Hanning window pushes this effect to a much lower level (no longer visible on this figure).

Exercise 32 (Amplitude distribution of filtered noise)   Generate a white binary sequence with a length of $N = 100,000$ and pass it through a filter with a long impulse response. Plot the histogram of the input and the filtered output.
Hint: Use the following instructions to generate the filter coefficients:

```
CutOff = 0.1        % Cut off frequency, relatively to fSample
[b,a] = butter(6,CutOff*2); % filter                          □
```

Observations Although the input amplitude distribution is a binary sequence, it can be seen that the output is almost normally distributed (see Figure 2-20). This is a direct consequence of the central limit theorem that states, loosely, that the sum of a large number of independent random variables tends to a normal distribution. A discrete time filter calculates the output by making a convolution between the input and the impulse response:

$$y(t) = \sum_{n=0}^{\infty} u(n)h(t-n).$$ (2-15)

So the output at time t consists of the sum of a large number of weighted input samples. Due to the central limit theorem, the output tends to be normally distributed, as is seen in Figure 2-20. It is difficult to generate a filtered noise sequence that is not approximately normally distributed.

2.6.2 What have we learned in Section 2.6? Further reading

Random noise excitations should be processed with care. The FFT of a random noise sequence does not converge towards a limit value for a growing length of the excitation. Well-

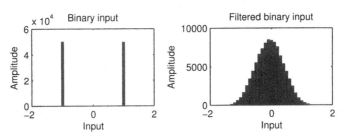

Figure 2-20 Histogram of a binary input, and the corresponding filtered output.

designed averaging techniques are needed to smooth the estimated power spectrum. At least a few realizations of the signal are needed. In practice this is done by breaking the original data record in a number of sub records. Even the availability of only a few sub records improves the smoothness already significantly (Bendat and Piersol, 1980; Pintelon and Schoukens, 2001). We also learned that it is possible to shape the power spectrum of random noise using a shaping filter. This is a simple operation, but it drives the amplitude distribution to become Gaussian. To the best of the authors knowledge, there does not exist a (simple) procedure to generate noise with a user defined power spectrum and amplitude distribution.

2.7 DIFFERENTIATION, INTEGRATION, AVERAGING, AND FILTERING OF PERIODIC SIGNALS

In some applications, the user needs to integrate or differentiate the measured signals. The aim of this exercise is to illustrate how the periodic nature of a signal can be used to calculate, for example, its derivative, and to improve the SNR by averaging over repeated periods and eliminating non excited frequencies.

Exercise 33 (Exploiting the periodic nature of signals: Differentiation, integration, averaging, and filtering) Generate a periodically repeated sawtooth function using the following MATLAB® instructions:

```
u0 = [0:N/2-1 N/2:-1:1];
    % create a sawtooth
u0 = u0-mean(u0); u0 = u0/std(u0);
    % eliminate the mean value and normalize the signal
u0 = kron(ones(1,M+1),u0);
    % repeat the signal over M + 1 periods, choose M = 10
[b,a] = butter(2,0.3); u0 = filter(b,a,u0); u0(1:N) = [];
    % reduce the high frequency components and eliminate the first period
u = u0+SNR*randn(size(u0));
    % add noise to the signal, put SNR = 0.001
```

■ Calculate the derivative of $u(t)$ and $u_0(t)$ by transforming the signal to the frequency domain and multiplying the spectra with $j\omega$. Take care of the frequencies above half the sampling frequency (see also Exercise 19.b)! Plot the signals together with the derivatives.

■ Reduce the influence of the noise by averaging (in the time domain) over the M measured periods. Plot the noise signal.

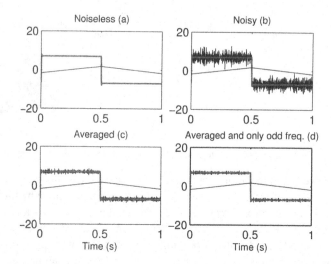

Figure 2-21   Calculation of the derivative of a periodic function using FFT techniques.
(a) The noiseless signal (black) and its derivative (gray). (b) Shows also the
measured derivative. (c) The measured derivative, averaged over $M = 10$
periods. (d) As (c) but the even frequencies put equal to zero.

- Verify the spectrum of $u_0(t)$ and observe that only the odd harmonics are present.
  This will not be changed by passing the signal through a linear system (e.g. integrat-
  ing, differentiating, ...). Hence we can put the even frequencies also equal to zero.
  Plot this "filtered" noise signal.

Observations (see Figure 2-21)

- The figure of the noiseless data shows the expected behavior. The derivative of a
  sawtooth is a staircase function. Notice that around the "discontinuity" a ringing
  phenomenon can be seen. This is due to the "Gibbs" phenomenon: A sum of sines
  cannot produce a discontinuous function.

- The derivative of the noise signal looks much noisier than the original signal. This is
  because we added white noise, and the differentiation action amplified the high frequency noise. It is possible to reduce the noise amplification by reducing the frequency band over which the derivative is calculated. A compromise between
  systematic errors (a part of the signal is removed) and noise sensitivity should be
  made.

- Averaging over M periods reduces the noise by $\sqrt{M}$ (*Figure 2-21, c*). Eliminating
  all even frequencies (one line in two) reduces the noise once more with a factor
  $\sqrt{2}$ (why?), (see *Figure 2-21 d*).

3

FRF Measurements

What you will learn: This chapter shows the reader how to measure the frequency response function (FRF) of a linear dynamic system with single or multiple inputs. The following topics are addressed:

- Direct impulse response function measurement (see Exercise 34).
- The response of a linear system on a sine excitation: transient and steady-state response (see Exercise 35).
- FRF measurements using broadband excitations using multisines (see Exercise 36) and random noise excitations (see Exercise 37).
- Understanding and reducing the impact of leakage on FRF measurements (see Exercises 38, 39, 40, 51, and 52).
- Dealing with process (output) noise in FRF measurements (see Exercises 42, 43, 44, 45, and 56).
- Dealing with disturbing noise on the input measurements in FRF measurements (see Exercises 46, 47, and 48).
- FRF measurements under feedback conditions (see Exercises 49, and 50).
- Measuring the FRF matrix of a multiple-input-multiple-output system (see Exercises 53, 54, 55, and 57).

3.1 INTRODUCTION

The aim of this chapter is to study how the nonparametric frequency response function (FRF) of a linear dynamic system can be measured. Making this nonparametric intermediate step in the system identification procedure offers multiple advantages. The user gets a good idea of the behavior of the system dynamics, and sees at the same time that it is also possible to measure the power spectrum of the disturbing noise. This can simplify significantly the initialization of the parametric identification step. Moreover, the user will get a good idea about the quality of the measurements in an early phase of the measurement campaign so that it is even possible to adapt the experiments to improve the result or to remove measurement problems.

In this chapter we consider first the perfect situation, where no disturbing noise nor nonlinear distortions are present. This allows us to study the intrinsic properties/problems of the basic algorithms, using impulse excitations, and periodic and random excitations. Next we study the impact of a disturbing process or measurement noise and learn how to deal with it. We will show the reader also to deal with disturbing noise on the input measurements. The impact of nonlinear distortions on FRF measurements is out of the scope of this chapter, it studied in detail in Chapters 5 and 6. Finally we illustrate how to measure the FRF of a system that is captured in a feedback loop. In this chapter it will not only be shown how the FRF can be extracted from the measurements, it will also be shown how a nonparametric noise analysis can be made so that uncertainty bounds can be given around the measurement results. Most of the exercises in this chapter are focused on single-input single-output (SISO) systems, but at the end of the chapter we also give a brief introduction to the problems related to the measurement of the frequency response matrix (FRM) of multiple-input multiple output (MIMO) systems.

3.2 DEFINITION OF THE FRF

Consider the linear time-invariant dynamic system, characterized by its impulse response $g_0(t)$ (Figure 3-1).

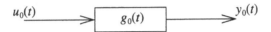

Figure 3-1 A linear time-invariant dynamic system with impulse response $g_0(t)$, excited by the input signal $u_0(t)$ and output $y_0(t)$.

In the continuous time domain, the system response is described by

$$y_0(t) = \int_{-\infty}^{\infty} g_0(t-\tau)u_0(\tau) \ d\tau. \tag{3-1}$$

For discrete time systems, the relation is

$$y_0(t) = \sum_{k=-\infty}^{\infty} g_0(t-k)u_0(k). \tag{3-2}$$

In the frequency domain, the relation for continuous and discrete time systems is given by

$$Y_0(\Omega) = G_0(\Omega)U_0(\Omega), \tag{3-3}$$

with U_0, Y_0, G_0 the continuous or discrete Fourier transform of u_0, y_0, g_0, respectively, $\Omega = j\omega$ for continuous time systems, and for discrete time systems $\Omega = e^{j\omega/\omega_s}$ where $\omega_s = 2\pi f_s$ with f_s the sample frequency. $G_0(\Omega)$ is the transfer function of the system. A nonparametric measurement of the transfer function at a discrete set of frequencies $G_0(\Omega_k)$, $k = 1, ..., F$ is called the frequency response function (FRF) of the system. In this chapter we will learn how to extract g_0, G_0 from measured data. For simplicity, we will make all simulations in the discrete time domain, but all methods can be generalized without any problem to measurements of continuous time systems.

Further reading: There are many books on system theory available. Amongst others, an introduction to continuous and discrete linear time invariant systems can be found in the book of Oppenheim *et al.* (1997).

3.3 FRF MEASUREMENTS WITHOUT DISTURBING NOISE

In this section we study the measurement of the FRF without considering the influence of disturbing processes or measurement noise. This allows us to focus on the errors that are induced due to the fact that the system is measured during a finite time. First we will study the behavior of the dynamic system in the time domain. A direct measurement of the impulse response will be made, and the transient and steady-state behavior of the system will be introduced. Next we will turn our attention to the frequency domain and study the popular methods for the measurement of FRFs.

3.3.1 Direct measurement of the impulse response

In this exercise we make a direct measurement of the impulse response. The input signal is an impulse ($u(t) = \delta(t)$ with $\delta(t = 0) = 1$, and zero elsewhere), the output is by definition the impulse response: $g_0(t) = G_0(\delta(t))$. The major advantage of this method is its simplicity, it is often used in practice for this reason. In many mechanical applications, a direct impulse response measurement can be made by a simple hammer excitation. The major disadvantage of this method is twofold: (i) a large peak value of the impulse is needed to get a sufficiently high signal-to-noise-ratio, the method can only be applied under low noise measurement conditions. (ii) Too large peak values often drive the system into a nonlinear operation mode.

Exercise 34 (Impulse response function measurements)   Create a second order discrete time system with a resonance of 10 dB using, for example, the MATLAB® instruction [b,a] = cheby1(2,10,2*fc/fs) with $f_s = 256$ Hz the sample frequency, and f_c the cut off frequency. Select $f_c/f_s = 0.1$. Excite this system with an impulse of amplitude 1, and measure its input and output in $N = 128$ points.

- Plot the input/output measurements in the time domain, and calculate the corresponding FFT spectrum. Normalize the FFTs by dividing by $\sqrt{N}$.

- Plot the time and frequency domain signals.

- Calculate the FRF of $G(z)$ directly from the transfer function $G(z) = \sum b_k z^{-k} / \sum a_k z^{-k}$ using the MATLAB® instruction freqz(b,a,ω), with $\omega = 2\pi f/f_s$.

- Plot the FRF and the output spectrum on one plot. Determine the scaling factor of the output spectrum such that it fits with the FRF. Compare the scaling factor to the amplitude of the input spectrum.

Observations (see Figure 3-2) The impulse response does not start before the impulse is applied. This is the defining characteristic of causal systems. The impulse response oscillates with a frequency that corresponds to the resonance frequency of the system, and it decreases exponentially toward zero (verify this by making a log plot). The FFT of the output shows the FRF of the transfer function of the discrete system, sampled at the frequencies $kf_s/(N)$, with f_s normalized to 256 Hz. The FFT results are compared to the FRF that is calculated directly from the transfer function, and a perfect agreement is found.

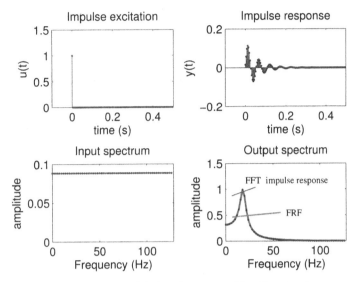

Figure 3-2   Direct measurement of the impulse response. Top: Time domain measurements. Bottom: Frequency domain measurements.

3.3.2 Transient and steady-state response for periodic excitations

Instead of applying an impulse to probe the system at all frequencies simultaneously, it is also possible to measure the FRF frequency per frequency using a sine excitation. The major advantage is that such measurements will have a much higher SNR, but at a cost of a longer measurement time, and this for two reasons. It is not only necessary to measure at least one period at each probing frequency, it is also required to wait till the transients have disappeared, and this is done each time that the frequency is changed. The transient effect is studied in this section.

Exercise 35 (Study of the sine response of a linear system: transients and steady-state)   Consider the discrete time system of the previous exercise and excite it at 16 Hz with a cosine with amplitude 1. Select the sample frequency $f_s = 256$ Hz. Measure $N = 128$ points of the input/output signals which corresponds to 8 periods of the excitation. Use the last measured period to estimate the steady-state response (all transients are assumed to be negligible in this interval), and subtract it from the rest of the record to make the transient visible.

- Plot the output and the estimated transient in the interval $[0, 0.25]$ s.

- Plot also the input and output spectrum. Use a scale factor of $1/N$ in the FFTs.

- Study the impact of the length of the measurement on the estimates of the steady-state and transient response. How can you verify that the record was long enough to reach the steady-state ?

□

Observations (see Figure 3-3) From the left figure it is seen that the transient decays exponentially. It can be concluded that the steady-state is reached, only when the transient is sufficiently small at the end of the measurement interval. In the right figure, the spectrum of the input (a pure cosine, no leakage) appears as an impulse. At the output, the skirts which are due to the transients are clearly visible.

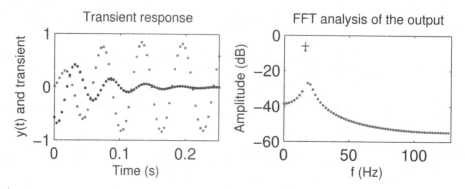

Figure 3-3   Left: Output (gray) and transient response (black). Right: The input (black +) and output (gray dots) FFT.

Exercise 36 (Study of a multisine response of a linear system: transients and steady-state)   In this exercise we repeat the previous exercise but this time for a multisine excitation. As was illustrated in Chapter 2, a multisine excites all frequencies simultaneously (similar to an impulse), but the phases are chosen to avoid the impulsive nature so that a more persistent excitation of the system is obtained; it is "continuously" excited during the whole experiment. The major advantage of these signals is that the same power is injected into the system for a much lower peak value of the excitation.

Create a multisine with a period of $NPeriod = 128$ points, exciting the system up to 100 Hz, with frequency components having an equal amplitude and random phase (see Chapter 2, Exercise 20). Normalize the peak value of the signal to 1 so that it is possible to compare the results with those obtained with an impulse excitation. Calculate the rms value of this signal and compare it to that of an impulse. Apply it to the system of Exercise 34, and measure the input and output over $M = 3$ periods of the excitation signal. Estimate the steady-state , and next the transient. Verify if the steady-state estimate is reliable.

Make the following series of plots on a first figure:

■ the measured output,

■ the estimated transient,

■ the estimated transient (absolute value) on a logarithmic amplitude scale.

Calculate the spectrum of the input signal (1 period). Segment the output in 3 sections, containing each exactly one period (see also Chapter 2, Exercise 15 to select correctly the begin and end point of the data for an FFT analysis). Calculate for each segment the FRF by making a division of the output spectrum by the input spectrum. Make a plot of:

■ the amplitude spectrum of one period of the input (scale the FFT with $\sqrt{NPeriod}$),

■ the FRF at each segment, together with the exact FRF.

□

Discussion (see Figure 3-4) The response of the system on three successive periods of the input is drawn at the left of this figure. Notice that it is hard to see deviations from period-

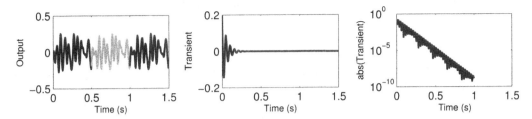

Figure 3-4   Top: 3 periods of the multisine response of the system. Middle: Transient. Bottom: Amplitude of the transient on a logarithmic amplitude scale.

icity in this figure. In the middle figure, the transient is made visible by subtracting the third output period from the first and second period. From the logarithmic plot (right plot), it can be clearly seen that the transient decays exponentially, the local successive maxima are on a straight line. Note also that the estimated transient amplitude is below 1e-10 at the start of the third period that was used to calculate the steady-state response. This shows that the transient error on the steady-state estimate will be smaller than this value. In Figure 3-5, the flat amplitude spectrum of the multisine excitation can be seen, all frequencies are equally excited as was the case for an impulse. Observe that the amplitude of the spectrum is almost 5 times larger than what was obtained with the impulse excitation in Section 34 for the same settings (see Figure 3-2). This will result in a much higher SNR in the presence of disturbing noise. In the next three subplots, it can be seen that the errors on the FRF decrease fast. This is because the impact of the transient that creates these errors decreases exponentially. Figure 3-5 b–d shows the measured FRF and the error. Observe that the error drops fast because the transient drops exponentially with a growing period number.

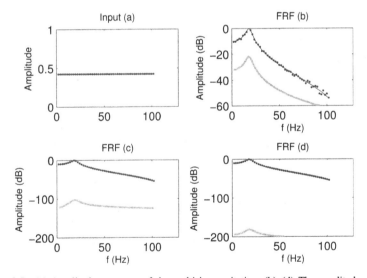

Figure 3-5   (a) Amplitude spectrum of the multisine excitation. (b)–(d) The amplitude of the FRF estimate for the first, second, and third period. Thin black line: G_0. Dots: Measured FRF. Gray line: $|G - G_0|$.

3.3.3 FRF measurements using random excitations

Noise is a very popular excitation signal to measure the FRF. The basic reason for this popularity is the apparent ease of generation: a random sequence is passed through a digital filter to select the frequency band of interest. The drawback of this signal is that it is not periodic, and hence the FRF measurements will suffer from leakage. It is impossible to reach a periodic steady-state response like in Exercise 36, because the excitation is not periodic. To reduce the leakage effects, the data are divided in M blocks (or subrecords) and dedicated averaging methods are used to combine all these data in a single estimate with a lower noise sensitivity.

Exercise 37 (FRF measurement using a noise excitation and a rectangular window) Generate a random noise sequence $r(t)$ with length $128 \times M$, with M the number of realizations. Filter this sequence by a generator filter with bandwidth $0.3 f_s$ and $f_s = 128$ Hz, using a digital filter `[b,a] = butter(2,2*0.3)` and apply the resulting signal $u(t)$ to the system in Section 34. Measure the output $y(t)$.

Segment the input/output record in M subrecords, calculate the FFT on each of these segments (is the normalization of the FFTs of importance?), and calculate the FRF at frequency ω_r as

$$\hat{G}(j\omega_r) = \frac{\frac{1}{M}\sum_{s=1}^{M} Y^{[s]}(r)\overline{U}^{[s]}(r)}{\frac{1}{M}\sum_{s=1}^{M} U^{[s]}(r)\overline{U}^{[s]}(r)} = \frac{\hat{S}_{YU}(\omega_r)}{\hat{S}_{UU}(\omega_r)}, \tag{3-4}$$

with $X^{[s]}$ the FFT spectrum of segment s. Repeat this exercise for $M = 1, 4, 16, 256$. Plot $|\hat{G}|$ and $\hat{S}_{UU}$ (scale the FFTs with $\sqrt{128}$).

Observations In Figure 3-6, the measured FRF is shown, together with the exact value G_0 and the error $|G - G_0|$. Note that in this exercise there is no disturbing noise added, the errors are completely due to leakage. It can be clearly seen that initially the error drops fast with growing M. For larger values of M, the error reduction becomes proportional to $1/\sqrt{M}$ which means that the error disappears slowly. For large values (in this case $M > 256$), it can be seen that at the resonance frequency the error no longer decreases and a systematic error becomes visible. To probe the relationship between M and the error, the realized amplitude spectrum of the excitation $\hat{S}_{UU}$ should be plotted (see Figure 3-7). In this figure it is clearly visible that for a small number of realizations, $\hat{S}_{UU}$ has many frequencies with low power (dips). At those lines the measurement is extremely sensitive to the leakage errors. For larger values of M, these dips disappear. At the same time the errors are also smoothed in the averaging process.

Exercise 38 (Revealing the nature of the leakage effect in FRF measurements)
In this exercise it will be illustrated that the leakage effect has intrinsically a highly structured nature although it looks like noise in Figure 3-6. Consider N measured input and output data points, sampled from a system G_0 that is excited with noise. It can be shown that the relation between the FFTs U_0, and Y_0 of these records is

$$Y_0 = G_0 U_0 + T, \tag{3-5}$$

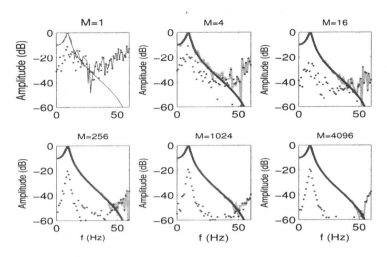

Figure 3-6 FRF measurement using a noise excitation and a rectangular window, averaged over M realizations. Black line: Exact FRF. Gray line: Measured FRF. Dots: The magnitude of the complex error $|G - G_0|$.

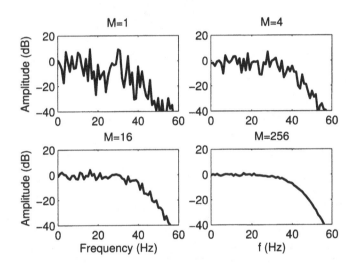

Figure 3-7 Power spectrum of the random noise excitation, averaged over M realizations.

with T a transient term that has the same mathematical structure as the system transfer function G_0. The measured FRF $\hat{G}$ is then

$$\hat{G} = \frac{Y_0}{U_0} = G_0 + \frac{T}{U_0}. \tag{3-6}$$

The leakage effect is nothing other than T/U_0, which has a random nature because U_0 is a complex normally distributed random signal. The aim of this exercise is to illustrate this transient behavior of leakage.

Create a second-order discrete time system with a resonance of 10 dB using, for example, the MATLAB® instruction [b,a] = cheby1(2,10,2*fc/fs) with f_s = 256 Hz the sample frequency, and f_c the cutoff frequency. Select f_c/f_s = 0.1. Excite this system with a white Gaussian random noise excitation $u_0(t)$ with a length $N + N_{\text{Trans}}$, with $N = 10{,}000$ and $N_{Trans} = 1000$. Measure its input and output. Remove the first N_{Trans} points to eliminate the effect of the initial conditions.

■ Process the remaining data in blocks of N_{Block}, with $N_{\text{Block}} = [125\ 250\ 500]$ in the time domain. Then transform the data to the frequency domain (block by block), and calculate for each block the output starting from the input using the relation:

$$\text{yMod = real(ifft(G.*U))} \tag{3-7}$$

(see also Chapter 2), with G the FRF of G_0 calculated at the FFT frequencies:

$$\text{G = freqz(b,a,2*pi*f/fs)} \tag{3-8}$$

■ Calculate for each block the error between the modeled output and the actual output of the system and plot these for the different block lengths.

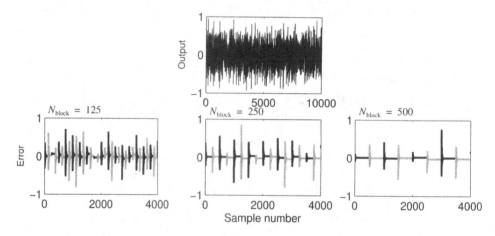

Figure 3-8 Illustration of the transient nature of the leakage effect. The output of a dynamic system is calculated block per block using the FFT IFFT and the FRF of the system. Observe that at the beginning of each block a transient that is due to the excitation of the previous block is appearing. This is disturbing the FRF measurement in two directions: This transient is missing in the analysis of the previous block, and it is disturbing the analysis of the next block.

Discussion (see Figure 3-8) From the error plot in this figure, the transient nature of the error becomes clearly visible. This transient is the response of the system to the input of the previous block (for example block number $k - 1$) and acts as a disturbance for the FRF measurement in the actual block (block number k). Moreover, the transient that is connected to block k is not measured in block k but is disturbing block $k + 1$. This shows that in a single block two errors are combined: the additional transient of the previous block, and the missing transient that is in the next block.

Observe that the level of the transients does not depend upon the block length. The longer the block, the smaller the relative power of the transients compared to that of the total signal in the block. This shows that the variance (power) of the error will drop in $1/N_{\text{Block}}$.

In the next exercise, an alternative will be proposed to reduce the impact of the transients. From this exercise it is seen that the errors are due to problems with the analysis at the borders of the blocks. By replacing the rectangular window (no weighting of the time domain data in the measured window) in this exercise by a window that goes to zero at the ends, it seems possible to reduce the errors. This is the basic idea behind windowing methods, for example a Hanning window, or a diff window (see Exercise 2, 22.a). A more advanced method that exploits explicitly the smooth behaviour of the transfer function and the transient (= leakage) is explained in Section 3.7.

Exercise 39 (FRF measurement using a noise excitation and a Hanning window)

In the previous exercise it turns out that leakage disturbs the FRF measurements. The impact of leakage can be reduced by multiplying the data with a window in the time domain. The basic idea is to use a weighting of the data that goes to zero at the beginning and the end of the time window. A very popular window is the Hanning window:

$$w_{\text{Hann}}(t) = 0.5(1 - \cos t2\pi/N). \tag{3-9}$$

where $t = 0, 1, ..., N - 1$. This window is multiplied with the signal for which the FFT is calculated: $X_{Hann}(k) = \text{FFT}(x(t)w(t))$.

- Repeat Exercise 37, using the Hanning window for $N = 1024$ and $M = 4, 16, 256, 1024, 4096$. Use the MATLAB® instruction: `hanning(N,'periodic')`.

- Compare next the measured FRF for the rectangular window (see Exercise 37) and the Hanning window, averaged over $M = 128$ realizations, and a block length of $N = 128, 256, 512, 1024, 2048, 4096$.

□

Observations For the Hanning window, the same observations as in the previous exercise can be made, but this time the errors are significantly smaller at most frequencies (Figure 3-9). For a low number of averages ($M = 1$) the stochastic leakage errors dominate. By increasing the number of averages, the stochastic errors are averaged to zero and the remaining errors are dominated by the bias errors of the leakage effect. It can be seen that both errors are smaller for the Hanning window than for the rectangular window. Note that the systematic errors are dominant around the resonance frequency. The level of these errors strongly depends on the length of the window.

This is shown in Figure 3-10. Here the evolution of the errors is shown for an increasing record length for the Hanning and the rectangular window. It can be seen that the errors of the Hanning window are again well below those of the rectangular window, and the size drops faster for growing N for the Hanning window than it does for the rectangular window (Schoukens et al., 2006) and Table 3-1.

TABLE 3-1 Comparison of the rectangular, Hanning, and Diff window

Window	Systematic Error $(M \to \infty)$	Variance
$w_{\text{Rect}}(k) = 1$	$O(N^{-1})$	$O(M^{-1}N^{-1})$
$w_{\text{Hann}}(k) = 0.5(1 - \cos k2\pi/N)$	$O(N^{-2})$	$O(M^{-1}N^{-2})$
$w_{\text{Diff}}(k) = 1 - e^{j\frac{2\pi}{N}k}$	$O(N^{-2})$	$O(M^{-1}N^{-2})$

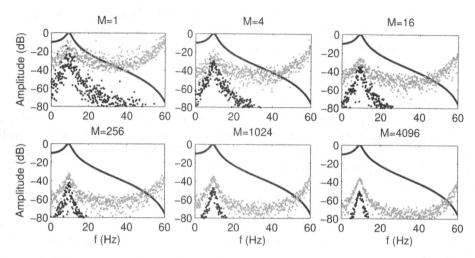

Figure 3-9 FRF measurement using a noise excitation and a Hanning window, averaged over M realizations. Black line: G_0 Gray dots and black dots: Amplitude complex errors of the rectangular and Hanning window, respectively.

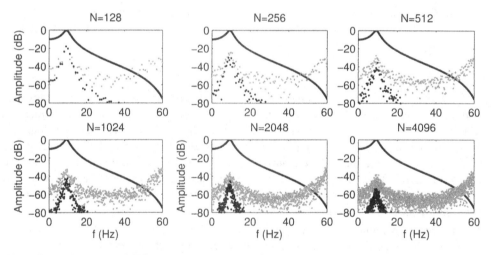

Figure 3-10 Comparison of FRF measurements with the rectangular and Hanning window for $M = 128$ realizations, and varying length N of the sub-blocks. Full line: G_0 Gray dots and black dots: Amplitude complex errors of the rectangular and Hanning window, respectively.

Exercise 40 (FRF measurement using a noise excitation and a diff window) In the previous exercise it was shown that the leakage errors can be reduced by multiplying the data with a window in the time domain. This can also be interpreted as filter operation (convolution) of the data in the frequency domain. The simplest filter is the diff operation (diff(X) in MATLAB®):

$$X_{\text{diff}}(k + 0.5) = X(k + 1) - X(k), \tag{3-10}$$

where the index $k + 0.5$ denotes symbolically that the corresponding frequency is the middle frequency of bin k and $k + 1$. For the Hanning window the filter operation corresponds within a factor -0.25 to a double difference $X_{\text{difdif}}(k) = -2X(k) + X(k+1) + X(k-1)$ so that $X_{\text{Hann}}(k) = -0.25(-2X(k) + X(k+1) + X(k-1))$.

- Repeat Exercise 37, using the diff window. Apply the window this time in the frequency domain using the 'diff' operation. Beware of the frequency shift of half a bin when comparing the result with the exact value of the FRF.

- Compare next the measured FRF for the rectangular window (Exercise 37) and the diff window, averaged over $M = 128$ realizations, and a block length of $N = 128, 512, 2048, 8192$.

□

Observations For the diff window, the same observations as in the previous exercise can be made (Figure 3-11). Note that the systematic errors are again dominating around the resonance frequency, and averaging does not significantly reduce these errors. The level of the errors strongly depends on the length of the window.

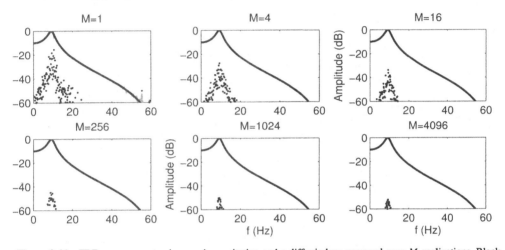

Figure 3-11 FRF measurement using a noise excitation and a diff window, averaged over M realizations. Black line: Exact FRF. Gray line: Measured FRF. Dots: The magnitude of the complex error $|G - G_0|$.

This is shown in Figure 3-12. Here the evolution of the errors is shown for an increasing record length for the diff and the rectangular window. It can be seen that the errors of the diff window are well below those of the rectangular window, and the size drops faster for growing N for the diff window than it does for the rectangular window (see also Table 3-1 and the paper (Schoukens *et al.*, 2006).

Exercise 41 (FRF measurements using a burst excitation)   In all the previous exercises, the only source of errors was the leakage effect because no disturbing noise was added. The leakage errors were due to the fact that it was not possible to measure in steady-state an integer number of periods, because the excitation was nonperiodic. An alternative is to use a burst excitation, that starts at the beginning of the measurement window, and falls back to zero well before the end of the window. This makes it possible to include the transient effects completely in the window (the begin and end transients), thus eliminating all the leakage effects. This is illustrated in the next exercise.

Consider a second-order discrete time system with a resonance of 20 dB, using for example the MATLAB® instruction [b,a] = cheby1(2,20,2*fc/fs) with $f_c/f_s = 0.1$, put $f_s = 256$ Hz. Excite this system with a burst white random sequence (rms

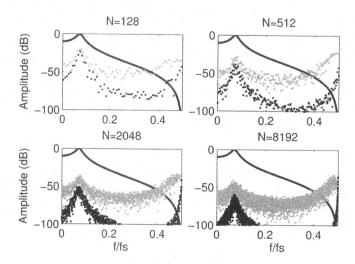

Figure 3-12 Comparison of FRF measurements with the rectangular and diff window for $M = 128$ realizations, and varying length N of the sub-blocks. Full line: G_0. Gray dots and black dots: Amplitude complex errors of the rectangular and diff window, respectively.

value of 1) that excites the system in $t = 0, 1, ..., N_{burst} - 1$, with $N_{burst} = 64, 128, 192$. Measure $N = 256$ samples of the input and output in an interval of 1 s. Calculate the FRF by simple division of the output/input FFTs (has the scaling of the FFTs any importance?). Make the following plots:

- The measured input and output on top of each other,
- The measured FRF, compared to the exact value.

□

Observations (Figure3-13) In this figure, the output transient that remains, once the excitation is stopped, is shown. Notice that the error on the FRF increases if a larger part of this end transient is missing. This illustrates that leakage effects are essentially a problem of unmeasured transients.

3.3.4 What have we learned in Section 3.3? Further reading

The response of a linear time invariant system consists of two parts: the transient and the steady-state response. The transients decay exponentially for systems described by ordinary differential equations (e.g., if no diffusion processes are present).

A periodic input results in a periodic output once the transient decayed. In many measurement applications the users wait till the transients are gone to start the measurement of the FRF making explicitly use of the periodicity. The major reason to do so it to avoid leakage errors; however, this is done at a cost of lost measurement time. Another advantage of measuring under steady-state conditions is that for well-designed periodic excitation signals it becomes possible to extract from a single measurement a nonparametric noise model (see for example Section 3.5), and also a nonlinear distortion analysis can be made (see Chapter 5, Exercise 81).

For random excitations it is much more difficult to avoid leakage. The classic solution to reduce the problem is to average the results over multiple realizations of the input and to

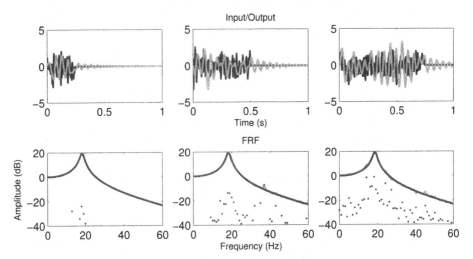

Figure 3-13 Measurement of the FRF with a burst random excitation for different burst lengths of 64, 128, 192, respectively. Top: The input (black) and output (gray). Bottom: The FRF, exact (black), measured (gray), error (dots).

apply a window before the FFTs are calculated. In Section 3.7 we will introduce a very recently developed method, called the local polynomial method. At a cost of an increased computing time, it is possible to reduce the leakage error with several order of magnitude.

The reader can find more information on the classical methods to measure the FRF using random excitation in the book of Bendat and Piersol (1980) that focuses on frequency domain methods. Alternatively the impulse response can by measured in the time domain using correlation methods, as is explained in Godfrey (1969, 1980). More information on the use of windowing methods and a detailed theoretical analysis is available in Antoni and Schoukens (2007, 2009) and the references therein. In Schoukens *et al.* (2000) it is shown that replacing a sine excitation by a multifrequency excitation like a multisine can reduce the measurement time significantly, especialy for high SNR measurements.

3.4 FRF MEASUREMENTS IN THE PRESENCE OF DISTURBING OUTPUT NOISE

In this section we repeat the study of nonparametric FRF measurements, but this time we consider the impact of process (output) noise on the measurements.

3.4.1 Impulse response function measurements in the presence of output noise

The major advantage of a direct impulse response function measurement is its simplicity; it is often used in practice for this reason. The major disadvantage is the noise sensitivity, the technique can only be applied in experimental conditions with a high signal-to-noise-ratio. This will be clearly illustrated in the next series of exercises. First the impulse response is measured directly, while in the next section more advanced excitation signals are used again. From these results it will become clear that the SNR increases significantly when the impulse excitation is replaced by an excitation that is applied during the whole experiment.

Exercise 42 (Impulse response function measurements in the presence of output noise)   In this exercise we repeat Exercise 34, where the FRF of a linear dynamic system is obtained by a direct measurement of the impulse response.

Create a second-order discrete time system with a resonance of 10 dB using, for example, the MATLAB® instruction `[b,a] = cheby1(2,10,2*fc/fs)` with $f_s = 256\,\mathrm{Hz}$ the sample frequency, and f_c the cutoff frequency. Select $f_c/f_s = 0.1$. Excite this system with an impulse of amplitude 1, and measure its input and output in $N = 128$ points. Disturb the output with white random Gaussian noise $N(0, \sigma^2 = 0.02^2)$. Next repeat this experiment 100 times and average the results. Calculate also the standard deviation of the measured FRF as a function of the frequency.

- Plot the input/output measurements in time domain, and calculate the corresponding FFT spectrum. Normalize the FFTs by dividing it by the record length N.

- Plot the time and frequency domain signals.

- Calculate the FRF of $G(z)$ directly from the transfer function $G(z) = \sum b_k z^{-k} / \sum a_k z^{-k}$ using the MATLAB® instruction `freqz(b,a,ω)`, and $\omega = 2\pi f/f_s$, with ω evaluated at the FFT frequencies.

- Plot the time and frequency domain signals.

- Plot the FRF and the output spectrum on one plot. Determine the scaling factor of the output spectrum such that it fits as close as possible to the FRF. Compare the scaling factor to the amplitude of the input spectrum.

□

Observations (see Figure 3-14) Besides the remarks made before in Exercise 34, the following additional observations can be made. From the output time domain measurements (Figure 3-14), it can be seen that the impulse response fades exponentially to zero, which is not the case for the noise. Increasing the record length does not result in a proportional increase of information about the system because the noise dominates completely at the end of the window. After a while, only noise is added to the record which will deteriorate the SNR. The bad SNR results also in a very poor estimate of the FRF (Figure 3-14, bottom left). A possibility to improve the SNR is to repeat the experiment periodically, and to average the successive measurements in the time domain. At the bottom right of Figure 3-14 the same result is shown after $M = 100$ averages. Observe that the SNR increased with $\sqrt{M} = 10$ (20 dB). From the repeated measurements it is also possible to estimate the standard deviation of the noise as a function of the frequency.

3.4.2 FRF measurement in the presence of output noise using noise and multisine excitations

In this section we replace, again, the impulse excitation by noise or multisine excitations so that the system is excited throughout the whole experiment. This allows us to put more power into the system for the same maximum excitation levels. Two classes of excitation signals are used: random noise excitation and multisines. In order to minimize the leakage errors, a diff window is used for the random noise excitations (see Exercise 37 and Exercise 39). For the multisines, the golden rule of thumb is to measure an integer number of periods, to break this long record in subrecords, each containing a single period, and to calculate the FFT using a rectangular window (sometimes called "no window"). The FRF is estimated as

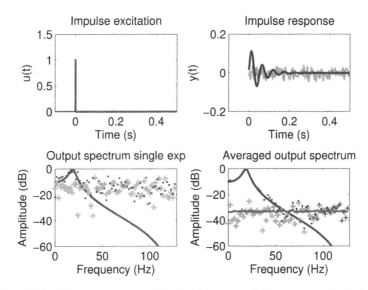

Figure 3-14 Direct measurement of the impulse response in the presence of output noise. Top: Time domain measurements. Black: The exact signal. Gray: the disturbing noise. Bottom: FFT spectrum of the output. Number of averages 1 (left), and 100 (right). Black line: Exact value (= FRF). Black dots: The measurements. Gray crosses: The error. Blue line: The standard deviation of the errors.

$$\hat{U}(k) = \frac{1}{M}\sum_{l=1}^{M} U^{[l]}(k), \ \hat{Y}(k) = \frac{1}{M}\sum_{l=1}^{M} Y^{[l]}(k), \text{ with } \hat{G}(k) = \frac{\hat{Y}(k)}{\hat{U}(k)}. \tag{3-11}$$

In these equations, $X^{[l]}$ is the FFT of the l th subrecord.

In the exercise the different behavior for random noise and multisine excitations is analyzed.

Exercise 43 (Measurement of the FRF using a random noise sequence and a random phase multisine in the presence of output noise) Generate a white Gaussian random noise sequence $r(t)$ (bandwidth is $0.5f_s$). Select the measurement length equal to $1024*M$ data points, with M the number of realizations. Choose $f_s = 128$ Hz. Scale the rms value to 0.33 to get a peak value that is comparable to the amplitude of the impulse that is used in Exercise 42. Apply these signals to the system of Exercise 42. Disturb the output with white random Gaussian noise $N(0, \sigma^2 = 0.02^2)$. Measure the output $y(t)$. Do the same for a random phase multisine (period length $N = 1024$) that excites the full frequency band, and consider M periods. Add to both signals $N_{\text{Trans}} = 1024$ points that will be used to eliminate the initial transients. Repeat the exercise for $M = 1, 4, 16, 64$.

- Plot the input and output signal plus the disturbing noise for the first 1024 input/output samples.

- Process the data, using a block length of 1024 data points so that exactly one period fits into a block. Calculate the FRF G using the appropriate method for each type of excitation signal. Use a diff window for the noise excitation, and a rectangular window for the multisine (see Exercises 36, 37, and 39 in this chapter). Is the scaling factor of the FFT's important?

■ Plot, as a function of the frequency, the exact FRF G_0, the complex error $G - G_0$ for the noise excitation measurements, and the error for the multisine excitation. Repeat this for all values of M.

□

Observations From Figure 3-15 it is seen that the system is continuously excited. The response does not fade away as was the case for impulse measurements. The collected information will increase with a growing data length (proportional to the square root of the number of samples), which is not the case for an impulse excitation where the output fades away. Note that the peak value of the excitation is similar to that of the impulse excitation of Exercise 42.

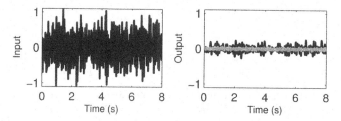

Figure 3-15   Measured input or excitation (left) and output or response (right). Black: Exact signal. Gray: Output noise.

This is also confirmed in Figure 3-16 where the errors of the FRF measurements are shown for both excitations for an increasing number of averages M. Notice that for $M = 1$ the noise excitation result is very poor compared to the multisine measurements. This is due to the fact that the amplitude spectrum of a single noise realization of finite length shows large dips resulting in a very poor SNR at those frequencies (see Exercise 37 and Figure 3-7). Averaging over an increasing number of realizations removes this problem. It can be seen from the figure that the errors for both excitations have a very similar behavior for $M \geq 16$. The average error level decreases as $1 / \sqrt{M}$.

Exercise 44 (Analysis of the noise errors on FRF measurements)   In this exercise, a more detailed analysis of the errors that were observed in Exercise 43 is made. The standard deviation will be calculated directly from the data (sample variance), and indirectly through a theoretical variance analysis.

Repeat Exercise 43, but fix the number of blocks to $M = 16$, and repeat the whole exercise $N_{repeat} = 10$ times. Estimate the FRF from each repeated set. Calculate the mean value and the standard deviation of the measured FRF over the 10 repetitions. Next calculate the theoretical variance of the FRF from a single realization (e.g., the last one) using the following formulas.

Noise excitation: Calculate first the coherence $\gamma^2(f)$ as a function of the frequency:

$$\gamma^2(f) = \frac{|\hat{S}_{YU}(f)|^2}{\hat{S}_{UU}(f)\hat{S}_{YY}(f)} \approx \frac{\left|\dfrac{1}{M}\sum_{k=1}^{M} Y^{[k]}(f)\bar{U}^{[k]}(f)\right|^2}{\left(\dfrac{1}{M}\sum_{k=1}^{M} U^{[k]}(f)\bar{U}^{[k]}(f)\right)\left(\dfrac{1}{M}\sum_{k=1}^{M} Y^{[k]}(f)\bar{Y}^{[k]}(f)\right)}, \tag{3-12}$$

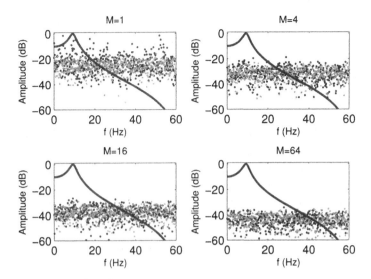

Figure 3-16 Evolution of
the error $|G - G_0|$ as a function of the number M of averaged blocks or periods for
a noise excitation (black dots) or a multisine (gray dots). Black line: exact FRF
$|G_0|$.

with $U^{[k]}(f)$, $Y^{[k]}(f)$ the windowed FFT spectra of block k. The variance estimate is then given by

$$\sigma_G^2(f) \approx |G_0(f)|^2 \frac{1 - \gamma^2(f)}{\gamma^2(f)}. \tag{3-13}$$

Replace in (3-13) the exact FRF $G_0(f)$ by its measured value.

Multisine excitation (no window is applied!): In this case the variance $\sigma_Y^2(f)$ is first estimated:

$$\hat{\sigma}_Y^2(f) = \frac{1}{M-1} \sum_{k=1}^{M} |Y^{[k]}(f) - \hat{Y}(f)|^2 \quad \text{with} \quad \hat{Y}(f) = \frac{1}{M} \sum_{k=1}^{M} Y^{[k]}(f). \tag{3-14}$$

Next the variance on the FRF averaged over M blocks is obtained as

$$\sigma_{G_{\text{averaged}}}^2(f) = \frac{1}{M} |G_0(f)|^2 \frac{\sigma_Y^2(f)}{|Y_0(f)|^2} \approx \frac{1}{M} \frac{\hat{\sigma}_Y^2(f)}{|U_0(f)|^2}. \tag{3-15}$$

where the additional factor M accounts for the variance reduction due to the averaging over M realizations. So the standard deviation of the averaged FRF over the total data set (M blocks and N_{repeat} realizations) is given by

$$\sigma_{G_{\text{averaged}}}^2(f) = \frac{1}{N_{\text{repeat}}M} \sigma_G^2(f). \tag{3-16}$$

The reason for the (different) scale factors in (3-15) and (3-16) is that with the expression in (3-13) the variance at the level of a single block is estimated. By averaging over M or $N_{\text{repeat}}M$ blocks, the variance on the average is reduced by the number of averaged measurements.

Make the following plots:

- The measured FRF and standard deviation together with the errors, for both excitations.

- Estimate the theoretical variance from the last realization (M blocks), using (3-13).

- Compare the sample variance with the estimated theoretical variance.

Observations From the top of Figure 3-17 it can be seen that the errors of the averaged FRF are well described by the sample standard deviation. This clearly indicates that for these measurements the stochastic errors are dominating, at this level no systematic errors can be observed. A further increase (either increasing M or N_{repeat}) of the number of averages would reveal that, for the noise excitation, systematic errors would pop up around the resonance frequency. These errors decrease fast proportional to squared block length N^{-2}. On the bottom

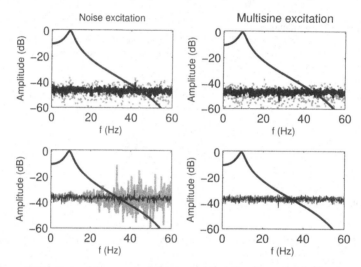

Figure 3-17 Study of the errors and the standard deviation for a noise (left) and a multisine (right) excitation. Top: FRF (black line). Thin black line: Measured standard deviation on the averaged FRF. Gray dots: $|G - G_0|$.. Bottom: Thin black line: Measured standard deviation for a single realization Bottom gray line: Estimated theoretical standard deviation.

part of the figure, it can be observed that the theoretical variance, calculated from a single realization, coincides well with the sample variance, after properly scaling for the number of processed blocks per realization (M in this case). Notice that for the noise excitation, the theoretical value calculated on the basis of (3-13) is not very reliable. This is due to the fact that for a small number of realizations, the estimate of the coherence is very poor. The expression in (3-15) behaves much better under these conditions because, in this case, no measured values appear in the denominator.

Exercise 45 (Impact of the block (period) length on the uncertainty) In   practice the block length is set in order to get a sufficient frequency resolution in the measurements: $\Delta f = f_s / N$. In this exercise we study the impact of the block length N on the behavior of the errors.

Repeat the previous exercise for a fixed number of blocks $M = 64$ and vary the block or period length $N = 128, 512, 2048, 8192$. Excite the full frequency band each time, so that the number of sines in the multisine equals $N/2$. Calculate for each value of N the theoretical value of the standard deviation, for both the random noise excitation and the multisine. Compare the results.

Observations (See Figure 3-18) It can be clearly seen that the variance level is independent from the length of the block or period. Increasing this length increases the frequency resolution of the measurement, but the uncertainty does not drop. This is because the power per excited frequency line remains constant in this experiment. For short data records the leakage errors dominate (see also Exercise 37), especially around the resonance frequency. For increasing block lengths it can also be seen that the impact of the leakage errors is decaying proportionally to N^{-1} (see also Table 3-1).

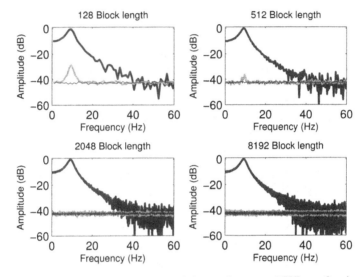

Figure 3-18   Evolution of the standard deviation on the averaged FRF as a function of the block length. Bold black line: Measured FRF (here shown for the noise excitation). Gray line: Standard deviation for noise excitation. Thin black line: Standard deviation for multisine excitation.

3.4.3 What have we learned in Section 3.4? Further reading

In this section we studied the impact of disturbing noise at the output of a system on the FRF measurement. The same methods as in the previous section can be used under these conditions, but besides the leakage errors we now face also the impact of the disturbing noise. Tools are provided to estimate the variance of the measured FRF. The coherence (see equations (3-12) and (3-13)) can be used for this purpose. For measurements with periodic excitations under steady-state conditions it is also possible to make a direct noise analysis using equations (3-14)–(3-16). From (3-15) it is seen that the variance on the FRF is inversely proportional to the power at a given frequency. Within a given measurement time, it is possible to increase this power by reducing the number of excited frequencies in the multisine. This can be done by putting a number of harmonics equal to zero, or by reducing the period length of the multisine. It is clear that the price to be paid is a reduced resolution of the FRF measurement. This is a degree of freedom that can be used during the experiment design. We refer the reader to Bendat and Piersol (1980) for more information on the use of the coherence, and to Pintelon and Schoukens (2001) for more information on the design of multisine signals.

3.5 FRF MEASUREMENTS IN THE PRESENCE OF INPUT AND OUTPUT NOISE

In this section, we mainly repeat the exercises from the previous section where only disturbing (process) noise on the output was considered. This time also input disturbances are added to the measurements of the input, assuming that the input and output noise disturbances $n_u(t)$, $n_y(t)$ are mutually independent; that is, knowledge about $n_u(t)$ does not give any information about $n_y(t)$:

$$u(t) = u_0(t) + n_u(t), \, y(t) = y_0(t) + n_y(t). \tag{3-17}$$

The input noise level in the exercises is selected to have a very poor SNR of 6 dB at the input. This will allow us to illustrate clearly the additional problems that are created by input disturbances. In the previous section it turned out that the disturbing output noise increases the uncertainty of the measured FRF, but it does not create systematic errors. This is not so for input noise disturbances in combination with a random excitation, where systematic errors will also occur.

Exercise 46 (FRF measurement in the presence of input/output disturbances using a multisine excitation)   Repeat Exercise 43, but change the rms value of the excitation to 1, and add this time also disturbing noise n_u to the input, with n_u normally distributed $N(0, \sigma_u^2 = 0.5^2)$. Select for the number of averages $M = [8 \ 32 \ 128 \ 512]$.

Observations   From Figure 3-19 it can be seen that the errors behave completely differently depending upon the applied excitation signal (noise or multisine) and the corresponding method that is used to estimate the FRF. Increasing the number of averages reduces the errors of the multisine measurements, completely similar to the previous output noise experiment. However, for the noise excitation, the situation is completely different. The errors converge to a level that is independent of M ; these are systematic errors. This is due to the presence of the input noise and the appearance of quadratic terms in the denominator of the estimator (3-4). This is completely similar to what is illustrated on the resistor example in Chapter 1, Exercises 12 and 13. For these situations, the use of multisines avoids a lot of problems and increases the quality of the measurements considerably. For multisine excitations (or in general, periodic excitations) the signals are first averaged before making the division in equation (3-11), so that the impact of the noise disappears (averaged towards zero) for a growing number of averages.

What will happen to the systematic errors of the noise excitation experiments if the SNR of the input or output measurements varies?

Exercise 47 (Measuring the FRF in the presence of input and output noise: Analysis of the errors)   In this exercise, a more detailed analysis of the errors is made, similar to Exercise 44. Repeat this exercise for the noise settings specified in Exercise 46 and the number of blocks $M = 16$. Calculate the theoretical standard deviation for the noise excitation with formulas (3-12) and (3-13). For the periodic (multisine) excitation, the following extended formulas should be used:

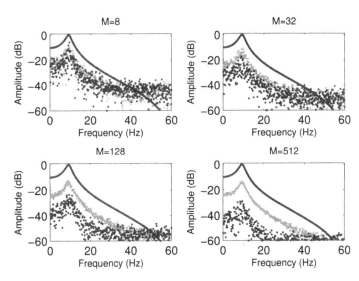

Figure 3-19 Evolution of the error $G - G_0$ of the averaged FRF G, as a function of the number M of averaged blocks or periods. For a noise excitation (gray dots) or a multisine (black dots). Black line: Exact FRF G_0.

$$\sigma_G^2(k) \approx \frac{|\hat{G}(j\omega_k)|^2}{M}(\hat{\sigma}_Y^2(k)/|\hat{Y}(k)|^2 + \hat{\sigma}_U^2(k)/|\hat{U}(k)|^2 - 2\text{Re}(\hat{\sigma}_{YU}^2(k)/(\hat{Y}(k)\overline{\hat{U}(k)}))) ,\qquad (3\text{-}18)$$

with
$$\hat{\sigma}_U^2(k) = \frac{1}{M-1}\sum_{l=1}^{M}|U^{[l]}(k) - \hat{U}(k)|^2, \ \hat{\sigma}_Y^2(k) = \frac{1}{M-1}\sum_{l=1}^{M}|Y^{[l]}(k) - \hat{Y}(k)|^2$$

$$\hat{\sigma}_{YU}^2(k) = \frac{1}{M-1}\sum_{l=1}^{M}(Y^{[l]}(k) - \hat{Y}(k))\overline{(U^{[l]}(k) - \hat{U}(k))} . \qquad (3\text{-}19)$$

Just as in the previous exercise, the variances σ_G^2 should be normalized by the total number of processed blocks $N_{\text{repeat}}M$ in order to get the variance of the mean value over all these measurements.

Observations From the results in Figure 3-20, it is again clearly seen that the multisine excitation/method gives much better results. For the noise excitation, the errors are much larger than the standard deviation, which is completely due to the presence of the systematic errors. For the multisine measurements, the standard deviation is in good agreement with the observed errors. On the bottom of the plot, the theoretical standard deviations, calculated with (3-13) for noise excitations, and (3-18) for periodic excitations, are compared to the sample variances. In both cases there is a good agreement. The scattering of the theoretical variance values at the higher frequencies is due to the fact that in equation (3-18) the measured values of the input and output are used instead of the theoretical values because the latter are not known. For higher frequencies, the SNR of the output measurements becomes very low because the measured system has a low pass nature.

Exercise 48 (Measuring the FRF in the presence of input and output noise: Impact of the block (period) length on the uncertainty) In this exercise, the impact of the block (period) length on the uncertainty is analyzed, similar to Exercise 45.

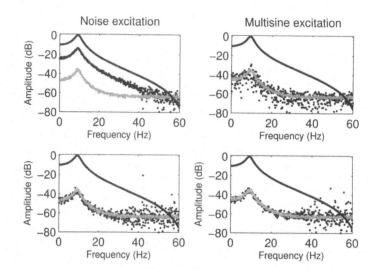

Figure 3-20 Study of the errors $|G - G_0|$ and the standard deviation for a noise (left) and a multisine (right) excitation. Top: FRF (black line); Gray line: Measured standard deviation on the averaged FRF. Black dots: The errors. Bottom: Gray line: Measured standard deviation on the mean value. Black dots: Standard deviation obtained from the theoretical expressions.

Repeat the exercise of Section Exercise 45, but add noise to the input as specified in Exercise 46.

Observations   From Figure 3-21 it can again be concluded that the variance level is independent of the length of the block or period. Increasing this length increases the frequency resolution of the measurement, but the uncertainty does not drop. This is because the power per excited frequency line remains constant in this experiment. Observe that the uncertainty for the noise excitation measurements does not disappear by increasing the block length.

3.5.1 What have we learned in Section 3.5? Further reading

In this section we learned that disturbing noise on the input results in a bias on the FRF measurement that does not disappear with a growing number of averages. This is also in agreement with the conclusions of Section 1.6 in Chapter 1. Depending upon the periodic and nonperiodic nature of the excitation, different methods are available to reduce this problem.

Using periodic excitations allows the input and output DFTs to be averaged before making the final division to get the FRF. This increases the SNR of the averaged input measurement and makes the bias to decrease as an $O(M^{-1})$.

This method does not work for random excitations because in that case the averages of the DFTs converge to zero. A first alternative solution is to use nonlinear averaging techniques, for example logarithmic averaging methods as discussed in Guillaume *et al.* (1992). Guillaume gives detailed analytical expressions for the bias level as a function of the SNR of the input measurements and shows that the bias decreases exponentially with the SNR of the input. It turns out that for an input SNR of 6 dB the bias is already smaller than 2‰. A second alternative is to use the indirect method that is explained in the next section, Exercise 50. In that case the FRF is measured from an external, exactly known, reference signal to the input and the output of the system. The final estimate is then obtained as the division of both FRFs.

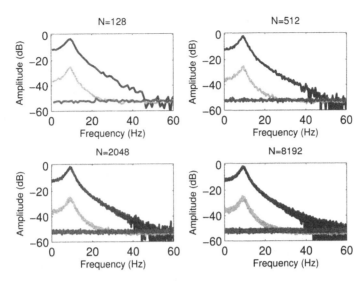

Figure 3-21 Evolution of the standard deviation as a function of the block length. Black: Measured FRF with noise excitation. Light gray: Standard deviation for noise excitation. Dark gray: Standard deviation for multisine excitation. $M = 64$ blocks are averaged.

3.6 FRF MEASUREMENTS OF SYSTEMS CAPTURED IN A FEEDBACK LOOP

The goal of this exercise is to illustrate the problems that appear when measuring the FRF of a system that is captured in a feedback loop. For such a measurement the process noise is fed back to the input of the system. Hence the input is no longer independent of the noise as it was in all the previous examples. This creates systematic errors for the noise excited measurements, while the periodic processing of multisine measurements does not suffer from this problem at all. The systematic errors can also be avoided if the external reference signal is available using an indirect measurement of the FRF.

3.6.1 Measurements in the loop: The direct method

From experimental point, it is often very tempting to make a direct measurement of the input and the output of the system, even when it is captured in a feedback loop. In many cases the exact feedback configuration is not known, or the user might even not aware that the system is operating under feedback conditions. This will lead to a systematic error on the FRF measurement as is shown in the next exercise.

Exercise 49 (Direct measurement of the FRF under feedback conditions)   Consider the setup in Figure 3-22: a system $y = G_0 u + v$ is captured in a feedback loop:

The system is given by: filter(b,a,u), with

$$a = \begin{bmatrix} 1 & -0.6323 & 0.90978 \end{bmatrix}, \text{ and } b = \begin{bmatrix} 0 & 0.12775 & 0.063873 \end{bmatrix}, \quad (3\text{-}20)$$

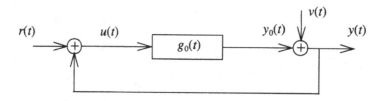

Figure 3-22   System captured in a unit feedback loop.

which can be generated in MATLAB® as

```
[bFF,aFF] = cheby1(2,20,2*0.25); bFF(1) = 0; bFF = bFF*2;
```

The process noise v is selected to be white normally distributed noise with $\sigma_v = 0.2$. Make two experiments. In the first experiment, the excitation $r(t)$ is a white normal distributed random excitation with standard deviations σ_r, and length $N*M$, with $M = 256$, the number of blocks, and $N = 1024$, the block length. In the second experiment a multisine with the same parameters is used. Add to both signals $N_{Trans} = 1024$ points that will be used to allow for the initial transients to disappear: the first N_{Trans} are neglected when processing the data. Select 4 different excitation levels $\sigma_r = 0, 0.2, 0.4, 0.8$, and measure each time the FRF between u and y, using formula (3-4) for the noise excitation and (3-11) for the multisine excitation. Apply a diff window for the random noise excitation, and a rectangular window for the multisine excitation (why?).

■ Plot for each excitation level the exact FRF, together with the measured FRF for the noise and the multisine. Make a plot for the amplitude and the phase.

                                                                                                    □

Discussion In Figures 3-23 and 3-24, the measured FRFs are shown. It can be seen that without an external reference signal ($r = 0$), the inverse controller characteristic C^{-1} is measured ($C = 1$ in this example). When the power of the reference excitation grows, the measured FRF shifts from C^{-1} to G_0 for the noise excitation. For the multisine measurements, the expected value (obtained by averaging over many periods of the multisine), equals G_0. This reveals one of the very strong advantages of periodic excitations.

3.6.2 Use of an external reference signal: The indirect method

In the previous exercise it turned out that the FRF that is measured under feedback conditions using a nonperiodic excitation like noise is prone to systematic errors. Instead of switching to periodic excitations, it is possible to remove these errors by using the indirect method. Instead of measuring directly the FRF from u to y, we measure two FRFs: the first from r to u, and the second from r to y, resulting in G_{ur} and G_{yr}. The estimate of the FRF from u to y is then finally obtained as

$$ G(k) = \frac{G_{yr}(k)}{G_{ur}(k)} = \frac{\left(\dfrac{S_{YR}(k)}{S_{RR}(k)}\right)}{\left(\dfrac{S_{UR}(k)}{S_{RR}(k)}\right)} = \frac{S_{YR}(k)}{S_{UR}(k)}. \tag{3-21} $$

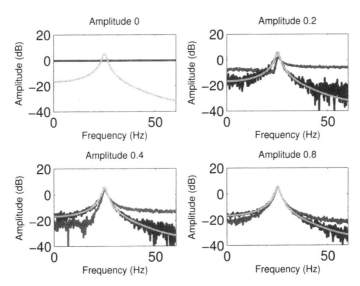

Figure 3-23 Measurement of an FRF of a system captured in a closed loop, for a fixed process noise level (0.2 Vrms) and a varying input level. Light gray: Exact FRF. Dark gray: Noise excitation. Black: Multisine excitation.

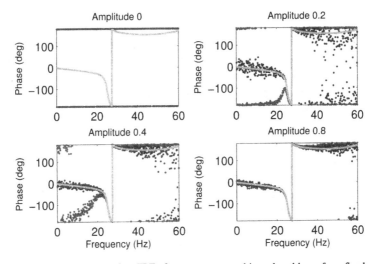

Figure 3-24 Measurement of an FRF of a system captured in a closed loop, for a fixed process noise level (0.2 Vrms) and a varying input level. Light gray: Exact FRF. Dark gray: Noise excitation. Black: Multisine excitation.

Exercise 50 (The indirect method) ■ Repeat Exercise 49, but use the indirect method (3-21) to process the measurements from the noise excitation. Plot the amplitude of the measured FRF and the amplitude of the complex error for the 4 excitation levels.

□

Discussion In Figure 3-25 the measured FRFs are shown. It can be seen that the systematic errors for the noise excitation are removed using the indirect method. The quality is now comparable to the results of the multisine excitation. This illustrates that it is always a

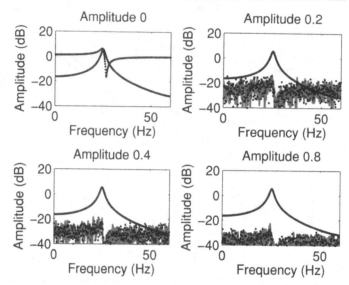

Figure 3-25   Measurement of an FRF of a system captured in a closed loop using the indirect method, for a fixed process noise level (0.2 Vrms) and a varying input level. Black line: Exact FRF. Gray line: Errors indirect method. Black dots: Errors multisine excitation.

good idea to store the reference signal r together with the measured input and output because it allows to get better results. A disadvantage of the indirect method compared to the multisine is that some information will be lost if there are nonlinearities present in the actuator or in the controller because the linear approximations can not capture the nonlinear behavior of the relations between r and u, y. However, as long as the plant is linear, the nonlinearities will not introduce systematic errors on the FRF from u to y.

3.6.3 What have we learned in Section 3.6? Further reading

Measuring under feedback conditions should be done with care. The process noise is fed back to the input and this creates a nonzero cross-correlation between the input and the process noise. This results eventually in a systematic error on the FRF if no special actions are taken. This problem is known for a long time, it was discussed by Wellstead (1977, 1981). Wellstead showed that the expected value of the direct method is given by

$$\mathbb{E}\{\hat{G}_{\text{direct}}(k)\} = \frac{G_{FF}(k)S_{rr}(k) - G_{FB}^*(k)S_{vv}}{S_{rr}(k) + |G_{FB}(k)|^2 S_{vv}(k)}, \tag{3-22}$$

with G_{FF}, and G_{FB} respectively, the FRF of the feed forward and feed back, and S_{rr}, S_{vv} the power spectrum of the external reference signal and the process noise. From this expression it is seen that the G_{FF} is retrieved when the reference signal dominates over the process noise, but the inverse feedback $1/G_{FB}$ is measured when the process noise dominates. Usually a mixture of both FRF's is obtained, resulting in a biased view on G_{FF}.

Exact expressions of the bias are given in Heath (2001). The indirect method avoids this bias by combining two direct measurements, respectively from the reference signal to the input and to the output of the system. Alternatively, the bias can be removed by averaging the DFT's over multiple periods of a periodic excitation.

3.7 FRF MEASUREMENTS USING ADVANCED SIGNAL PROCESSING TECHNIQUES: THE LPM

In Exercises 39 and 40 we studied how to deal with leakage problems. The best solution is to measure the steady-state response to a periodic excitation (e.g. a multisine), but this is not always possible. Waiting till the transients are gone is a loss of measurement time, while in other applications it is not possible to apply periodic excitations. Instead nonperiodic or random excitations are used. The windowing methods were developed during around the 1960s because they could be calculated very efficiently. However, nowadays the computers are much faster than 50 years ago which opens the possibility to use solutions with a better leakage rejection at a cost of making more calculations. In the next two exercises we will propose such a method. Compared to the windowing methods, the calculation time grows with a factor thousand, but with the computers that we have available nowadays it is still possible to calculate the FRF a thousands frequencies per second which is still very acceptable in many applications. At the same time the leakage errors are reduced to a level where they do no longer hurt for most applications. The basic idea is to observe that leakage errors are due to transient effects (see Exercise 38) which results in the exact relation (3-5). In this exercise we will make a polynomial approximation of the transfer function G_0 and the transient T that are both smooth functions of the frequency. In the exercise we use a 2nd degree polynomial with complex coefficients. To keep the approximation error small, the approximation is only made in a small frequency band around the central frequency on which the FRF is estimated. Next this frequency window is shifted over the entire frequency band of interest. At the edges it is not possible to center the frequency band around the frequency of interest. Special actions should be taken there, using a non-symmetric selection of the frequencies but this is outside the scope of this exercise.

3.7.1 Measuring the FRF using the local polynomial method

Exercise 51 (The local polynomial method) Generate a random noise sequence $u_0(t)$ with length $128 \times 16 + N_{Trans}$. Apply the resulting signal $u(t)$ to a second-order discrete time system with a resonance of 10 dB using, for example, the MATLAB® instruction [b,a] = cheby1(2,10,2*fc/fs) with $f_s = 128$ Hz the sample frequency, and f_c the cutoff frequency. Select $f_c/f_s = 0.1$.

- Apply the random noise excitation sequence to the system, and measure its input and output. Eliminate the first N_{Trans} points of the simulated input and output so that the initial transients of the simulation are eliminated.

- Create a set of linear equations:

$$Y(k + r) = G(k + r)U_0(k + r) + T(k + r) \qquad (3\text{-}23)$$

around the central frequency with $r = [-n, ..., -1, 0, 1, ..., n]$ and $n = 3$ by using a local polynomial approximation around the central frequency for G_0, T

$$G_0(k+r) = G_0(k) + a_1 r + a_2 r^2,$$
$$T(k+r) = T(k) + t_1 r + t_2 r^2.$$

(3-24)

- Solve this set of equations in least square sense with respect to the coefficients $[G_0(k), a_1, a_2, T(k), t_1, t_2]$ (e.g. using the \ operator of MATLAB®) and retrieve $\hat{G}_{Poly}(k) = \hat{G}_0(k)$ at the kth-FFT line.

- Repeat this at all frequencies $k = \left[n, n+1, ..., \frac{N}{2} - n \right]$, with $N = 128 \times 16$.

- Estimate the FRF G_{Hann} also with the Hanning method (see Exercise 39) using sub-records with a length of 128 points.

- Plot the amplitude of G_0, and the complex errors $G_0 - G_{Poly}$ and $G_0 - G_{Hann}$.

□

Discussion (see Figure 3-26) From the figure it can be seen that for the actual settings the errors of the LPM (local polynomial method) are much smaller than those for the Hanning window. Actually it can be shown that the errors drop as an $O(N^{-3})$, while it was mentioned before that for the Hanning method the errors are an $O((N/M)^{-1})$. Observe also that the resolution of the LPM is 16 times higher in this case. Unlike the Hanning method, the LPM does not split the original long data record in shorter subrecords. We encourage the reader to change the parameters (the degree of the polynomial approximation, the width of the window) to learn more about the properties of the algorithm.

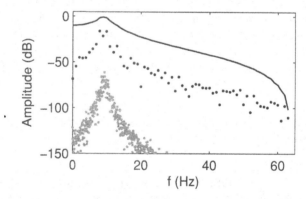

Figure 3-26 Comparison of the FRF measured with the LPM method and the Hanning method. Full line: G_0 Black dots: Errors of the Hanning windowing method. Gray dots: Errors of the LPM method.

3.7.2 Estimating a nonparametric noise model using the local polynomial method

Besides the estimate of the FRF, the user needs also an estimate of the power spectrum of the disturbing noise in order to calculate uncertainty bounds (see also Exercise 44). A first possibility is to estimate the power spectrum of the output noise using estimates of the second order cross- and auto-power spectra:

$$S_{VV}(k) = S_{YY}(k) - |S_{YU}(k)|^2 / S_{UU}(k).$$

(3-25)

However, the quality of this estimate is closely related to that of the windowing methods discussed before. For the LPM and alternative estimate can be obtained by analyzing the residuals of the least squares estimate in the previous solution:

$$E(k + r) = Y(k + r) - (\tilde{G}(k + r)U_0(k + r) + \tilde{T}(k + r)), \qquad (3\text{-}26)$$

with $\tilde{G}, \tilde{T}$ the estimated polynomial models evaluated in the frequency band centered around k for $r = [-n, ..., -1, 0, 1, ..., n]$.

Exercise 52 (Estimation of the power spectrum of the disturbing noise)

■ Estimate the variance of the error at frequency k for the settings of the previous exercise using

$$\hat{\sigma}^2(k) = \frac{1}{2n + 1 - 6} \sum_{r = -n}^{n} |E(k + r)|^2. \qquad (3\text{-}27)$$

The denominator in this expression accounts for the degrees of freedom in the residuals of the least squares fit.

■ Plot the amplitude of the FFT of the output (normalized by $1/\sqrt{N}$) and the estimated variance.

$\square$

Observations (see Figure 3-27) From this plot it is easy to get an idea of the quality of the method. Because there is no disturbing noise added to the simulation output, the observed errors are due to leakage errors. As it can be seen they became very low in this example. We encourage the user to change the settings of the algorithm and observe the behavior of the algorithm also in the presence of disturbed output measurements.

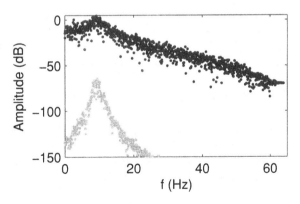

Figure 3-27 Estimation of the power spectrum of the disturbances: Bold dots: Measured output. Gray dots: Estimated variance of the output disturbances using the LPM method (remark: $\mathrm{dB}(\sigma^2) \equiv 10\log_{10}(\sigma^2) = 20\log_{10}(\sigma)$).

3.7.3 What have we learned in Section 3.7? Further reading

A major problem of the classical FRF measurement techniques is the sensitivity to leakage. Classically, leakage errors are reduced using windowing methods, as illustrated in Section 3.3. However, nowadays we can do much better using the local polynomial method at a cost of an increased number of calculations (typically a factor 1000), but even then we still can process a few thousand frequencies per second. The detailed study and illustrations of the properties of the local polynomial method is outside the scope of this book. The reader is referred to Schoukens *et al.* (2009b) for a first introduction to the topic. A detailed theoretical analysis of the properties is given in Pintelon *et al.* (2010a, 2010b).

3.8 FREQUENCY RESPONSE MATRIX MEASUREMENTS FOR MIMO SYSTEMS

3.8.1 Introduction

Till now we considered in this chapter single-input single-output (SISO) systems. In this section we will introduce the reader to the measurement of the frequency response function matrix (FRM) of multiple-input multiple-output (MIMO) systems. For simplicity we will focus on systems with 2 inputs and 2 outputs. This allows to illustrate the additional problems that appear when moving from SISO to MIMO for a minimal increase of the complexity. The scalar FRF response function (3-3) for a SISO system should be generalized to a frequency response matrix for a MIMO system that relates the multiple inputs to the multiple outputs:

$$
\begin{bmatrix} Y_{[1,1]}(k) \\ Y_{[2,1]}(k) \end{bmatrix} = \begin{bmatrix} G_{11}(k) & G_{12}(k) \\ G_{21}(k) & G_{22}(k) \end{bmatrix} \begin{bmatrix} U_{[1,1]}(k) \\ U_{[2,1]}(k) \end{bmatrix}.
\tag{3-28}
$$

with $X_{[p,q]}(k)$ the DFT of the pth input or output of experiment q. Equation (3-28) shows that from a single experiment we get at each frequency two linear equations, while we have to measure 4 unknowns: ($G_{11}(k)$, $G_{12}(k)$, $G_{21}(k)$, $G_{22}(k)$). For that reason at least two experiments should be combined to have as many equations as unknowns.

3.8.2 Measuring the FRM using a rectangular window

Questions that will be addressed in this section are: How to design the experiments? How to estimate the (co-)variance on the FRM estimate? In the first exercise we show how to solve the problem with multisines, next we discuss how random excitations can be used, and finaly we estimate also the covariance matrix of an FRM.

Exercise 53 (Measuring the FRM using multisine excitations)   In this exercise we will use multisine (periodic) excitations to measure the FRM so that we can focus the attention completely on the new MIMO aspects without having to deal with leakage problems. To do so the following series of steps can be made (among many other possibilities).

■ Create the following MIMO system:

G_{11}: cheby1(2,10,0.2)

G_{12}: cheby1(2,20,0.5)

G_{21}: butter(4,0.6)

G_{22}: `cheby1(2,15,0.8)`

■ Generate a multisine $u(t)$ with a flat amplitude spectrum up to $0.4f_s$. Select $f_s = 128$ Hz, and set the number of points in a period $N = 128$.

■ Generate two experiments:

$$\begin{aligned} y_{[1,1]}(t) &= G_{11}(q)u(t) + G_{12}(q)u(t) \\ y_{[2,1]}(t) &= G_{21}(q)u(t) + G_{22}(q)u(t) \end{aligned} \tag{3-29}$$

and

$$\begin{aligned} y_{[1,2]}(t) &= G_{11}(q)u(t) - G_{12}(q)u(t) \\ y_{[2,2]}(t) &= G_{21}(q)u(t) - G_{22}(q)u(t) \end{aligned} \tag{3-30}$$

using the `filter` instruction of MATLAB® for each subsystem. Wait $NTrans = 64$ points to remove the initial transients and select next precisely one period of the input and output signals.

■ Calculate the FFT of the input and outputs: U and $Y_{[p,q]}$ with $p, q = 1, 2$.

■ At each frequency k the following set of equations is solved:

$$\begin{bmatrix} Y_{[1,1]}(k) & Y_{[1,2]}(k) \\ Y_{[2,1]}(k) & Y_{[2,2]}(k) \end{bmatrix} = \begin{bmatrix} G_{11}(k) & G_{12}(k) \\ G_{21}(k) & G_{22}(k) \end{bmatrix} \begin{bmatrix} U_{[1,1]}(k) & U_{[1,2]}(k) \\ U_{[2,1]}(k) & U_{[2,2]}(k) \end{bmatrix} \tag{3-31}$$

or

$$Y_k = G_k U_k \tag{3-32}$$

using the \ operator of MATLAB®: `GkEst = Yk/Uk` with `GkEst` the estimate at frequency k of the FRM, consisting of the FRF's $(G_{11}(k), G_{12}(k), G_{21}(k), G_{22}(k))$.

■ Plot the amplitude characteristic of the estimated FRFs.

Discussion (see Figure 3-28) In the figure it can be seen that in this exercise the FRFs are very well retrieved. The relative errors are less than -50 dB. These errors are due to remaining initial transient effects and can be further reduced by increasing the waiting time *NTransient* before the measurements are made. In this exercise we made only one possible choice for the design of the multisine excitations. A detailed discussion on the possible choices and the impact on the quality of the measurements is made in the paper Dobrowiecki et al. (2006).

Exercise 54 (Measuring the FRM using noise excitations) In this exercise we will measure again the FRM of the system that was used in the previous exercise in the absence of disturbing noise, but this time using a noise excitation. In Exercise 37 we solved this problem on SISO systems using (3-4). The spectra were averaged over multiple realizations in order to reduce the sensitivity to leakage and noise errors. The auto- and cross-spectra were defined. In this exercise we follow exactly the same procedure, but this time we have that S_{YU} and S_{UU} are matrices.

■ Consider the MIMO system defined in Exercise 53.

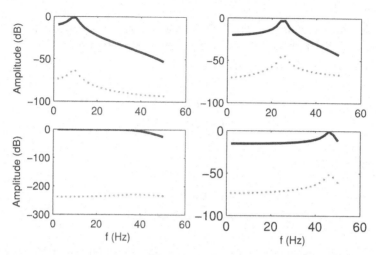

Figure 3-28 Measuring FRM of a 2×2 MIMO system using a multisine excitation. The FRF of $G_{11}(k)$, $G_{12}(k)$, $G_{21}(k)$, $G_{22}(k)$ is shown (from the left to the right and top to the bottom). Full line: The exact value. Gray dotted line: The amplitude of the complex error.

■ Generate two filtered white noise sequences (generator filter is `butter(4,2*0.4)`) of length $512 \times M + NTrans$, with $M = 50$ the number of realizations and apply these signals to the MIMO system and measure the output. Drop the first $NTrans = 1024$ points to eliminate the initial simulation transient effects.

■ Segment the input/output records in M sub records, calculate the FFT on each of these segments, and calculate the cross- and auto-spectra:

$$\hat{S}_{YU}(k) = \frac{1}{M} \sum_{s=1}^{M} \begin{bmatrix} Y_{[1,s]}(k) \\ Y_{[2,s]}(k) \end{bmatrix} \begin{bmatrix} \overline{U}_{[1,s]}(k) & \overline{U}_{[2,s]}(k) \end{bmatrix}, \qquad (3\text{-}33)$$

and

$$\hat{S}_{UU}(k) = \frac{1}{M} \sum_{s=1}^{M} \begin{bmatrix} U_{[1,s]}(k) \\ U_{[2,s]}(k) \end{bmatrix} \begin{bmatrix} \overline{U}_{[1,s]}(k) & \overline{U}_{[2,s]}(k) \end{bmatrix}. \qquad (3\text{-}34)$$

■ At each frequency k the following set of equations is solved:

$$\hat{S}_{YU}(k) = G_k \hat{S}_{UU}(k) \qquad (3\text{-}35)$$

using the \ operator of MATLAB®: `GkEst = SYU(k)/SUU(k)` with `GkEst` the estimate at frequency k of the FRM, consisting of the FRF's $(G_{11}(k), G_{12}(k), G_{21}(k), G_{22}(k))$.
Plot the amplitude characteristic of the estimated FRFs. □

Discussion (see Figure 3-29) ■ In the figure it can be seen that also in this exercise the FRF's are well retrieved, but this time the relative errors are in the order of -30 dB which is significantly larger than those in Figure 3-28 even if more data points were processed. These errors are due to leakage effects as discussed in Exercise 37. Averaging over more realizations will make the estimates smoother, but the leakage errors create also systematic errors so that the error does not vanish to zero. The only possibility to reduce the error is to use longer sub-blocks. The reader can try this by increasing the length of the sub-blocks from 512 to 2048 or even higher. The errors will drop in $1/\sqrt{N_{Block}}$. In the figure we show also the variance of the estimated FRM (see also Exercise 55). It can be seen that the errors are scattered around this estimated value.

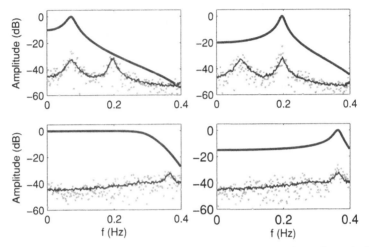

Figure 3-29 Measuring FRM of a 2×2 MIMO system using a random noise excitation. The FRF of $G_{11}(k)$, $G_{12}(k)$, $G_{21}(k)$, $G_{22}(k)$ is shown (from the left to the right and top to the bottom). Full line: The exact value. Gray dotted line: The amplitude of the complex error. Thin black line: The variance of the FRM (remark: $dB(\sigma^2) \equiv 10\log_{10}(\sigma^2) = 20\log_{10}(\sigma)$).

Exercise 55 (Estimate the variance of the measured FRM). The covariance matrix of the FRM is defined as the covariance matrix of the column vector $[G_{11}(k), G_{12}(k), G_{21}(k), G_{22}(k)]^T$. It is calculated from the measurements using

$$C_{FRM}(k) = \frac{1}{M-n_u}\hat{S}_{UU}(k)^{-1} \otimes (\hat{S}_{YY}(k) - \hat{S}_{YU}(k)\hat{S}_{UU}(k)^{-1}\hat{S}_{YU}(k)^H), \qquad (3-36)$$

with $n_u = 2$ the number of inputs. $A \otimes B$ is the Kronnecker product and is calculated in MATLAB® as kron(A,B). It replaces each entry $A_{[i,j]}$ of A by the matrix $A_{i,j}B$.

■ Plot the estimated variance of the FRFs on top of the results of Exercise 54. The variances of the FRFs are at the main diagonal of $C_{FRM}(k)$.

Discussion See Exercise 54

Exercise 56 (Comparison of the actual and theoretical variance of the estimated FRM) In this exercise we verify the quality of the estimated variance of the FRM (3-36) by comparing it with the variance that is obtained from repeated simulations.

■ Repeat Exercise 54 $NRep = 100$ times using each time a different realization of the random inputs. Calculate for each realization the estimated FRM and the estimate for the covariance matrix. Calculate from these results the variance of the estimated FRM. Average also the estimated variance for each of the FRFs at each frequency. Plot both results on top of each other. Compare the averaged FRM with the exact value and plot the amplitude of the complex difference.

□

Discussion (see Figure 3-30) From the figure it is clear that the variance that is estimated from the repeated simulations and the averaged "theoretical" variance coincide very well. It can also be observed that the error between the exact value of the FRM and the averaged estimate is well below the plotted standard deviation. This shows that the stochastic contribution to the leakage error on a single realization dominates at most frequencies the systematic leakage error. Only after averaging the result over a larger number of realizations, the systematic error will be the dominant one This can be seen in the figure by comparing the error of the averaged FRM with the standard deviation. If both become equal, the systematic errors dominate, while the stochastic errors dominate at those frequencies where the errors is far below the standard deviation.

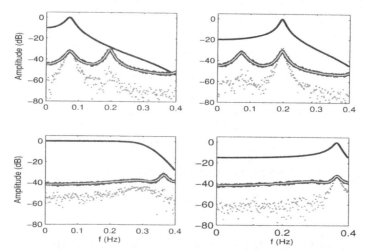

Figure 3-30   Measuring FRM of a 2×2 MIMO system using a random noise excitation and a rectangular window. The FRF of $G_{11}(k)$, $G_{12}(k)$, $G_{21}(k)$, $G_{22}(k)$ is shown (from the left to the right and top to the bottom). Full line: The exact value of the FRF. White line: The measured standard deviation. Dark gray line: The averaged calculated variance (remark: $dB(\sigma^2) \equiv 10\log_{10}(\sigma^2) = 20\log_{10}(\sigma)$). Light gray dots: The amplitude of the complex error on the averaged FRM.

3.8.3 Measuring the FRM using random noise excitations and a Hanning window

In Exercise 54 and 55 we measured the FRM using random noise excitations in combination with a rectangular window. No noise was added in order to focus on the leakage errors. In Exercise 39 it was shown that the impact of leakage on the FRF measurement could be reduced by using a Hanning window. In this exercise we will repeat this, but now for the FRM.

Exercise 57 (Measuring the FRM using noise excitations and a Hanning window)

■ Consider the setup of Exercise 54. Process the data in the same way, but use a Hanning window when calculating the FFTs of the subrecords (see also Exercise 39).

- Repeat the simulations $NRep = 100$ times, and process the data as in Exercise 56.
- Plot the results.

□

Discussion (see Figure 3-31) When comparing these results to those in the previous exercise obtained with the rectangular window, it can be seen that the skirts are much smaller now. However at the resonance frequencies, the reduction of the error is much smaller which is in agreement with the experience obtained before for the FRF measurements.

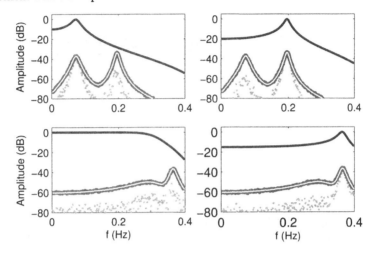

Figure 3-31 Measuring FRM of a 2×2 MIMO system using a random noise excitation and a Hanning window. The FRF of $G_{11}(k)$, $G_{12}(k)$, $G_{21}(k)$, $G_{22}(k)$ is shown (from the left to the right and top to the bottom). Full line: The exact value of the FRF. White line: The measured standard deviation. Dark gray line: The averaged calculated variance (remark: $dB(\sigma^2) \equiv 10\log_{10}(\sigma^2) = 20\log_{10}(\sigma)$). Light gray line: The amplitude of the complex error on the averaged FRM.

3.8.4 What have we learned in Section 3.8? Further reading

Measuring the FRM is a more tedious task than measuring the FRF. The essential difference is that the impact of the different inputs on each output should be separated from each other. This requires to combine multiple subexperiments. The design of these experiments is critical, failing to make a good choice will result in extremely large uncertainties on the FRM. For multiple input systems, it is not enough to tune the power spectrum of the signal, also the relation between the excitations at the different inputs of the system should be welltuned. In the literature, the design of orthogonal excitation signals has been discussed. If the number of inputs can be written as 2^n, Hademard matrices can be used, as proposed by Guillaume *et al.* (1997). Dobrowiecki *et al.* (2006) generalized this solution later on to orthogonal multisines for an arbitrary number of inputs.

4

Identification of Linear Dynamic Systems

What you will learn: This chapter learns the reader how to identify a parametric model for a linear dynamic single-input single-output system, starting from experimental data. Time and frequency domain methods will be discussed and compared. The following topics are addressed:

- A first introduction to basic system identification aspects in the time domain (Exercise 58) and the frequency domain (Exercise 59). The numerical conditioning of the problem (Exercise 60).

- Simulation versus prediction errors (Exercise 62).

- Shaping model errors in the time and the frequency domain (see Exercises 64 and 65).

- Study of the time domain prediction error framework using parametric noise models: identification of parametric plant and noise models; one step ahead prediction; identification under feedback conditions; model uncertainty (see Exercises 66, 67, 68, 69, and 70).

- Frequency domain system identification using nonparametric noise models and periodic excitations; its comparison with the time domain prediction error framework (see Exercises 71, 72, and 73).

- Frequency domain system identification using nonparametric noise models and random excitations; its comparison with the time domain prediction error framework (see Exercise 74).

- Illustration of a typical identification run, using a time and a frequency domain identification toolbox (see Exercises 75 and 76).

4.1 INTRODUCTION

Consider the linear dynamic system G_0 in Figure 4-1. The aim of this chapter is to find a parametric model for the system G_0 starting from the measured input $u(t)$ and output $y(t)$. Due to process and measurement disturbances, the measured values u, y differ from the true values u_0, y_0, that are related by the true system $y_0(t) = G_0(q)u_0(t)$. In Figure 4-1 we added all possible disturbances: the input to the system is disturbed by generator noise $n_g(t)$, the output of the system is disturbed by process noise $n_p(t)$. On top of that, the measured input

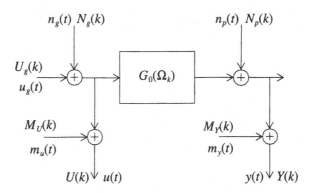

Figure 4-1 Frequency domain representation of the measurement process, with all possible disturbances. Note that the system can be captured in a feedback loop.

and output is also disturbed by measurement noise $m_u(t)$, $m_y(t)$. All these noise sources are not always present in every application, and their impact will depend upon the selected framework. In this chapter we will consider two situations with respect to the nature of the excitation signal that can be nonperiodic or periodic.

■ Nonperiodic excitations: We consider the input signal to be an arbitrary or random signal. In that case the generator noise $n_g(t)$ is a part of the input that is applied to the system, it is not a disturbance in the measurement $u(t)$. The measurement noise $m_u(t)$ acts always as a disturbance but it does not affect the evolution of the system. We have that $u(t) = u_0(t) + m_u(t)$ with $u_0(t) = u_g(t) + n_g(t)$.

■ Periodic excitations: We consider the input $u_0(t)$ to be periodic and all nonperiodic signals n_g and m_u, n_p, m_y act as disturbances but their impact will be different. We have that $u_0(t) = u_g(t)$. Only the 2nd group of disturbances m_u, n_p, m_y increases the variance of the plant estimates. The generator noise n_g does not increase the variance for well-designed identification methods, nor does it contribute to the knowledge of the plant as it did under the "nonperiodic" situation.

Initially we will assume that the input is exactly known, only the output is disturbed as shown in Figure 4-2. All output disturbances are combined in the noise source $v(t)$. This is the typi-

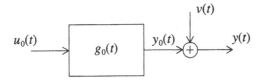

Figure 4-2 Time domain representation of the simplified problem.

cal setting of the classical time domain identification framework. A thorough study of this problem can be found in the books of Ljung (1999) or Söderström and Stoica (1989). The full problem, as shown in Figure 4-1 is discussed in detail in the book of Pintelon and Schoukens (2001) where a frequency domain approach is proposed. In this chapter you will learn how to solve both problems, and we illustrate what are the equivalences and differences between the time and frequency domain approach. At the end of this chapter, you will be able to select and use the proper method to solve your problem.

4.2 IDENTIFICATION METHODS THAT ARE LINEAR-IN-THE-PARAMETERS. THE NOISELESS SETUP

4.2.1 Introduction

The aim of this section is to give a better understanding of the basic problems that should be addressed when identifying a linear dynamic system: What is the basic idea? What is the impact of the initial state (transients) of the system? What is the relation between transients in the time domain and leakage effects in the frequency domain? Also numerical issues and the impact of model errors is studied. In order to keep the focus as much as possible on these aspects, we simplify the identification method in this section to a formulation that is linear-in-the-parameters, avoiding the use of numerical optimization procedures to find the parameter estimates that minimize the cost function. These methods are optimal for a very specific behavior of the disturbing noise. In Sections 4.3, 4.4, and 4.5 we will address the general situation that considers disturbing process and measurement noise with an arbitrary power spectrum.

4.2.2 Introduction to time domain and frequency domain identification

A. Time Domain Identification

Consider a single-input–single-output linear dynamic system:

$$y_0(t) = G_0(q)u_0(t), \tag{4-1}$$

with

$$G_0(q) = \frac{B(q)}{A(q)} = \frac{\sum_{k=0}^{n_b} b_k q^{-k}}{\sum_{k=0}^{n_a} a_k q^{-k}}, \text{ with } q^{-1} \text{ the backward shift operator: } u_0(t-1) = q^{-1}u_0(t). \tag{4-2}$$

Such a system puts a relation on the input and output signal:

$$y_0(t) = G_0(q)u_0(t) = \frac{B(q)}{A(q)}u_0(t), \tag{4-3}$$

or

$$A(q)y_0(t) = B(q)u_0(t) \Leftrightarrow A(q)y_0(t) - B(q)u_0(t) = 0. \tag{4-4}$$

which can be rewritten as

$$y_0(t) = b_0 u_0(t) + b_1 u_0(t-1) + \cdots + b_{n_b} u_0(t-n_b) - a_1 y_0(t-1) - \cdots - a_{n_a} y_0(t-n_a), \tag{4-5}$$

with a_0 normalized to $a_0 = 1$. Writing (4-5) for $t = 0, 1, ..., N-1$, a set of linear equations in $\theta_0^T = [a_1 ... a_{n_a} \ b_0 ... b_{n_b}]$ is obtained:

$$K(u_0, y_0)\theta_0 = \tilde{y}_0, \tag{4-6}$$

with

$$K(u_0, y_0) = \begin{bmatrix} -y_0(-1) & -y_0(-2) & \dots & -y_0(-n_a) & u_0(0) & \dots & u_0(-n_b) \\ -y_0(0) & -y_0(-1) & \dots & -y_0(-n_a+1) & u_0(1) & \dots & u_0(-n_b+1) \\ & & \dots & & & \dots & \\ -y_0(N-2) & -y_0(N-3) & \dots & -y_0(N-1-n_a) & u_0(N-1) & \dots & u_0(N-1-n_b) \end{bmatrix}$$

$$\tilde{y}_0 = \begin{bmatrix} y_0(0) \\ y_0(1) \\ \dots \\ y_0(N-1) \end{bmatrix}. \tag{4-7}$$

Exercise 58 (Identification in the time domain) Goal: Identify a transfer function model of a linear dynamic system.

Calculate the output of the linear dynamic system $y_0(t) = G_0(q)u_0(t)$ that is driven by filtered Gaussian noise $u_0(t) = G_{Gen}(q)r(t)$ with length $N + N_{Trans}$. The first N_{Trans} data points of the simulation are eliminated to get rid of the initial transients of the simulation. From the remaining N data points, the system parameters θ_0 are identified, by minimizing the least squares cost:

$$V = e^T(\theta)e(\theta), \quad \text{and} \quad e(\theta) = \tilde{y}_0 - K(u_0, y_0)\theta. \tag{4-8}$$

Notice that this kind of estimation problems is solved in Exercise 6 in Chapter 1 and can be solved using the \ operator of MATLAB®.

■ Generate the system G_0 [b0,a0] = cheby1(3,5,2*0.1)

■ Generate   the   generator   filter
  G_{Gen}: [bGen,aGen] = cheby1(5,1,2*0.45).

■ Select $N = 5000$ and $N_{Trans} = 1000$.

■ What is the minimal order n_a, n_b of the model needed to describe the exact system?

■ Estimate the model parameters $\hat{\theta}$ two times:
  i) The first time $[y_0(0), y_0(1), \dots, y_0(N-1)]^T$ is used as right-hand side in (4-6). The first rows of the corresponding matrix $K(u_0, y_0)$ contain the unknown (not measured) values of $u(k), y(k)$, $k = -1, -2, \dots$. Replace these unknown values by zero.
  ii) The second time $[y_0(n), y_0(n+1), \dots, y_0(N-1)]^T$ with $n = \max(n_a, n_b)$ is used as right-hand side in (4-6). Now the first rows of the matrix $K(u_0, y_0)$ do no longer contain unknown data $u_0(-1), y_0(-1), \dots$ so that all values $u_0(k), y_0(k)$ in the matrix K are known, no values $y_0(k < 0)$, $u_0(k < 0)$ are needed.

■ Plot for each of the solutions the amplitude of the complex error $|G_0(j\omega) - G(jw, \hat{\theta})|$ as a function of the frequency.

Observations (see Figure 4-3) Although the models are estimated from undisturbed data (no disturbing noise is present), it turns out that there are model errors if the initial state effects are not removed. These effects are similar to the transient effects that were observed in Chapter 3, Exercise 38, and they will appear as an $O(N^{-1})$ when the length of the data N is

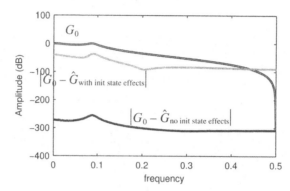

Figure 4-3 Identification of a third order Chebyshev filter using the ARX model implemented in the time domain. There is no disturbing noise added to the output. Gray: Initial state effects not removed. Black: The impact of the initial states is removed.

growing to infinity. This can be easily checked by repeating this exercise on records of growing length. Notice also that the initial state effects are random effects for a random input. They vary from one realization to the other.

It is easy to remove the initial state effects in this case, as explained in the exercise. Instead of starting at time $t = 0$, the matrix $K(u_0, y_0)$ is formed starting at time $t = n$ with $n = \max(n_a, n_b)$. In that case no unknown data points are entering in the matrix K. The relative error on this solution is in the order of -250 or even -300 dB, corresponding to the 12 to 15 digits of calculation precision which is indeed the MATLAB® precision.

B. Frequency Domain Identification

Consider again the single-input–single-output linear dynamic system of Section A, but rewrite the equations in the frequency domain, using the DFT $U_0(k)$, $Y_0(k)$ of $u_0(t)$, $y_0(t)$, and

$$Y_0(l) = G_0(j\omega_l)U_0(l) + T(l). \tag{4-9}$$

(see also Chapter 3, Exercise 38) with $G_0(j\omega_l)$ the transfer function of the system at frequency $2\pi l f_0 = 2\pi l f_s / N = l 2\pi / N$ (f_s normalized to 1):

$$G_0(z_l = e^{jl\frac{2\pi}{N}}) = \frac{B(z_l)}{A(z_l)} = \frac{\sum_{k=0}^{n_b} b_k z_l^{-k}}{\sum_{k=0}^{n_a} a_k z_l^{-k}}, \tag{4-10}$$

and $z_l = e^{jl2\pi/N}$ the discrete domain frequency variable at frequency lf_0. $T(l)$ is due to the begin- and end effects (leakage), and has a completely similar behavior as the initial state effects in the previous time domain exercise:

$$T(z_l) = \frac{I(z_l)}{A(z_l)} = \frac{\sum_{k=0}^{n_i} i_k z_l^{-k}}{\sum_{k=0}^{n_a} a_k z_l^{-k}}, \quad \text{with } n_i = \max(n_a, n_b) - 1. \tag{4-11}$$

The relation on the input and output DFT coefficients becomes

$$A(z_l)Y_0(l) = B(z_l)U_0(l) + T(z_l) \Leftrightarrow A(z_l)Y_0(l) - B(z_l)U_0(l) - I(z_l) = 0, \tag{4-12}$$

Normalizing again $a_0 = 1$, a set of linear equations is found:

$$K\theta_0 = [K_G(U_0, Y_0) \; K_I]\theta_0 = \tilde{Y}_0, \tag{4-13}$$

with $\theta_0^T = [a_1 \ldots a_{n_a} \; b_0 \ldots b_{n_b} \; i_0 \ldots i_{n_i}]$, and

$$K_G(U_0, Y_0) = \begin{bmatrix} -z_0^{-1}Y_0(0) & -z_0^{-2}Y_0(0) & \ldots & -z_0^{-n_a}Y_0(0) & U_0(0) & \ldots & z_0^{-n_b}U_0(0) \\ -z_1^{-1}Y_0(1) & -z_1^{-2}Y_0(1) & \ldots & -z_1^{-n_a}Y_0(1) & U_0(1) & \ldots & z_1^{-n_b}U_0(1) \\ & & \ldots & & & \ldots & \\ -z_{N/2}^{-1}Y_0(N/2) & -z_{N/2}^{-2}Y_0(N/2) & \ldots & -z_{N/2}^{-n_a}Y_0(N/2) & U_0(N/2) & \ldots & z_{N/2}^{-n_b}U_{N/2}(N/2) \end{bmatrix},$$

$$K_I = \begin{bmatrix} 1 & z_0^{-1} & \ldots & z_0^{-n_i} \\ 1 & z_1^{-1} & \ldots & z_1^{-n_i} \\ \ldots & \ldots & \ldots & \ldots \\ 1 & z_{N/2}^{-1} & \ldots & z_{N/2}^{-n_i} \end{bmatrix} \quad \text{and} \quad \tilde{Y}_0 = \begin{bmatrix} Y_0(0) \\ Y_0(1) \\ \ldots \\ Y_0(N/2) \end{bmatrix}. \tag{4-14}$$

Notice that these equations are complex. A real equivalent is found by defining

$$K_{\text{real}} = \begin{bmatrix} \text{real}(K) \\ \text{imag}(K) \end{bmatrix} \quad \text{and} \quad \tilde{Y}_{\text{real}} = \begin{bmatrix} \text{real}(\tilde{Y}_0) \\ \text{imag}(\tilde{Y}_0) \end{bmatrix}, \quad \text{with } K_{\text{real}}\theta_0 = \tilde{Y}_{\text{real}}. \tag{4-15}$$

Exercise 59 (Identification in the frequency domain)   Goal: Identify a transfer function model of a linear dynamic system in the frequency domain.
Consider the data from Exercise 58 and transform these data to the frequency domain using the FFT. Identify the system parameters θ_0, by minimizing the least squares cost:

$$V = e^T(\theta)e(\theta) \quad \text{and} \quad e(\theta) = \tilde{Y}_{\text{real}} - K_{\text{real}}(U_0, Y_0)\theta. \tag{4-16}$$

Notice that this kind of estimation problems is solved in Exercise 6, using the \ operator.

■ What is the minimal order n_a, n_b of the model needed to describe the exact system?
■ Estimate the model parameters $\hat{\theta}$ two times:
   (i) the first time formulating all equations without the transient term T so that the leakage effects are not included in the model,
   (ii) the second time using the full equations, including also the transient term.
■ Plot for each of the solutions the amplitude of the complex error $|G_0(j\omega) - G(j\omega, \hat{\theta})|$ as a function of the frequency.

Observations (see Figure 4-4)   A completely similar behavior as for the previous exercise is observed. If the full model equations are used, including the transient term to eliminate the begin and end effects, a solution that is exactly equal to the time domain solution is found. Actually, formulating the problem in the time or frequency domain does not change the solution, there is a full equivalence between the time and the frequency domain. If the transient term is not included, the solutions are different, but the average behavior is completely similar. In the frequency domain, leakage effects become visible. These can be interpreted as the

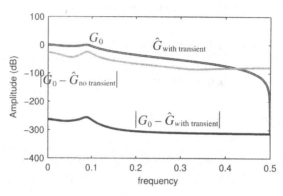

Figure 4-4   Identification of a third order Chebyshev filter using the ARX model implemented in the frequency domain. There is no disturbing noise added to the output. The error is plotted for models with and without "leakage" removal using an additional transient term.

influence of the begin and end states of the system, completely similar to the initial state effects in the time domain. Again these effects disappear as an $O(N^{-1})$ (check this by calculating the solution for different record lengths).

4.2.3 Numerical conditioning

A. Introduction
In this exercise, we study the numerical condition problems that appear when solving the estimation problem of the previous exercises. To do so, a linear least squares problem has to be solved. In practice, these calculations can become numerically ill conditioned, leading to poor solutions because of the increased sensitivity to calculation errors. This results in poor models, even in the absence of disturbing noise. As a user it is important to be aware of these problems, because these are often misinterpreted and not recognized as such. Using dedicated algorithms, it is possible to improve the numerical conditioning so that more complex systems can be identified. The next exercise illustrates on undisturbed data what parameters strongly affect the numerical conditioning. How to deal with these numerical problems is out of the scoop of this book.

Exercise 60 (Numerical conditioning)   Goal: Study the dominant parameters that affect the numerical conditioning.
The setup of this exercise is similar to that of Exercise 58. Calculate the output of the linear dynamic system $y_0(t) = G_0(q)u_0(t)$ that is driven by filtered Gaussian noise $u_0(t) = G_{Gen}(q)r(t)$ with length $N + N_{Trans}$. The first N_{Trans} data points of the simulation are eliminated to get rid of the initial transients of the simulation. From the remaining N data points, the system parameters θ_0 are identified, by minimizing the least squares cost in (4-8).

- Generate the bandpass system
  G_0: `[b0,a0] = cheby1(OrderG0,20,2*BW)`
  scan for the `OrderG0` the values [1:15]
  and for the bandwidth BW = 0.25 + [-0.2 0.2]/s, s = [1:100]
- Generate the corresponding generator filter for each bandwidth:
  G_{Gen}: `[bGen,aGen] = cheby1(5,5,2*BW);`
- Select $N = 5000$ and $N_{Trans} = 1000$.

- Put n_a, n_b equal to the model order 2 x `OrderG0` (the factor 2 accounts for the fact that G_0 is a bandpass filter).

- Estimate the model parameters $\hat{\theta}$, eliminating the initial state effects.

- Calculate for each problem the condition number of K using the MATLAB® instruction κ = `cond(K)`.

- Make a `contour` plot of the condition number as a function of the relative bandwidth: $0.2/(0.25 \times s)$, and the system order. Use logarithmic scales for the relative bandwidth and for the condition number κ.

Observations (see Figure 4-5) On this figure, the evolution of the condition number κ

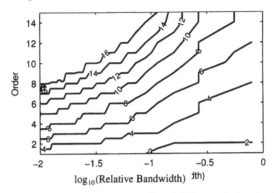

Figure 4-5   Study of the minimum number of digits to calculate the ARX estimate of a Chebyshev filter with a ripple of 20 dB. The order and the relative bandwidth.

can be seen. It is clear that the numerical conditioning deteriorates (condition number κ becomes larger) if the order of the system increases and/or if the relative bandwidth becomes smaller. The number of digits needed is about $\log_{10}(\kappa)$. MATLAB® calculates with 15 digits, so the upper limit for the allowed complexity of the system can be found immediately from the figure. The reader should be aware that it is possible to identify more complex systems, but then more robust numerical procedures than those used in this exercise are needed.

4.2.4 Simulation and prediction errors

A. Introduction

Once an estimated model is available, it can be used to calculate the output of the system from a known input. During these calculations delayed inputs and outputs are needed. To implement these calculations, two possibilities exist. Either delayed simulated or measured output values are used. The behavior of both choices can be very different and will be illustrated in this exercise. Consider the estimated model with the leading coefficient of $A(q)$ normalized to one: $a_0 = 1$.

$$G(q, \hat{\theta}) = \frac{\sum_{k=0}^{n_b} \hat{b}_k z_l^{-k}}{\sum_{k=0}^{n_a} \hat{a}_k z_l^{-k}} = \frac{B(q, \hat{\theta})}{1 + \tilde{A}(q, \hat{\theta})}. \tag{4-17}$$

From this equation it follows immediately that

$$y(t) = B(q, \hat{\theta})u_0(t) - \tilde{A}(q, \hat{\theta})y(t), \tag{4-18}$$

where $\tilde{A}(q, \hat{\theta})y(t) = \sum_{k=1}^{n_a} \hat{a}_k y(t-k)$ depends only on past values of $y(t)$. If these are re-placed by the measured output $y(t) = y_0(t)$, we call $\hat{y}(t) = B(q, \hat{\theta})u_0(t) - \tilde{A}(q, \hat{\theta})y_0(t)$ a pre-diction, and the output is called the "predicted" output. Alternatively, the past values can be replaced by the simulated values, and we call the output $\hat{y}(t) = B(q, \hat{\theta})u_0(t) - \tilde{A}(q, \hat{\theta})\hat{y}(t)$ a 'simulated' output. We will generalize and formalize these concepts later in Section 4.3.3 that introduces the reader to the prediction error framework for system identification. We will illustrate the different behavior of "simulation" and "prediction" in the next exercises. It will be shown that prediction is much more robust than the simulation, for example, a shorter simulation transient (initial state effects disappear much faster), much lower sensitiv-ity to model errors. Of course the price that has to be paid for this advantage is that the mea-sured outputs should be available. We first make the comparison in the absence of model errors. Later on we come back to the situation with model errors. Finally, the link between both approaches and the underlying disturbing noise model will be made clear.

Exercise 61 (Simulation and one-step-ahead prediction)   Goal: Calculation of the output using the known input and an estimated model.

In this exercise we assume that an estimated model is available. Also the input $u_0(t)$ is known. Next the output will be calculated twice, a first time using the simulation method, next using also the past measured output values (prediction method).

- Consider the setup of Exercise 58 and 59. Select the estimated model parameters of the method that includes the initial state effects (time domain) or the leakage effects (frequency domain method).

- Calculate the simulation and prediction error. Notice that the simulated output can be easily calculated using the MATLAB® instruction `ySim = filter(bEst, aEst, u0)`.

- Plot both errors together with the output.

Observations (see Figure 4-6) ■ In this figure it can be seen that both the simulation

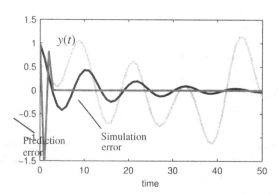

Figure 4-6   Comparison of the behavior of the prediction and the simulation error.

and prediction error converge to zero (the estimated model equals the exact model within the MATLAB® calculation precision). However, the initial state effect of the predicted output vanishes much faster than it does for the simulated output. Indeed, from (4-18), it can be seen that for the prediction method, the impact of the initial states is completely removed after $\max(n_a, n_b)$ samples, so it is set by the order of the system. For the simulation method, the transient time is set by the time constants of the plant.

4.2.5 Influence of model errors

A. Introduction: Impact of Model Errors

In the previous exercises, we identified a linear dynamic system using a model that was 'rich' enough to describe the modelled system exactly, no model errors are present. In practice this is often not the case. Model errors that appear frequently are nonlinear distortions, and unmodelled dynamics. The impact of nonlinear distortions is studied intensively in Chapter 5. In this exercise we focus on unmodelled dynamics. Theoretically, the best solution is to increase the model complexity by adding poles and zeros to the model, or equivalently by increasing the model order. However, in some applications this is an undesirable solution, and the user prefers to continue with a too simple model. In the next exercises we illustrate the impact of this choice on the identified model, the simulation and the prediction error. Next we also illustrate that by focusing the power on the frequency band of interest, it is possible to get a good match between the model and the plant in that frequency band.

Exercise 62 (Identify a too-simple model)   Goal: Illustrate the effect of using a too-low model order.

Consider the setup of Exercise 58, but use the following settings:

- Define the system G_0: [b0,a0] = cheby1(5,5,2*0.1)
- Define the generator filter G_{Gen}: [bGen,aGen] = cheby1(5,1,2*0.15*3)
- Select $N = 5000$ and $N_{\text{Trans}} = 1000$.
- The exact system G_0 is a fifth order system. Put $n_a = n_b = 4$ for the model, so that model errors will be present.
- Estimate the model parameters $\hat{\theta}$, eliminating the initial state effects (see Exercise 58).
- Plot the amplitude of both transfer functions $|G_0(j\omega)|$, $|G(j\omega, \hat{\theta})|$ as a function of the frequency.

Observations (see Figure 4-7) Just as in Exercises 58 and 59, there is no disturbing

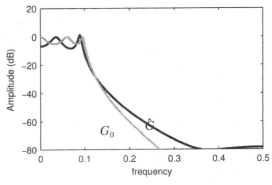

Figure 4-7   Fit of a too-simple fourth order model on a fifth order system.

noise in this exercise. But because the model order is too low to cover the full complexity of the plant, the identified model is different from G_0. These differences are called model errors. As can be seen in the figure, quite large errors appear in this case.

Exercise 63 (Sensitivity of the simulation and prediction error to model errors)

Goal: The aim of this exercise is to check what happens with the simulation and prediction errors in the presence of model errors.

Consider the results of the previous Exercise 62, and use the estimated model parameters to predict and to simulate the output of the system (see also Exercise 61). Make a plot of both errors, together with the output $y_0(t)$ of the system.

Observations (see Figure 4-8) In this figure it can be seen that also in this case the tran-

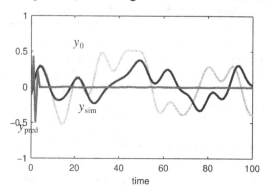

Figure 4-8 Comparison of the prediction error (light gray) and simulation error (black) of the too simple model of Figure 4-7.

sient of the prediction error vanishes much faster than that of the simulation error. However, on top of that it turns out that the simulation error is much more sensitive to the model errors than the prediction error does. The latter one becomes remarkably small, even in the presence of the large model errors as they were observed in the previous exercise. This result can be generalized: Prediction errors are less sensitive to model errors than simulation errors. The reason for that is the fact that the prediction method uses much more information, because it uses the measured past outputs, while the simulation method does not because it relies completely on the (wrong) model to simulate the past output data.

Conclusion "Prediction" is much more robust to model errors than "simulation," but it can only be used if the measured output data are available. This is the typical situation in control, where the next output sample $y_{pred}(t)$ is predicted on the basis of the past measurements of the output and the known inputs up to time t. In simulation studies this is not possible, in that case we want to compute the system output $y_{sim}(t)$ in new situations without making the experiment and so the measured past output are not available. In that case only the simulation setup can be used.

B. Introduction: Focusing the Identification Procedure on a User Selected Frequency Band

In this exercise we will show that even with a too simple model it is possible to focus the model on a user selected frequency band: the model is no longer required to fit the data everywhere, only in the user selected frequency band a good match is desired. We selected in this exercise the frequency band around the first resonance of the system. There are two possibilities to get this focus. In the time domain, the measured signals u_0, y_0 can be prefiltered. This does not affect the linear relation between both signals, but it will change the estimated model. In the frequency domain, the focus is obtained by selecting only those frequencies in the cost function that are in the frequency band of interest. Both approaches are illustrated in the next two exercises.

Exercise 64 (Shaping the model errors in the time domain: Prefiltering)

Goal: Focus the model on the user selected frequency band [0.04 0.08] (f_s = 1) using a prefilter.

Consider the setup of Exercise 62

- Generate $u_0(t)$, $y_0(t)$ as described before in Exercise 62.
- Generate the prefilter $L(q)$: `[bL,aL]  =  butter(5,[0.04,0.08]*2)` to focus on the frequency band of interest.
- Select N = 5000 and N_{Trans} = 1000
- Generate the prefiltered signals $u_L(t) = L(q)u_0(t)$, $y_L(t) = L(q)y_0(t)$.
- Eliminate the first N_{Trans} points to avoid the transients of the simulation.
- The exact system G_0 is a fifth-order system. Put $n_a = n_b = 2$ for the model, so that model errors will be present, but a single resonance can be covered.
- Estimate the model parameters $\hat{\theta}$, eliminating the initial state effects (see Exercise 58).
- Plot $|G_0|$, $|\hat{G}|$, and the amplitude of the complex error $|G_0(\omega) - G(w, \hat{\theta})|$ as a function of the frequency.

Observations (see Figure 4-9) There are still model errors visible, the model order is

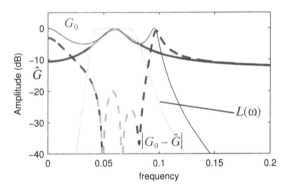

Figure 4-9 Improving the quality of a too simple model by using a weighting filter $L(\omega)$ implemented as a prefilter on the time domain signals. The gray region is the frequency band of interest.

still too low to cover the full complexity of the plant, the identified model is different from G_0. However, in the user selected frequency band, (the pass-band of the prefilter $L(\omega)$ in Figure 4-9), we get a much smaller model error. Outside this band the errors are even larger than in Exercise 62, because we are using an even simpler model of order 2 instead of order 4. This illustrates that the user can strongly influence the behavior of the model errors.

Conclusion The user can balance the complexity of the model, the model errors, and the noise sensitivity. By using a prefilter, it is possible to tune the model errors, keeping them small in a user-defined frequency band.

Exercise 65 (Shaping the model errors in the frequency domain: frequency weighting)

Goal: focus the model on the user selected frequency band L = [0.04 0.08] (f_s = 1) by selecting only the frequencies of interest in the cost function.

Consider the setup of Exercise 62

- Generate $u_0(t)$, $y_0(t)$ as described before.

- Select $N = 5000$ and $N_{\text{Trans}} = 1000$.

- Eliminate the first N_{Trans} points to avoid the transients of the simulation.

- The exact system G_0 is a fifth-order system. Put $n_a = n_b = 2$ for the model, so that model errors will be present, but a single resonance can be covered.

- Estimate the model parameters $\hat{\theta}$ in the frequency domain, eliminating the leakage effects (see Exercise 59). In the cost function (4-16), only the frequencies of interest are considered $L = \{l : 0.04 \le l / N \le 0.08\}$:

$$K = e^H(\theta)e(\theta) = \sum_{l \in L} \bar{e}(l, \theta)e(l, \theta), \quad \text{with} \quad e(l, \theta) = G_0(2\pi l) - G(2\pi l, \hat{\theta}). \qquad (4\text{-}19)$$

- Plot the amplitude of the complex error $|G_0(\omega) - G(w, \hat{\theta})|$ as a function of the frequency.

Observations (see Figure 4-10) The results look very similar to those of Exercise 64

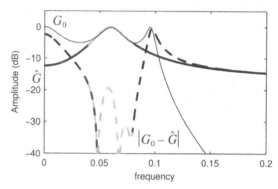

Figure 4-10 Improving the quality of a too simple model by selecting the frequencies that contribute to the cost. The gray region is the frequency band of interest.

where the frequency weighting was implicitly obtained using the prefiltering of the input and output signals. The major difference between both approaches is the increased flexibility for the user with the frequency domain method. An arbitrary frequency grid can be selected. It is no problem at all to eliminate a number of very narrow bands with spurious components (e.g., the frequency and harmonics of the mains), or using very complex patterns selecting for example only the odd frequencies to eliminate even nonlinear distortions. It would be quite involved to realize such requirements in the time domain using prefiltering.

4.2.6 What have we learned in Section 4.2? Further reading

In this section we presented a number of basic problems that the user faces when identifying a model for dynamic systems. In order to present the basic concepts as clear as possible, we started with an oversimplified situation where no disturbing noise is considered. Even then a number of fundamental issues and choices becomes visible. In the noiseless case the identification problem can be formulated as a linear least squares problem using an ARX method (see Section 4.2.2) that can be solved in the time or in the frequency domain. Both solutions are completely equivalent and equal up to the MATLAB® calculation precision (Ljung, 1999; Söderström and Stoica, 1989; Pintelon and Schoukens, 2001). It turned out that special care had to be taken to deal with the initial conditions. In the frequency domain also the end conditions should be accounted for (Pintelon *et al.*, 1997; Pintelon and Schoukens, 2001). The solution is disturbed by unmodeled transient effects if no explicit model term is added. These errors disappear as an $O(N^{-1/2})$.

The numerical conditioning of the equations to be solved becomes a problem either when the system becomes more complex (increasing number of poles and zeros), or when the relative bandwidth of the system becomes very small (see Exercise 60). Well-selected robust numerical methods should be used to deal with this class of systems (Bultheel *et al.*, 2005; Pintelon and Schoukens, 2001; Pintelon and Kollár, 2005).

Once a model is available, there are mainly two different possibilities to calculate the output. In the simulation mode, the output is calculated using past input values only. The measured output is not used. The alternative is to predict the output one step ahead using both the past input and output measurements (Ljung, 1999; Söderström and Stoica, 1989). While the simulation mode is very useful to simulate the systems behavior for new unobserved inputs (no new experiments need to be made), the prediction mode is often used in control design where only the short time behavior of the system should be predicted starting from the observed past inputs and outputs. Prediction errors are much more robust: The simulation transient decays much faster (Exercise 61). Simulation errors are more sensitive to model errors than prediction errors (Exercise 63). Model errors can be tuned (i) by selecting a well choose power spectrum of the input, (ii) by choosing a focusing filter, or (iii) by selecting the frequency band of interest (Exercises 64 and 65) (Ljung, 1999; Söderström and Stoica, 1989; Pintelon and Schoukens, 2001).

4.3 TIME DOMAIN IDENTIFICATION USING PARAMETRIC NOISE MODELS

4.3.1 Introduction

The aim of this section is to identify a linear dynamic system in the presence of disturbing noise. In Section 4.2 a basic introduction to the identification problem was given using a formulation that is linear-in-the-parameters. The major advantage of this approach is that the parameters are obtained as the solution of a linear set of equations. In this section we will illustrate that this choice is often not the optimal one, because this solution can be very sensitive to disturbing noise. More involved weighted least squares cost functions will be needed that also account for the noise behavior. The least squares cost function of the previous exercises is replaced by a weighted least squares cost where the choice of the weighting function will be set by the noise model. This has a double impact on the identification problem: (i) in general a nonlinear optimization problem needs to be addressed to find the estimate, while for the simple solution only a set of linear equations needed to be solved; (ii) also a noise model needs to be selected and identified. Two major possibilities exist for the latter choice, either a parametric or a non parametric noise model can be used.

The parametric approach models the noise as filtered white noise, and the noise model parameters are identified simultaneously with the plant parameters, which increases the complexity of the numerical optimization methods. This leads to the classical prediction error framework that is almost always formulated and solved in the time domain (see Figure 4-2). This approach is followed in Section 4.3.

Alternatively, a nonparametric noise model can be used. This is mostly done in the frequency domain, by estimating at each frequency the variance of the noise in a nonparametric preprocessing step. Next it is used as a nonparametric weighting in the weighted least squares cost function. So only the plant model has to be estimated in the parametric model identification step. This results in a simpler optimization problem to be solved and a smaller risk to get stuck in a local minimum. This approach is illustrated in Section 4.4 (for periodic excitations) and 4.5 (for random excitations). In Section 4.7 we study how to deal with (correlated) noise on the input and output measurements assuming that periodic excitations are used.

4.3.2 Parametric noise models

In this section we learn how to use parametric noise models. First we introduce the concept of a noise model, showing that it is possible to predict the noise behavior. It will become clear when these methods are working well, and under what conditions they fail. Next these results are used to understand the full identification problem where a plant and a noise model are identified. Three different classes of disturbing noise models will be used which cover a wide set of practical problems. It will also be illustrated that each noise model corresponds to an identification method that is optimal for the specific situation. Mixing the noise model and the identification method leads to suboptimal results, with a larger uncertainty, and under some conditions even biased models are obtained.

In the first exercise it will be shown that it is possible to predict future values of a colored noise sequence. Two methods will be illustrated. In the first, very simple method, the predicted value of the noise is chosen equal to the last known value $\hat{v}(t) = v(t-1)$. In the second method, an autoregressive model $A(q, \theta)v(t) = e(t)$, with $e(t)$ white noise, is identified, normalizing the leading coefficient equal to $a_0 = 1$. Next the noise value is predicted by putting $e(t) = 0$, and solving for $v(t)$ as a function of the past values $v(t-k)$ leading to

$$\hat{v}(t) = -\tilde{A}(q, \hat{\theta})v(t) = -\sum_{k=1}^{n_a} a_k v(t-k). \tag{4-20}$$

This result is generalized in Section 4.3.3 to more complex noise model with

$$v(t) = \frac{B(q, \theta)}{A(q, \theta)} e(t). \tag{4-21}$$

Exercise 66 (One-step-ahead prediction of a noise sequence) Goal: Illustrate the prediction of a noise sequence.

Consider a colored noise sequence $v(t) = H_0(q)e(t)$ with $e(t)$ white Gaussian noise.

- Generate the system H_0: `[bGen,aGen] = butter(5,2*BW)`
- Put BW = [0.01 0.05 0.1 0.2 0.3 0.4] respectively
- Select N = 5000 and N_{Trans} = 1000.
- Generate $N + N_{\text{Trans}}$ points,
  `v = filter(bGen,aGen,randn(1,N+NTrans))`
  and eliminate the first N_{Trans} points to avoid transient effects.
- Make a first prediction using the last known value $\hat{v}(t) = v(t-1)$ as prediction.
- Make a second prediction by estimating an AR-model. Use the ARX program of Exercise 58 and put the order $n_b = 0$ (no input terms). Choose $n_a = 5$ and $n_a = 50$. Calculate for both models the predicted noise value by
  $\hat{v}(t) = -\tilde{A}(q, \hat{\theta})v(t) = -\sum_{k=1}^{n_a} \hat{a}_k v(t-k)$, with the normalization $a_0 = 1$.
- Plot the noise sequence, and both predicted sequences on one plot. Plot the prediction error on another plot. Repeat this for all values BW of the bandwidth of the noise, and for both model orders.

Observations (see Figure 4-11) From both figures it can be seen that it is possible to make a reasonable good prediction for the noise. The simple method works better for noise

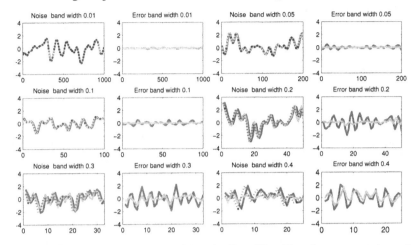

Figure 4-11   One-step-ahead prediction of the noise. Dotted line: The noise sequence to be predicted. Dark gray line: keep the last value of the noise. Light gray line: One-step-ahead prediction using an autoregressive model with order $n_a = 5$. .

with a low bandwidth with respect to the sample frequency. For a higher bandwidth, the method fails completely. The auto-regressive AR-model does a much better job. But also here we see that the error increases with increasing bandwidth. Increasing the order of the model results again in a better prediction (see Figure 4-12). It can be shown that in general, the opti-

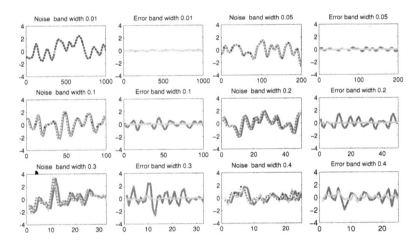

Figure 4-12   One-step-ahead prediction of the noise. Dotted line: The noise sequence to be predicted. Dark gray line: Keep the last value of the noise. Light gray line: One-step-ahead prediction using an auto-regressive model with order $n_a = 50$.

mal one step ahead prediction of the filtered noise $v(t)$ is

$$\hat{v}(t|t-1) = \sum_{k=1}^{\infty} h(k)e(t-k)$$
$$= (1 - H_0^{-1}(q))v(t),$$

(4-22)

with $h(t)$ the impulse response of $H_0(q)$. In practice, the coefficients $H_0(q)$ has to be estimated from a finite data record. In this exercise a specific choice $H_0(q) = 1/A(q)$ is made to keep the exercise as simple as possible, but more general noise models are frequently used, for example in Section 4.3.3, as a part of the introduction to the prediction error framework system identification approach.

4.3.3 Identification of parametric plant and noise models

In the first set of exercises (Section 4.2) we identified the model of a system in the absence of disturbing noise. In the second series of exercises (Section 4.3.2), the disturbances were modeled as filtered white noise. In the next step, we will identify a plant model in the presence of disturbing noise, using the observations $y(t)$ and the known input signal $u_0(t)$:

$$y(t) = G_0(q)u_0(t) + H_0(q)e(t) = y_0(t) + v(t). \tag{4-23}$$

Using the prediction $\hat{v}(t|t-1)$ of the noise disturbance, the one-step-ahead prediction of the output becomes

$$\hat{y}(t|t-1) = G_0(q)u_0(t) + \hat{v}(t|t-1). \tag{4-24}$$

Replacing $\hat{v}(t|t-1) = (1 - H_0^{-1}(q))v(t) = (1 - H_0^{-1}(q))(y(t) - G_0(q)u_0(t))$ [see (4-22)], leads eventually to the optimal prediction of the output:

$$\hat{y}(t|t-1) = H_0^{-1}(q)G_0(q)u_0(t) + (1 - H_0^{-1}(q))y(t). \tag{4-25}$$

The model parameters in $G_0(q, \theta)$, $H_0(q, \theta)$ can be estimated by minimizing the cost function:

$$\frac{1}{2N}\sum_{t=0}^{N} (y(t) - \hat{y}(t|t-1))^2 = \frac{1}{2N}\sum_{t=0}^{N} (H_0^{-1}(q)(y(t) - G_0(q, \theta)u_0(t)))^2, \tag{4-26}$$

which is known as the prediction error method. From (4-25) and (4-26) it can be seen that the optimal predictor depends on the properties of the disturbing noise, and hence different estimation schemes will result. These are known as:

■ ARX model: The disturbing noise and the system have a common denominator:

$$y(t) = \frac{B(q)}{A(q)}u_0(t) + \frac{1}{A(q)}e(t) = G_0(q)u_0(t) + v(t). \tag{4-27}$$

Notice that this is the model that was used in the exercises of Section 4.2 and it leads to a problem that is linear-in-the-parameters (check this in equation (4-26)). If this model is corresponding to the reality (for example, the disturbing noise enters the system close to its input), we have that the noise $v(t)$ carries also information about the poles of the system.

■ ARMAX model: This is a slightly generalized model, the disturbing noise and the system have a common denominator but an additional filtering of the noise with $C(q)$ is added.

$$y(t) = \frac{B(q)}{A(q)}u_0(t) + \frac{C(q)}{A(q)}e(t) = G_0(q)u_0(t) + v(t). \tag{4-28}$$

■ OE model (output error model): The output of the plant is disturbed by white noise

$$y(t) = \frac{B(q)}{A(q)}u_0(t) + e(t) = G_0(q)u_0(t) + v(t). \tag{4-29}$$

This corresponds to white noise disturbances added to the output of the system.

■ Box–Jenkins model (BJ model): There is no relation between the plant model and the noise model. This is the most general model:

$$y(t) = \frac{B(q)}{A(q)}u_0(t) + \frac{C(q)}{D(q)}e(t) = G_0(q)u_0(t) + v(t). \tag{4-30}$$

All these models, besides the ARX model, result in a numerical optimization problem that is nonlinear in the parameters. Numerical optimization methods are needed, and the implementation of these methods is out of the scope of this book. Instead we advice the reader to use for example the MATLAB® System Identification Toolbox that is based on the prediction error theory (Ljung, 1999).

Exercise 67 (Identification in the time domain using parametric noise models)
Goal: Identify a transfer function model of a linear dynamic system in the presence of disturbing noise.
Calculate the output of the linear dynamic system $y(t) = G_0(q)u_0(t) + H_0(q)e(t)$ that is driven by filtered Gaussian noise $u_0(t) = G_{Gen}(q)r(t)$ for the input, with $r(t)$ and $e(t)$ white Gaussian noise with standard deviation 1 and 0.1 respectively. Generate records with a length $N + N_{Trans}$. The first N_{Trans} data points of the simulation are eliminated to get rid of the initial transients of the simulation. From the remaining N data points, the plant and noise parameters θ_0 are identified, by minimizing the least squares cost (4-26), using the MATLAB® System Identification Toolbox.

■ Generate the system G_0: [b0,a0] = cheby1(2,10,2*0.36)
■ Generate the generator filter G_{Gen}: [bGen,aGen] = butter(3,2*0.4)
■ Select one of the following sets of noise filters coefficients for the noise filter H_0:
   - ARX noise: bARX = 1; aARX = a0
   - OE noise: $H_0 = 1$, or bOE = 1; aOE = 1;
   - BJ noise: [bBJ,aBJ] = butter(2,2*0.133);
■ Select $N = 5000$ and $N_{Trans} = 1000$.
■ Estimate the plant and noise model parameters $\hat{\theta}$ for each of these three disturbances, using the ARX, OE, and BJ model.
■ Repeat this 20 times, and calculate the rms value of the complex error $|G_0(j\omega) - G(j\omega, \hat{\theta})|$ as a function of the frequency.

Observations (see Figure 4-13) The results for the different combinations of noise models are shown. Also the power spectrum of the disturbing noise is shown. From this figure it can be concluded that each model is optimal if it corresponds to the true noise model structure, leading to the following optimal pairs: ARX-ARX; OE-OE; BJ-BJ. However, it can also be seen that using the BJ model gives a result that is very close to that of the ARX model and OE model for the ARX noise and OE noise, respectively. Only a slight increase of the

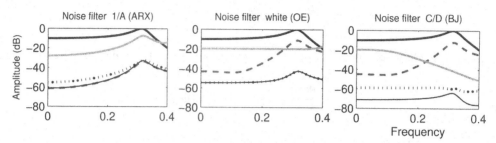

Figure 4-13 Comparison of the different time domain models on three different disturbing noise situations. Black line: G_0. Gray line: Power spectrum disturbing noise. Root mean square value $|G_0(j\omega) - G(j\omega, \hat{\theta})|$ for: ---: ARX model, ...: OE model, thin black line: BJ model.

rms error can be observed because more parameters are used in the BJ model than in the more specific noise models. So the BJ model could be selected as the general purpose tool. However, as mentioned before, the optimization of the parameters in this model is more prone to local minima, and needs on the average (much) more computation time. This is because two completely independent models have to be identified simultaneously.

4.3.4 Identification under feedback conditions

In this section we consider a very special but important problem: identification of a system under feedback conditions. The goal is to identify a system G_0 where the input u_0 does not only depend upon the user excitation $r(t)$, but also on past values of the output $y(t)$ that are fed back to the input. Notice that also the disturbance $v(t)$ is fed back to the input which creates a correlation between the input signal and the noise. Many identification methods like those based on ARX, ARMAX, and OE models will result in general in a biased estimate. Only the BJ model will still give consistent estimates under the condition that both the plant and the noise model are complex enough to capture the true systems. This is illustrated in the next exercise.

Exercise 68 (Identification Under Feedback Conditions Using Time Domain Methods)   Consider the setup in Figure 4-14. A system $y = G_0 u_0 + v$ is captured in a unit feedback loop with controller $C = 1$:

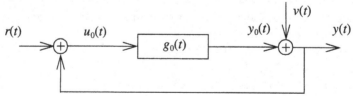

Figure 4-14   System captured in a unit feedback loop.

The system G_0 is given by the filter:
    `[bFF,aFF] = cheby1(1,20,0.5); bFF(1) = 0; bFF = bFF*10;`
The reference signal $r(t)$ is white Gaussian noise with standard deviation 1, filtered by the generator filter:
    `[bGen,aGen] = butter(3,2*0.4);`
The process noise v is filtered Gaussian noise with standard deviation $\sigma_v = 0.1, 0.5, 1$ respectively, the noise filter is

```
[bNoise,aNoise] = butter(OrderNoise,2*0.133);
```
with OrderNoise = 1.

- Calculate the relation between $y(t)$ and $r(t)$, $v(t)$, and generate the output $y(t)$.

- Select $N = 5000$ and $N_{\text{Trans}} = 1000$ (see Exercise 58).

- Estimate the plant model from $u_0(t)$, $y(t)$ using the ARX, OE, and BJ model for each of the values of σ_v. What is the order of the plant model to be used? What is the order of the noise model in the BJ model?

- Repeat this 100 times, and calculate the rms value of the complex error $|G_0(j\omega) - G(j\omega, \hat{\theta})|$ as a function of the frequency. Calculate also the bias. Plot the results for the three models and for the three levels of the disturbing noise.

- *Hint*: u(t) = filter(br,ar,r), with br = bFF; ar = aFF+bFF. The signal r is the reference signal. The contribution of $v(t) = H(q)e(t)$ to $y(t)$ is given by   filter(bv,av,e),   with   bv = conv(aFF,bNoise); av = conv(aFF+bFF,aNoise). The convolution of the polynomial coefficients calculates the coefficients of the product of the polynomials.

Observations (see Figure 4-15 and Figure 4-16) The first figure shows the rms error for

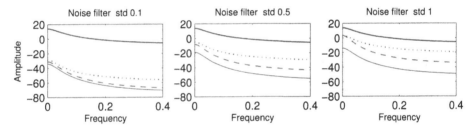

Figure 4-15   Comparison of the rms value of the error for different time domain models for identification under feedback conditions. Black line: G_0. Gray line: Power spectrum disturbing noise. ---: ARX model. ...: OE model. Thin black line: BJ model.

three different noise levels. For small noise levels, all models behave quite similarly, although the BJ model turns out to be slightly better than the others. This difference is increasing for increasing disturbing noise levels almost proportional to the noise level.

In Figure 4-16 the bias error is shown for $N = 5000$ and $N = 50,000$. Here it can be seen that there is a significant bias on the ARX and OE model. Notice that the bias grows proportional to the variance of the noise (and not to the standard deviation!). The observed error on the BJ model is much smaller. A more detailed analysis would show that this error is not significantly different from zero, and hence no bias can be detected for the BJ model. This can be checked by increasing the length N of the simulation. In that case the error for the BJ will decrease in $1/\sqrt{N}$ because the stochastic error dominate (and these are an $O(1/\sqrt{N})$), while the bias of the ARX and OE model will remain the same. This can be checked in Figure 4-16 (right) where $N = 50,000$ data points are processed in each simulation.

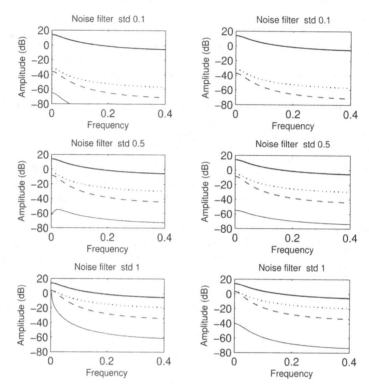

Figure 4-16 Comparison of the bias error for different time domain models for identification under feedback conditions. Left: $N = 5000$. Right $N = 50000$. Black line: G_0. Gray line: power spectrum disturbing nois.; ---: ARX model. ...: OE model. Thin black line: BJ model.

4.3.5 Estimating the variability of the estimated model

An estimate without an uncertainty interval has no value. At any time, the user should have an idea about the reliability of the results. Classically this is done by providing the covariance matrix on the parameters:

$$C_\theta = \mathrm{E}\{(\hat{\theta} - \mathrm{E}\{\hat{\theta}\})(\hat{\theta} - \mathrm{E}\{\hat{\theta}\})^T\}. \tag{4-31}$$

From this knowledge, it is possible to calculate an approximate value of the variance of any related quantity using linearization. This is done in most packages, hidden for the user. In this exercise we will illustrate the use of these tools to compare the estimated standard deviation on the amplitude and phase characteristic of the estimated transfer function model. These will be compared to the results obtained from the simulations.

    Exercise 69 (Generating uncertainty bounds for estimated models) Goal: Calculate the estimated variance of the amplitude and phase of the estimated transfer function model, starting from the estimated covariance matrix.

    Calculate the output of the linear dynamic system $y(t) = G_0(q)u_0(t) + H_0(q)e(t)$ that is driven by filtered Gaussian noise $u_0(t) = G_{\mathrm{Gen}}(q)r(t)$ for the input, with $r(t)$ and $e(t)$ white

Gaussian noise with standard deviation 1 and 0.1, respectively. Generate records with a length $N + N_{\text{Trans}}$. The first N_{Trans} data points of the simulation are eliminated to get rid of the initial transients of the simulation. From the remaining N data points, the plant and noise parameters θ_0 are identified, using the BJ model.

- Generate the system G_0:
  ```
  b0 = [0.2 0.6 0.3], a0 = [1 -1.2 0.9]
  ```
- Generate the excitation filter G_{Gen}:
  ```
  [bGen,aGen] = butter(3,2*0.2*1.5)
  ```
- Define the noise generating filter:
  ```
  bNoise = [1  0.7  0.7]; aNoise = [1 -0.8  0.3]
  ```
- Select $N = 1000$ and $N_{\text{Trans}} = 500$, and generate u_0, $y = y_0 + v$. Eliminate the first N_{Trans} data points.
- Estimate the plant and noise model parameters $\hat\theta$ using the BJ model from u_0, y.
- Repeat this 1000 times, and calculate the standard deviation of $|G(j\omega, \hat\theta)|$ and $\angle G(j\omega, \hat\theta)$ as a function of the frequency. Calculate also the standard deviation of the amplitude and phase characteristic, starting from the estimated covariance matrix (use the routines of the toolbox).

Observations (see Figure 4-17) The figure compares the estimated and observed stan-

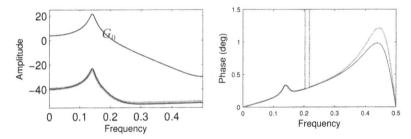

Figure 4-17   Comparison of the standard deviation on the estimates obtained from 1000 repeated simulations (gray), and from the estimated covariance matrix obtained from a single realization (black).

dard deviations. A very good agreement can be observed. Since the estimates are asymptotically normal distributed (see Exercise 3, Chapter 1), it is also possible to calculate for example the 95% uncertainty bounds, using the two sigma interval. From these simulations we can also conclude that it is possible to generate uncertainty bounds starting from only one realization. At the end of each estimation process, we end up we the numerical values of the estimated parameters and their covariance matrix.

Remarks: (i) The spike in the phase plot is due to small variations in the position of the 2π jump of the phase. As such, it does not indicate a large phase uncertainty, it is only an artifact that is due to the modulus 2π nature of the phase. (ii) The reader should be aware that the estimated covariance matrix will slightly vary from one realization to the other, and hence also the values of the theoretical standard deviations on the amplitude and phase will slightly change.

4.3.6 Focusing the BJ model on a user defined frequency band using prefiltering

In this section we study how the BJ model can be focused on a user-defined frequency band. A first (and sound) possibility is to design an appropriate excitation signal that puts most of its power in this band. However, for a given data set it is still possible to focus the identification model by applying a prefilter G_p to both the input and output signal. Such an operation does not change the relations:

$$y(t) = G_0(q)u_0(t) + v(t) \Rightarrow G_p(q)y(t) = G_0(q)G_p(q)u_0(t) + G_p(q)v(t).$$

So the estimation can be done on the prefiltered signals $y_p(t) = G_p(q)y(t)$, and $u_p(t) = G_p(q)u_0(t) + v_p(t)$. The reader should notice that the prefilter operation also changes the disturbing noise spectrum from $H_0 \rightarrow G_p H_0$. When a full BJ identification is performed, the noise model order should be increased to account for this effect. Eventually the estimate is almost not influenced by the prefiltering step, at least in theory. The only difference is the larger number of parameters in the noise model. However, in practice, the BJ model should be initialized, which is done using nonoptimal estimators that can be affected by the prefiltering operation. This can influence the convergence properties of the numerical optimization routine. On the other hand, if the noise model order is not adapted for the prefiltering, a sub-optimal result will be obtained because the noise model in use will be too simple. This results not only in a larger variability, also the uncertainty bounds that are calculated together with the estimate will be incorrect, because these rely heavily on the estimation of a correct noise model. All these aspects are illustrated in the next exercise.

Exercise 70 (Study of the behavior of the BJ model in combination with prefiltering)   Goal: identify a transfer function model of a linear dynamic system in the presence of disturbing noise using the BJ model, first without pre-filter, next using a pre-filter.

Calculate the output of the linear dynamic system $y(t) = G_0(q)u_0(t) + H_0(q)e(t)$ that is driven by filtered Gaussian noise $u_0(t) = G_{Gen}(q)r(t)$ for the input, and $e(t)$ white Gaussian noise with standard deviation 0.1. Generate records with a length $N + N_{Trans}$. The first N_{Trans} data points of the simulation are eliminated to get rid of the initial transients of the simulation. From the remaining N data points, the plant and noise parameters θ_0 are identified, using the BJ model.

- Generate the system G_0:
  ```
  [b0,a0] = cheby1(2,5,2*0.08); b0(2) = b0(2)*1.3;
  ```
- Generate the excitation filter G_{Gen}:
  ```
  [bGen,aGen] = butter(3,2*0.25); bGen(2) = 0.9*bGen(2);
  ```
- Define the noise generating filter:
  ```
  [bNoise,aNoise] = butter(2,2*0.2);
  bNoise = bNoise+0.1*aNoise;
  ```
- Generate the prefilter:
  ```
  [bf,af] = butter(3,2*0.01); bf = bf+0.01*af;
  ```
- Select $N = 1024$ and $N_{Trans} = 1000$, and generate $u_0, y_0, v(t)$. Generate also the prefiltered values $u_p(t) = G_p(q)u_0(t)$, $y_p(t) = G_p(q)y(t)$. Eliminate the first N_{Trans} data points.
- Estimate the plant and noise model parameters $\hat{\theta}$ using the BJ model using:
  - $y(t)$, $u_0(t)$ with the exact plant and noise model order (a),
  - $y_p(t)$, $u_p(t)$ with the same orders as in the previous step (b),
  - $y_p(t)$, $u_p(t)$ with an increased noise model order to account for the prefiltering (c).
- Repeat this 1000 times, and calculate the rms value of the complex error $|G_0(j\omega) - G(j\omega, \hat{\theta})|$ as a function of the frequency for the three situations. Calculate also the standard deviation of the amplitude characteristic of the estimated models, and compare to the theoretical uncertainty bound obtained from the estimated models.

Observations (see Figure 4-18 and Figure 4-19) The first figure shows the rms error

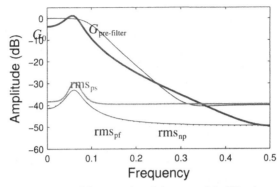

Figure 4-18   Comparison of the rms value of the error of the BJ estimates without and
with prefilter. The rms error is shown (i) without prefilter (np), (ii) with
prefilter and full noise model (pf), and (iii) with prefilter and simple noise
model (ps).

for three simulations. Also the prefilter characteristic is plotted. It can be seen that the results
of situation (a) and (c) are almost identical, as expected from the introduction. The rms error
on the prefiltered data with the too simple noise model (situation b) is significantly larger.
This is due to the nonoptimal frequency weighting by the too simple noise model.

In Figure 4-19 the standard deviation is shown for situation (b) and (c). The plot for (a)

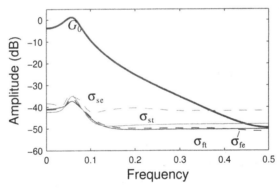

Figure 4-19   Comparison of the theoretic (t) and observed (e) uncertainty on the
amplitude of the estimated transfer function for the full (f) and simple (s)
noise model. σ_{se}, σ_{st} and σ_{fe}, σ_{ft} the estimated and theoretical standard
deviation for, respectively, the simple and the full model.

is identical to that of (c), and is not shown here. From this plot it is seen that for the (a) and (c)
situation, the observed standard deviation is well described by that calculated from the estimated model, using the data of the last run. For situation (b), this is no longer true. This is because the noise model is too simple, so that also the uncertainty bounds that are calculated
starting from this model are incorrect. This brings us to the conclusion that prefiltering can be
used to focus the BJ model at a cost of an increased variance on the estimated model. Also the
estimated variance on the estimates can be wrong.

4.3.7 What have we learned in Section 4.3? Further reading

The major new concept of this section is the observation that it is possible to predict colored noise. While white noise is unpredictable, it is possible to predict the value of colored noise with increasing precision for a decreasing bandwidth using a parametric noise model. In the full identification problem, a parametric plant and noise model are estimated simultaneously. In this section we focussed completely on time domain identification. An excellent and exhaustive introduction can be found in the books of Ljung (1999) and Söderström and Stoica (1989). Also for the parametric identification approach special care has to be taken for identification under feedback. Dedicated methods were developed to avoid systematic errors that are induced by the disturbing noise turning around in the loop (Van den Hof and Schrama, 1995). In this book we focus completely on SISO identification. The reader can find more information on MIMO identification under the following topics: subspace identification; state space identification; common denominator models; left fraction modelling; and so on.

4.4 IDENTIFICATION USING NONPARAMETRIC NOISE MODELS AND PERIODIC EXCITATIONS

In the Section 4.3, parametric noise models were used: the noise was modeled as filtered white noise, and the filter coefficients were explicitly estimated simultaneously with the plant model. In this section, we will use a nonparametric model for the noise. We will study two situations, the first assuming that a periodic excitation is used and that the measurements are made under steady state conditions (Section 4.4); secondly we will show how to deal with random excitations (Section 4.5).

4.4.1 Identification using nonparametric noise models and periodic excitations

In Exercise 47 Chapter 3 it was illustrated how to estimate the signals $\hat{U}(k)$, $\hat{Y}(k)$ and the noise levels $\hat{\sigma}_U^2(k)$, $\hat{\sigma}_Y^2(k)$, $\hat{\sigma}_{YU}^2(k)$ if a periodic excitation is used, and a series of repeated (successive) periods is measured. We will make use here of these results. For the periodic excitation we select a multisine signal (see Chapter 2). We advice the reader to make these exercises before continuing here, because in the next sections we assume that the reader is familiar with these techniques.

Once the averaged signal $\hat{U}(k)$, $\hat{Y}(k)$ and the noise (co-)variances $\hat{\sigma}_U^2(k)$, $\hat{\sigma}_Y^2(k)$, $\hat{\sigma}_{YU}^2(k)$ are obtained from the preprocessing step, the results are used to estimate the plant model by minimizing the following cost function (see also Figure 4-1):

$$V_F(\theta, Z) = \frac{1}{F}\sum_{k=1}^{F} \frac{\left|\hat{Y}(k) - G(\Omega_k, \theta)\hat{U}(k)\right|^2}{\hat{\sigma}_Y^2(k) + \hat{\sigma}_U^2(k)|G_0(\Omega_k, \theta)|^2 - 2\mathrm{Re}(\hat{\sigma}_{YU}^2(k)\overline{G(\Omega_k, \theta)}(\Omega_k, \theta))} \tag{4-32}$$

which can be rewritten as

$$V_F(\theta) = \frac{1}{F}\sum_{k=1}^{F} \frac{\left|A(\Omega_k, \theta)\hat{Y}(k) - B(\Omega_k, \theta)\hat{U}(k)\right|^2}{\hat{\sigma}_Y^2(k)|A(\Omega_k, \theta)|^2 + \hat{\sigma}_U^2(k)|B(\Omega_k, \theta)|^2 - 2\mathrm{Re}(\hat{\sigma}_{YU}^2(k)A(\Omega_k, \theta)\overline{B}(\Omega_k, \theta))}. \tag{4-33}$$

In these expressions $\hat{\sigma}_U^2(k)$, $\hat{\sigma}_Y^2(k)$, $\hat{\sigma}_{YU}^2(k)$ are the nonparametric noise models for the noise on the input, output and the covariance between both, as a function of the frequency. The frequency Ω_k is

$$\Omega_k = j\omega_k \text{ for continuous time systems,}$$

and $\Omega_k = e^{jk\frac{2\pi}{N}f_s}$ for discrete time systems (f_s is the sample frequency).   (4-34)

The reader can remark that this setting is more general than that of the classical prediction error framework in Section 4.3:

- Besides process and measurement noise on the output, we can deal here also with disturbing noise on the input. This leads to what is called the errors-in-variables framework. This generalized problem can be solved in the time or in the frequency domain. While it leads to much more complex methods in the time domain, the complexity of the problem remains the same in the frequency domain, for the settings of this section.

- In the frequency domain, the switch from a discrete time to a continuous time model requires only a change of the frequency variable. In the time domain, the identification of continuous time models requires again more advanced methods than those developed in the classical prediction error framework that is directed towards discrete time models.

We will not further elaborate on these aspects and refer the reader to the existing literature. In the next series of exercises we will learn how to apply the frequency domain identification methods using discrete time models and assuming that there is only disturbing noise on the output, the input is exactly known. Generalizing this to continuous time systems is straight forward by changing the frequency variable in the model. Also dealing with (correlated) input and output noise remains very simple under the condition that periodic excitations are used. In Section 4.7 we deal with these generalizations.

As a first exercise, we repeat Exercise 67, using this time periodic excitation signals, a nonparametric noise model, and a frequency domain method. The results will be compared to those obtained with the BJ model. The standard steps to use frequency domain identification methods in combination with periodic excitations and a nonparametric noise model are:

- Design a periodic excitation signal with a user specified amplitude spectrum (time domain equivalence: the power spectrum of the excitation), and with the desired resolution (time domain equivalence: the length of the experiment). These steps were illustrated in Chapter 2.

- Measure an integer number of periods, and check that the transients disappeared (see Exercises 35, Chapter 3).

- Calculate the sample mean and sample (co-)variances of the input and output DFT coefficients in the preprocessing step (see Exercise 46, Chapter 3).

- Optional: Select the frequencies that will be used during the identification step. This is the equivalence of prefiltering in the time domain (see Section 4.4.2).

- Start the identification process. Notice that no choice on the noise model should be made. In the time domain, a choice between the ARX, ARMAX, OE, or BJ model should be made.

Exercise 71 (Identification in the frequency domain using nonparametric noise models)   Goal: Identify a transfer function model of a linear dynamic system in the presence of disturbing noise, using a nonparametric noise model.

Calculate the output of the linear dynamic system $y(t) = G_0(q)u_0(t) + H_0(q)e(t)$ that is excited by a multisine signal, and disturbed by filtered Gaussian noise $H_0(q)e(t)$, with $e(t)$ white Gaussian noise with standard deviation 0.1. Generate records with a length $NM + N_{\text{Trans}}$, with M the number of periods and N the number of data points in one period. The first N_{Trans} data points of the simulation are eliminated to get rid of the initial transients of the simulation. From the remaining NM data points, a nonparametric noise model $\hat{\sigma}_Y(k)$ is extracted in the preprocessing step, and next the plant parameters θ_0 are identified, by minimizing the least squares cost (4-33), using the FDIDENT toolbox (Kollár, 1994) (see also Exercise 76).

- Generate the system G_0: `[b0,a0] = cheby1(2,10,2*0.36)`
- Generate the system G_{Gen}: `[bGen,aGen] = butter(3,2*0.4)`
- Select $N = 1024$, $M = 7$, and $N_{\text{Trans}} = 1024$.
- Generate a random phase multisine. The amplitudes are set by the amplitude characteristic the generator filter:

$$|U_{\text{MS}}(k)| = \left|G_{\text{gen}}\left(k\frac{2\pi}{N}\right)\right|, \quad k = 1, ..., N/2. \tag{4-35}$$

Scale the multisine so that its rms value is equal to that of a filtered Gaussian noise sequence $G_{\text{gen}}(q)r(t)$, with $\sigma_r = 1$.

- Generate $M = 7$ periods in steady state of $y(t) = G_0(q)u_0(t) + H_0(q)e(t)$ with $\sigma_e = 0.1$.
- Use for H_0 one of the following sets of noise filters coefficients:
  - ARX noise: `bARX = 1; aARX = a0`
  - OE noise: $H_0 = 1$, or `bOE = 1; aOE = 1;`
  - BJ noise: `[bBJ,aBJ] = butter(2,2*0.133);`
- Calculate $\hat{U}(k)$, $\hat{Y}(k)$, and $\hat{\sigma}_Y^2(k)$, used for frequency domain identification (we know that $\sigma_{\hat{U}}^2(k)$ and $\sigma_{\hat{Y}U}^2(k)$ equal zero).

$$\hat{U}(k) = \frac{1}{M}\sum_{l=1}^{M} U^{[l]}(k), \quad \hat{Y}(k) = \frac{1}{M}\sum_{l=1}^{M} Y^{[l]}(k), \tag{4-36}$$

$$\hat{\sigma}_Y^2(k) = \frac{1}{M}\frac{1}{M-1}\sum_{l=1}^{M} \left|Y^{[l]}(k) - \hat{Y}(k)\right|^2. \tag{4-37}$$

Notice that the extra division by M is made to get the (co-)variance on the mean values $\hat{U}(k)$, $\hat{Y}(k)$.

- Calculate also the mean values over the M periods in the time domain: $\hat{u}_0(t)$, $\hat{y}(t)$ (used for time domain identification).
- Time domain identification: Estimate the plant and noise model parameters $\hat{\theta}$ for each of these three disturbances, using respectively the ARX, OE, and BJ model for the AR, OE, and BJ noise source, starting from $\hat{u}_0(t)$, $\hat{y}(t)$.

Frequency domain identification FDIDENT: Estimate the plant parameters using the frequency domain estimator for each of the three noise filters. This can be done using the following code: Generate the long input and output data records, and split them in M subrecords of length N_{per}, next load the data in the toolbox data structure and process them (Section 4.7 shows how to use the graphical user interface).

```
u0 = reshape(u0,NPer,M);
U0 = fft(u0)/sqrt(N); U0=U0(Lines,:);
y = reshape(y,NPer,M); Y = fft(y)/sqrt(N); Y = Y(Lines,:);
% calculate the Fourier data
Fdat = fiddata(num2cell(Y,1), num2cell(U0,1),f(:));
% store the data in fiddata format
Fdat = varanal(Fdat);
% add variance analysis
nb = OrderG;        % order numerator
na = OrderG;        % order denominator
MFD = elis(Fdat,'z',nb,na,struct('fs',1));
```

■ Repeat this 20 times, and calculate the rms value of the complex error $|G_0(\omega) - G(w, \hat{\theta})|$ as a function of the frequency for all these estimates.

Observations (see Figure 4-20) From the results it can be seen that the rms value the er-

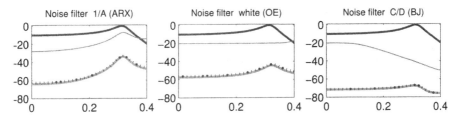

Figure 4-20 Comparison of time and frequency domain identification methods for three different noise types and the corresponding time domain model. Black line: G_0. Thin gray line: Power spectrum disturbing noise. ...: Rms error freq. dom. method. Bold gray line: Rms error time domain model (ARX, OE, BJ, respectively).

ror for the frequency domain identification method is almost equal to that of the optimal time domain method for each of the three noise situations. A small loss in efficiency of $(M-2)/(M-3)$ on the variance of the frequency domain method can be observed. This is due to the fact that in the parametric noise model less parameters are used than in the nonparametric one. Only in the ARX situation there can be a larger difference because in that case the noise model contributes to the knowledge of the plant model: the noise and plant model have the same poles. Of course this is only valid if the user can guarantee on the basis of prior knowledge that this is indeed the case, otherwise, a bias error will be introduced. This will be illustrated in Section 4.5.

So we can conclude that both methods are almost equivalent to each other for the experimental conditions of this exercise (no disturbances on the input, periodic excitation). Selecting between a parametric or a nonparametric noise model becomes a users choice (Can the user easily select the noise model structure and order? Can the user afford a loss of, e.g., 1.5 dB in efficiency? Does the Box–Jenkins model get stuck in a poor local minimum?).

Discussion In this exercise we knew from the setting that only $\hat{\sigma}_Y^2(k)$ differs from zero, and we could put $\sigma_U^2(k)$ and $\sigma_{YU}^2(k)$ equal zero. However in case $\sigma_U^2(k)$ is different from zero, the user is not allowed to put $\sigma_{YU}^2(k) = 0$, even if this is known from prior information. In order to keep the consistency of the estimate, it is required to calculate and use $\hat{\sigma}_{YU}^2(k)$ in the cost function in combination with

$$\hat{U}(k) = \frac{1}{M}\sum_{l=1}^{M} U^{[l]}(k), \quad \hat{Y}(k) = \frac{1}{M}\sum_{l=1}^{M} Y^{[l]}(k), \tag{4-38}$$

$$\hat{\sigma}_U^2(k) = \frac{1}{M}\frac{1}{M-1}\sum_{l=1}^{M} |U^{[l]}(k) - \hat{U}(k)|^2, \quad \hat{\sigma}_Y^2(k) = \frac{1}{M}\frac{1}{M-1}\sum_{l=1}^{M} |Y^{[l]}(k) - \hat{Y}(k)|^2,$$

$$\hat{\sigma}_{YU}^2(k) = \frac{1}{M}\frac{1}{M-1}\sum_{l=1}^{M} (Y^{[l]}(k) - \hat{Y}(k))\overline{(U^{[l]}(k) - \hat{U}(k))}. \tag{4-39}$$

This is because the sample co-variances calculated for a finite number of repetitions M will not be equal to zero (Pintelon and Schoukens, 2001).

4.4.2 Emphasizing a frequency band

The goal of this section is to emphasize a frequency band during the identification. A similar study was made for the Box–Jenkins model where it was be shown that this is not straightforward. When the full noise model was used, the prefiltering operation does not affect the final model properties, but it affected the initialization process. A similar result will be illustrated here if a nonparametric noise model is used. Prefiltering will not drastically change the properties of the estimate. However, in this case there is an alternative. In the frequency domain, it is possible to select arbitrarily what frequencies will be used in the cost function (4-32), offering also the freedom to the user to focus the estimate on a user defined frequency band. This is illustrated in the next exercise.

Exercise 72 (Emphasizing a frequency band) Goal: Emphasize a frequency band during the identification of a transfer function model of a linear dynamic system.
Calculate the output of the linear dynamic system

$$y(t) = G_0(q)u_0(t) + H_0(q)e(t) = y_0(t) + v(t) \tag{4-40}$$

that is driven by a multisine for the input, and $e(t)$ white Gaussian noise with standard deviation 0.1. Generate records with a length $MN + N_{\text{Trans}}$, with N the period of the multisine. The first N_{Trans} data points of the simulation are eliminated to get rid of the initial transients of the simulation. From the remaining data points, the plant and noise parameters θ_0 are identified, using the frequency domain method.

■ Generate the system G_0:
  ```
  [b0,a0] = cheby1(2,5,2*0.08); b0(2) = b0(2)*1.3;
  ```
■ Generate the excitation filter G_{Gen}:
  ```
  [bGen,aGen] = butter(3,2*0.25); bGen(2) = 0.9*bGen(2);
  ```
■ Define the noise generating filter:
  ```
  [bNoise,aNoise] = butter(2,2*0.2);
  bNoise = bNoise+0.1*aNoise;
  ```
■ Generate the prefilter:
  ```
  [bf,af] = butter(3,2*0.1); bf = bf+0.01*af;
  ```
■ Generate a zero mean random phase multisine with a flat amplitude up to the Nyquist frequency (not included). Scale the rms value to be equal to 1. Use $N = 1024$, $M = 7$, and $N_{\text{Trans}} = 1024$. Filter this signal with the generator filter to get u_0.

- Generate $u_0, y_0, v(t)$. Generate also the prefiltered values $u_p(t) = G_p(q)u_0(t)$, $y_p(t) = G_p(q)y(t)$. Eliminate the first N_{Trans} data points.

- Estimate the plant and noise model parameters $\hat{\theta}$ using the frequency domain-method with the settings of Exercise 71, for u_0, y, the prefiltered data u_p, y_p, and the restricted frequency band using only the FFT lines `[2:1:floor(0.1*N)]`.

- Repeat this 100 times, and calculate the rms value of the complex error $|G_0(j\omega) - G(j\omega, \hat{\theta})|$ as a function of the frequency for the three situations: no prefilter, prefilter, cut of the frequency band of interest.

Observations (see Figure 4-21) The figure shows the rms error for three simulations.

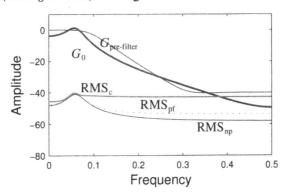

Figure 4-21   Comparison of the rms value of the error of the FD estimates without (np), with prefilter (pf), and with a restricted frequency band (c).

Also the prefilter characteristic is plotted. It can be seen that the results with and without pre-filtering are identical in the centre of the frequency band of interest, as expected from the introduction. Restricting the frequencies to the frequency band of interest results in a slightly larger uncertainty. This is because all information outside that band is not used in that case. Extending the selected frequency interval would reduce this loss.

The reader should notice that selecting the frequencies of interest is a very flexible method. In combination with a good selection of the excitation signal, the user gets a full freedom to tune the experiment to the frequencies of interest. Very complex frequency patterns can be used, avoiding for example to use the frequencies in the neighborhood of the mains frequency (and its higher harmonics). If also the excitation is tuned similarly using a multisine, no information loss will be present anymore.

4.4.3 Identification under Feedback Conditions

In this exercise, we repeat the feedback Exercise 68, but this time using a plant and noise system of order 2.

Exercise 73 (Comparison of the time and frequency domain identification under feedback)

- Consider the setup of Exercise 68.
- The system G_0 is given by the filter:
  ```
  [bFF,aFF] = cheby1(OrderG,20,0.5);bFF(1) = 0;
  bFF = bFF*2;  with OrderG = 2.
  ```
- The process noise v is filtered Gaussian noise with standard deviation $\sigma_v = 0.1, 0.5, 1$ respectively, the noise filter is

```
[bNoise,aNoise] = butter(OrderNoise,2*0.133);
```
with OrderNoise = 2.

■ Replace the excitation signal by a multisine as specified in Exercise 71 (amplitude spectrum, period length N = 1024).

■ Make the simulation as in Exercise 71, measuring the input (not the reference input to the feedback) of the system captured in the feedback loop, and the corresponding output of the feedback loop. Calculate the variances and covariance of the input and output, frequency per frequency and use this as the nonparametric noise model.

■ Identify the plant model using the BJ model and the frequency domain method using a nonparametric noise model.

■ Repeat this 100 times. Do the following:
- Calculate the rms value of the complex error $|G_0(j\omega) - G(j\omega, \hat{\theta})|$ as a function of the frequency for both methods.
- Calculate the mean value $G(j\omega, \hat{\theta})$ as a function of ω over the 100 realizations for both methods.
- Plot the results.

Observations (see Figure 4-22 and 4-23) ■ From these results it can be observed that

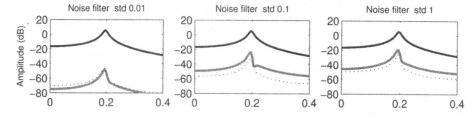

Figure 4-22   Comparison of rms-error of the time and frequency domain identification methods under feedback conditions. Black line: G_0 ; gray line: rms error of Box–Jenkins; ...: rms error freq. dom. method

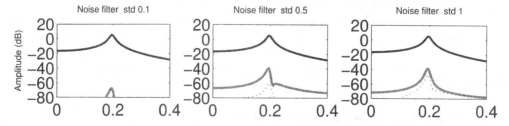

Figure 4-23   Comparison of bias error of the time and frequency domain identification methods under feedback conditions. Black line: G_0. Gray line: Bias error of Box–Jenkins. ...: Bias error frequency domain method.

both methods behave similarly for the simulation with low noise disturbances. For the other two noise levels, a (slight) difference can be observed between the time and frequency domain method, although both methods should be equivalent under these conditions. This difference is because the risk to get stuck in a local minimum is larger for the BJ model than for the frequency domain method, the latter has to solve a simpler optimization problem (only the plant model has to be identified). This can be verified by plotting all the estimated models on a single plot, making a few outliers visible for the BJ model (Figure 4-24).

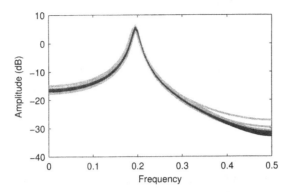

Figure 4-24   Check for the presence of outliers of the time and frequency domain identification methods under feedback conditions. Amplitude FRF of 100 simulations. Black line: Frequency domain. Gray curve: Box–Jenkins.

4.4.4 What have we learned in Section 4.4? Further reading

In Section 4.3 the parametric plant and noise model were simultaneous identified using arbitrary excitations, e.g. random or periodic excitations. If possible, we strongly advice the user to apply periodic excitations. This offers many potential advantages: Besides a full nonlinear analysis that can be made (see Chapter 5), it is also possible to extract a nonparametric noise model without any user interaction. This simplifies the identification process significantly because the simultaneously selection of the plant- and noise model order is a difficult task for unskilled users. For periodic excitations, it is easy to deal with (correlated) noise on the input and output data, and also direct identification under feedback conditions requires no special care. A detailed discussion is given in the book of Pintelon and Schoukens (2001).

4.5 FREQUENCY DOMAIN IDENTIFICATION USING NONPARAMETRIC NOISE MODELS AND RANDOM EXCITATIONS

In the previous section we restricted the excitation to be periodic in order to estimate in a preprocessing step the nonparametric noise model. Here we consider again the general situation where the system is excited with random excitations. To extract the noise model under these conditions, we preprocess the data first with the local polynomial method (see Exercise 51, Chapter 4). While we could tolerate disturbing noise on the input and the output data in Section 4.4, we impose here again that the input is exactly known as it was for the prediction error framework that was used in Section 4.3. So we will make this exercise under exactly the same conditions as defined for the classical time domain prediction error framework (see Section 4.3) and we process exactly the same data as in Exercise 67 to emphasize this.

Exercise 74 (Identification in the frequency domain using nonparametric noise models and a random excitation)   Goal: Identify a transfer function model of a linear dynamic system in the presence of disturbing noise making use of a nonparametric noise model. Calculate the output of the linear dynamic system $y(t) = G_0(q)u_0(t) + H_0(q)e(t)$ that is driven by filtered Gaussian noise $u_0(t) = G_{\text{Gen}}(q)r(t)$ for the input, with $r(t)$ and $e(t)$ white Gaussian noise with standard deviation 1 and 0.1, respectively. Generate records with a length $N + N_{Trans}$. The first N_{Trans} data points of the simulation are eliminated to get rid of the initial transients of the simulation. From the remaining N data points, the plant and noise parameters θ_0 are identified, by minimizing the least squares cost (4-26), using the MATLAB® System Identification toolbox of Ljung.

- Generate the system G_0: [b0,a0] = cheby1(2,10,2*0.36)
- Generate the system G_{Gen}: [bGen,aGen] = butter(3,2*0.4)
- Select the 'BJ noise' filter of Exercise 67:
  [bBJ,aBJ] = butter(2,2*0.133);
- Select $N = 5000$ and $N_{Trans} = 1000$.
- Nonparametric preprocessing: Estimate the FRF, the cleaned output (leakage errors eliminated), and its variance using the "Local Polynomial Method" as explained in Chapter 3, Exercise 51.
- Estimate the plant model using the BJ model with the time domain data, and use the frequency domain identification method of the previous section using the data obtained in the preprocessing step. Put the variance on the input and the covariance equal to zero.
- Repeat this at least 100 times, and calculate the rms value of the complex error $|G_0(\omega) - G(w, \hat{\theta})|$ as a function of the frequency for both methods. Plot the results.

Observations (see Figure 4-25) From these results it can be observed that both methods

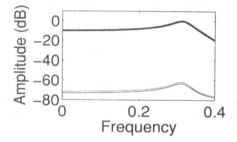

Figure 4-25 Comparison of rms error of the Box–Jenkins estimate (time domain, parametric noise model) and the frequency domain method (frequency domain, preprocessing local polynomial method, nonparametric noise model). Bold line: G_0. Thin black line: rms error Box–Jenkins estimate. Thin gray line: rms error frequency domain method.

behave very similarly although a small increase of 1.3 dB in the rms error can be observed for the nonparametric noise model. This loss is very typical for the use of nonparametric noise models. The major advantages of the nonparametric noise model are that (i) the user faces only the selection of the plant model structure instead of the simultaneous selection of the parametric plant and noise model structure; (ii) a simpler optimization problem needs to be solved, resulting in less convergence problems of the numerical optimization method. The major disadvantages are a small loss in efficiency, and it is impossible to express that the plant and noise model have common poles.

4.6 TIME DOMAIN IDENTIFICATION USING THE SYSTEM IDENTIFICATION TOOLBOX

In this section and Section 4.7, we familiarize the reader with the use of a time and frequency-domain system identification toolbox. Starting from the data, we show how to deal with the model structure selection and the parameter estimation and eventually the model validation. To process the data in the time domain, we use the MATLAB® System Identification toolbox of the Mathworks that is based on the prediction error framework using parametric noise models (Section 4.6). The toolbox is written by Lennart Ljung (Linkoping University, Sweden). We want to acknowledge Lennart for providing us with the best practices how to deal with these data in the toolbox. The second toolbox that we will illustrate in Section 4.7 is the frequency domain system identification toolbox, FDIDENT, written by Istvan Kollar (Budapest University of Technology and Economics). Also Istvan contributed to the preparation of these examples. The FDIDENT toolbox makes use of nonparametric noise models.

Exercise 75 (Using the time domain identification toolbox)   Goal: Make a complete identification run using the time domain identification MATLAB® toolbox, The system is excited with a filtered noise sequence.

- Generate the system G_0:
  ```
  [b0,a0] = cheby1(2,10,2*0.25); b0(2) = b0(2)*1.3;
  ```
- Define the noise generating filter:
  ```
  [bNoise,aNoise] = butter(1,2*0.2);
  bNoise = bNoise+0.1*aNoise;
  ```
- Generate a filtered random noise excitation:
  ```
  [bGen,aGen] = butter(3,0.3);
  u0 = filter(bGen,aGen,randn(1,N+NTrans));
  u0 = u0/std(u0);
  ```
  Use a data length $N = 1024 \times 7$, and $N_{\text{Trans}} = 1000$.
- Generate y_0, $v(t)$: $y(t) = G_0(q)u_0 + v(t)$ with $v(t) = 0.1G_{\text{noise}}(q)e(t)$ filtered white noise $e(t) \sim N(0, 1)$. Eliminate the first N_{Trans} data points to remove the transients of the simulation.
  ```
  y0 = filter(b0,a0,u0);
  yNoise = filter(bNoise,aNoise,randn(1,N+NTrans));
  yNoise = stdNoise*yNoise;
  y = y0+yNoise; u0(1:NTrans) = []; y(1:NTrans)=[];
  ```
  □

Discussion Processing the data. In the following, we discuss step by step the processing of the data.

A. Make an estimation and validation data set
In the first step the data are split in two parts: (i) an estimation set that is used to estimate the model parameters; (ii) a validation set that is used to verify the quality of the estimated model and to tune the selection of the model order.
```
tData = iddata(y(:),u0(:),1);
tDataEst = tData(1:N/2);
tDataVal = tData(N/2+1:end);
plot(tDataEst)
```
The results are shown in Figure 4-26 For the convenience of the reader with show the figures as they are plotted in the toolbox.

B. Make an initial guess of the delay
First we have to verify if there is a delay present in the system. Is there a direct term from the input to the output (e.g., the model is used to approximate band limited data), or is there one

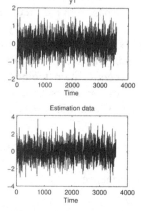

Figure 4-26   Split of the data in an estimation and a validation set.

or more samples delay between the input and the output (e.g., a zero-order-hold excitation). This is verified with the following instruction:

```
nk = delayest(tDataEst)
```

The result is $nk = 0$. No delay is detected which is in agreement with the presence of a term $b0(1) \neq 0$ in the definition of the system G_0 in this exercise.

C. Make an initial guess

A first initial guess of the plant model is obtained by the instruction:

```
m = pem(tDataEst,'nk',nk);
```

This routine estimates a state space representation with common poles for the plant and the noise model (see the "Users Manual" of the MATLAB® System Identification Toolbox).

This delivers a first estimate of the plant. The order is determined by the algorithm itself using a model selection rule that is based on the decrease of the singular values, indicating the rank of an internal matrix. Look at the model m in the work space of MATLAB®. In this case a state space model of order 3 is retrieved (this can vary depending upon the realization of the input!).

For the rest of this study we calculate also the other models of order 1 and 2 and store them in m1, m2, and m3.

```
m1 = pem(tDataEst,1,'nk',nk);  % order 1
m2 = pem(tDataEst,1,'nk',nk);  % order 2
m3 = pem(tDataEst,3,'nk',nk);  % order 3
```

These models are next evaluated:

```
compare(tDataVal,m1,m2,m3).
```

For each model the "Fit" and the FPE (final prediction error) are calculated and tabled in Table 4-1. The "Fit" is defined as:

$$Fit = 100\left(1 - \frac{\text{MSE}(y - y_{\text{sim}})}{\text{MSE}(y - \text{mean}(y))}\right). \tag{4-41}$$

The FPE is obtained from the cost function by multiplying it with a model complexity term (see also the AIC-model selection criterion in Chapter 1). The fit value is calculated on the validation data. It measures the capability of the model to simulate the output. From Exercise

61, it is known that it is more demanding to get a small simulation error than to get a small prediction error.

The results are also shown in Figure 4-27.

TABLE 4-1 Study of the simulation and prediction error of the estimated models

Model	Fit	FPE
m1	57.63%	0.0263241
m2	86.16%	0.0028511
m3	87.14%	0.0020932
m4	83.37%	0.00406543
mbj2	87,20%	0.00187675
mbj3	87,20%	0.00187869

The table shows that the first model m1 is poor, while the third model m3 is slightly better than the second one m2. In order to be sure not to miss important dynamics, we also test the model of order 4. The simulation error on the validation set starts to grow again, which points in the direction of over fitting (see Exercise 11 on model order selection in Chapter 1). We will continue with the model of order 3.

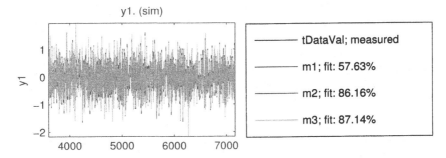

Figure 4-27   Comparison of models of order 1, 2, and 3. Left: Comparison of the simulated outputs on the validation data. Right: Quality of fit.

D. Analysis of the residuals

The validation of the model is continued by making an analysis of the residuals on the validation data. The difference between the simulation error and the measured output is analyzed using the command:

```
resid(tDataVal,m3)
```

This instruction tests if the residuals are uncorrelated (whiteness test), and if there is no cross-correlation left between the input and the output (unmodeled dynamics). Both are shortly discussed below.

(i) *Whiteness test*: For a perfectly tuned model, the simulation errors should be white. If there are no model errors, the dominating simulation errors are the noise disturbances. These are whitened by the noise model. White residuals correspond to uncorrelated residuals in the time domain, and that is verified in this test. In Figure 4-28 (top), the auto-correlation of the residuals are plotted, together with the 99% uncertainty interval around zero. It is seen that the a few samples is outside this interval (more than 1%). It can be concluded that the residuals are not completely white within the statistical precision of the test, but the quality of the model is already very good.

(ii) *Unmodeled dynamics*: We verify if the residuals are uncorrelated with the input. This test reveals if there are significant unmodelled dynamics left between the input and the residuals. These will become visible in a cross-correlation test between the input and the residuals. In Figure 4-28, it is seen that the same conclusions can be made as for the whiteness test.

Because both tests do not indicate the presence of large model errors, we can conclude that the estimated model is complex enough to explain the data. It might still be possible that

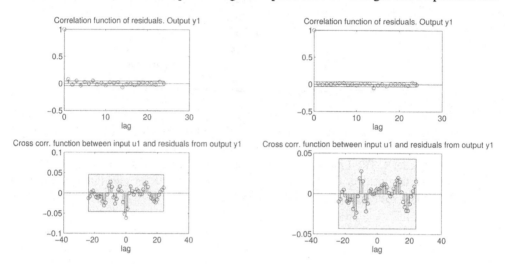

Figure 4-28   Correlation analysis of the residuals for the model m3 (left) and the BJ model mbj2 (right). Top: autocorrelation test. The light box indicates the 99% uncertainty interval. Bottom: Cross-correlation between input and output.

the simpler second order model would also be a valid choice. To analyze this we take a look at the poles and zeros of the third order model.

E. Analysis of the poles/zeros

In Figure 4-29 we plot the poles and zeros for the plant and noise model. Three stable poles are retrieved for the plant model. One pole is very close to a zero, suggesting that this is a candidate pole/zero pair that can be omitted from the model. Also the overlapping uncertainty bounds confirm this suggestion, although the reader should be careful to make a firm conclusion because also the covariance should be considered (do the poles and zeros move together over different realizations or not). The fit values in step C point also in the same direction: The difference between the fit-values for model m2 and m3 are indeed very small, although the prediction errors differ more.

On the right side of the figure, the poles and zeros of the noise model are shown. In this case a complex pole/zero pair is canceled. It can be observed that the canceled poles are complementary in the plant and noise model. This points to the fact that the underlying assumption of the initial model that the poles are shared is not valid. The zeros are used to separate the plant and noise model. Seemingly the previous model structure is not flexible enough to accommodate simultaneously a simple plant and noise model. Since the poles are common to both models, excess poles are needed that are canceled by the zeros in the model: one pole is canceled in the plant model, two poles are canceled in the noise model. This gives a strong in-

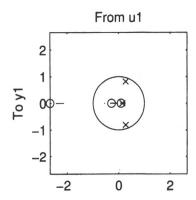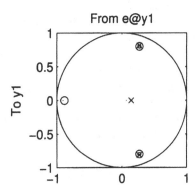

Figure 4-29 Poles (x) and zeros (o) of the estimated third order system $\hat{G}$ (left) and the noise model (right). One pole and zero coincide almost completely for the plant model. For the noise model two complex poles/zeros are coinciding..

dication that a more flexible BJ model could do a better job, because it does not link the poles of plant and noise model. The pole/zero analysis suggests to use a second order plant model and a first order noise model. This is verified in the next step.

F. Comparison with a Box–Jenkins model

Motivated by the conclusions of the previous step, we try to identify a Box–Jenkins model on the data. We make two trials, the first time using a plant model of order two, and the second time of order three. In both case the noise model order is set to 1. The models are estimated using the instructions:

```
mbj2 = bj(tDataVal,[3 1 1 2 0]);
mbj3 = bj(tDataVal,[4 1 1 30]);
compare(tDataVal,m3,mbj2,mbj3)
```

The fit value of both models is about the same and close to the m3-results: fit = 87.14% for m3, and 87.2% for the Box–Jenkins models (see Table 4-1). Next we make again an analysis of the residuals.

```
resid(tDataVal,mbj2);
```

The results are shown in Figure 4-28 (right). It can be seen that the mbj2 model is passing the correlation and the cross-correlation test better than the m2 model (left) did. This brings us to the conclusion that we can retain the Box–Jenkins model mbj2 as the best model.

As a final test we simulate the output of the validation set and compare it to the exact output y0 of the simulation. When we calculate the rms error for both models we find an error of 0.0070 for the m3 model and a slightly smaller value 0.0062 for the mbj2 model.

The model parameters of mbj2 model are close to the true values of the model parameters that are used in the simulation (scale the noise model parameter bNoise(1) to 1):

```
Discrete-time IDPOLY model:
 y(t) = [B(q)/F(q)]u(t) + [C(q)/D(q)]e(t)
B(q) = 0.0902 + 0.2493 q^-1 + 0.09116 q^-2
C(q) = 1 + 0.9726 q^-1
D(q) = 1 - 0.1579 q^-1
F(q) = 1 - 0.5377 q^-1 + 0.7357 q^-2
Estimated using BJ from data set z
Loss function 0.00186943 and FPE 0.00187675
```

G. Conclusion

At the end of this procedure we retrieve as the best model the `mbj2` model that uses a second-order plant model and a first order noise model. This model is slightly better than the initial `m3` model. This is mainly due to the more flexible model structure that allowed the plant and noise models to be decoupled. The retrieved model structure corresponds also with the exact models that were used to generate the simulation data.

4.7 FREQUENCY DOMAIN IDENTIFICATION USING THE TOOLBOX FDIDENT

The goal of this exercise is to illustrate more advanced aspects of the identification procedure using the frequency domain identification toolbox FDIDENT (Kollár, 1994). This toolbox is completely directed to the use of nonparametric noise models. A full identification run will be made, starting from the raw data to a final model. Some model selection tools and model tests will be discussed in more detail. Here we illustrate the use of periodic excitation signals. We refer the reader to Exercise 74 to deal with non-periodic excitations.

Exercise 76 (Using the frequency domain identification toolbox FDIDENT)
Goal: Make a complete identification run using the FDIDENT toolbox. The same system as in Exercise 75 is identified, but this time using a periodic excitation.

- Generate the system G_0:
  ```
  [b0,a0] = cheby1(2,10,2*0.25);b0(2) = b0(2)*1.3;
  ```
- Define the noise generating filter:
  ```
  [bNoise,aNoise] = butter(1,2*0.2);
  bNoise = bNoise+0.1*aNoise;
  ```
- Generate a zero mean random phase multisine with a flat amplitude exciting the spectral lines:
  ```
  Lines = [1:NPer/3].
  ```
  Scale the rms value to be equal to 1. Use a period length $N = 1024$, and generate $M = 7 + 1$ periods. The first period is used to eliminate all transient effects in the simulations ($N_{Trans} = 1024$).
- Generate $y_0, v(t)$: $y(t) = G_0(q)u_0 + v(t)$ with $v(t) = 0.1 G_{noise}(q)e(t)$ filtered white noise $e(t) \sim N(0, 1)$. Eliminate the first N_{Trans} data points.
- Generate a time domain object that can be imported by FDIDENT, normalizing the sample frequency equal to 1:
  ```
  ExpData = tiddata(y(:),u0(:),1);
  ```
- Start the GUI of FDIDENT (type `fdident` in the command window) and follow the menu in the GUI as explained below.
Observations The successive windows of the GUI are shown and shortly discussed.

H. Main window

The main window allows the data to be imported in the GUI, using either time or frequency domain data. Double click the 'Read Time Domain Data' to open the data importing window.

I. Importing and preprocessing the data

```
read time domain data
```
In a series of successive steps (see Figure 4-31), the data are
- loaded into the GUI (`Get Data`),

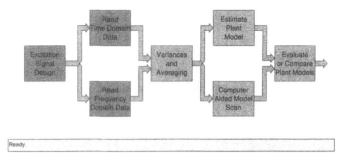

Figure 4-30   Opening window GUI-FDIDENT.

- the successive periods are separated (`Segmentation`),
- and converted to the frequency domain (`Con. to Freq`).
- A first possibility to select the frequencies of interest is offered (`Freq Select`).

The corresponding windows are shown.

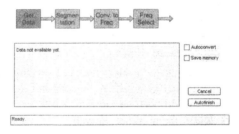

Figure 4-31   Opening window GUI-FDIDENT.

(i) Get the data: The data object `ExpData` that was created in the m-file is loaded.

(ii) Segmentation: Put `period length` to `1024` and click `Apply Periods`.

(iii) Convert to frequency domain: The frequency domain results are shown. The user can select, for example, the FRF, the input–output Fourier coefficients, etc., by making the appropriate choice under the `Type of Figure` instruction.

(iv) Frequency selection: In this window a first selection of the active frequencies to be used in the identification process can be used. We postpone in this exercise this choice, and `select all frequencies` in this step. Notice that the system was not excited above 0.33 Hz.

At the end of these 4 substeps, the Read Time Domain Data block is highlighted indicating that the data are ready to be processed in the next block of the main menu.

Figure 4-32   Load the time domain object `ExpData`.

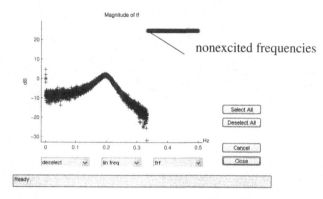

Figure 4-33 Frequency selection: all frequencies selected. The FRF is shown. Notice that above 3.3 Hz, no excitation was present which is indicated in the plot by the straight line.

J. Nonparametric noise analysis
```
variances and averaging
```
In this block, the data are averaged over the periods: the sample mean and sample variance are calculated. We advice to make the final frequency selection in this block in the window where the input and output data are shown.

In this window, it is very easy to select the excited frequency lines in the input window, using the `frequency selection` button. Selection of the not excited frequencies would not affect the MLE, but it would become more difficult to generate good initial estimates of the system parameters to start the nonlinear search.

Once the frequencies are selected, we are ready to start the identification step, as is visible in the main menu.

K. Parametric identification step
```
Estimate plant Model
```
or
```
Computer Aided Model Scan
```
All the information is now available to start the parametric model estimation step. In the toolbox, the sample maximum likelihood (SMLE) is used, minimizing the cost function (4-33). First a series of simplified cost functions is minimized to generate starting values (hidden for the user). The SMLE cost function for each of these parameters is calculated, and the best result (lowest cost function) is retained to start the nonlinear search. The user can select a single

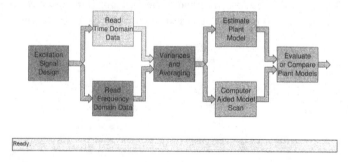

Figure 4-34 The time domain data are imported and transformed to the frequency domain.

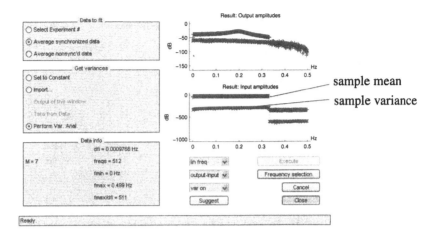

Figure 4-35   Sample mean and sample variances of the input and the output calculated in the
nonparametric preprocessing. Notice that the noise level for the input is 300 dB
below the actual input. This is the MATLAB® calculation precision,
corresponding to 15 digits (20 dB/digit).

model, or a whole bunch of models with different orders can be scanned. The last option is
chosen in this exercise because we will illustrate also the model selection procedures. Opening the `Computer Aided Model Scan` offers a number of user choices (see Figure 4-
37). In this window, the user has to select the nature of the model (for example discrete or
continuous time), the orders to be scanned (a selection - deselection tool is available). The
discussion of the other optional choices is out of the scope of this book, and the user is referred to the `help` functions of the GUI. The results are accessible in the `Evaluate or
Compare Plant Models` window.

Some of the available results are discussed in the next section.

L. Evaluation of the estimated models
`Evaluate or Compare Plant Models window`

(i) Comparing the estimated models

Once the estimates are available, it is tempting to select as "best" model the one corresponding to the lowest cost function value, but it will be shown that this is not the best or even
a good strategy (see also Exercise 11 in Chapter 1).

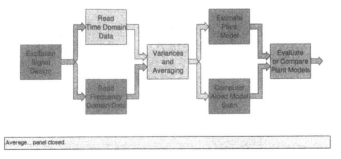

Figure 4-36   The nonparametric preprocessing step is finished. The data are ready to start the
parametric identification step.

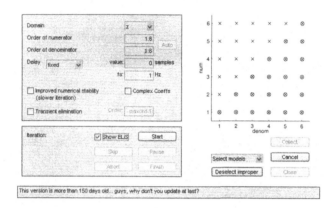

Figure 4-37   Preparation for the computer-aided model scan. The user has to select
- continuous or discrete time model,
- selected set of model orders to be scanned.

In this run, the lowest cost function was obtained for the model 5/5 (5 zeros, 5 poles). One could expect that the model 6/6 would do better than the 5/5 model, because the latter is a subset of the 6/6 class of models. Since the cost function for the 6/6 model is larger than that of the 5/5 model, it shows that the program got stuck in a local minimum for the 6/6.

Notice that the observed cost functions are close to the theoretical expected value (number of frequencies - $n_\theta/2$, with n_θ the number of free parameters in the model, for example for a 2/2 model, $n_\theta = 5$ (the model is invariant with respect to a scaling of all parameters). This is an indication that the models are reasonable, the remaining residuals can be explained by statistical properties of the noise. In the value of the cost function, there is no evidence of the presence of model errors. A cost function that is much larger than the theoretical value, is a strong indication of model errors. A cost function that is significantly smaller than the theoretic value is an indication for a wrong nonparametric noise model (e.g., the presence of a correlation of the noise over the frequencies).

In Figure 4-39, the 2/2 model is compared to the 5/5 one. The FRF and the amplitude of the complex errors $|G(j\omega_k) - G(j\omega_k, \hat{\theta})|$ is shown. These plots indicate that the behavior is quite similar. This points in the direction that 5/5 may be a too complex model for the data.

(ii) Selecting the best model using a model selection tool

In Chapter 1, Exercise 11, we learned that it is not always a good idea to choose the model corresponding to the lowest cost function. This can result in a higher model variability

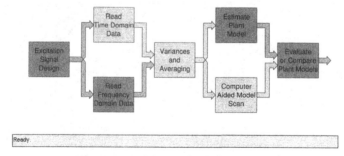

Figure 4-38   Evaluation and comparison of the estimated models: Evaluate   or
Compare Plant Models window.

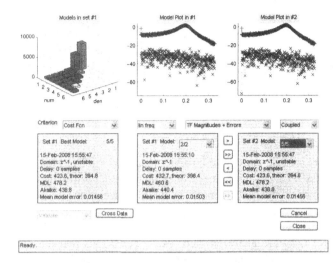

Figure 4-39 Evaluate or Compare Plant Models window
TF Magnitude + Erros is selected

due to an increased noise sensitivity of complex models (see Exercise 10). In the previous section it was observed that 5/5 may be too complex. In order to make a better choice, model selection tools are developed that balance the model complexity versus the model variability by adding a penalty factor for the complexity to the cost function (4-33) (see also Excercise 11). The Akaike information criterion (AIC) or the minimum description length (MDL) are two popular tools that start from the weighted least squares cost function V:

$$V_{\text{AIC}} = V\left(1 + \frac{2n_\theta}{N}\right), \quad V_{\text{MDL}} = V\left(1 + \frac{2n_\theta \log N}{N}\right). \tag{4-42}$$

It can be seen from (4-42) that for the same value of the cost function V, a more complex model results in a higher AIC or MDL criterion. In Figure 4-40, the MDL criterion selects the 2/2 model as the best one.

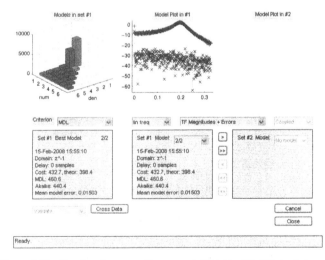

Figure 4-40 Evaluate or Compare Plant Models window
TF Magnitude + Errors is selected; MDL criterion is selected

(iii) Residual analysis

The residuals are that part of the data that the model could not reproduce. Since we have access in the frequency domain to good estimates of the FRF $G(j\omega_k)$, a lot of information can be gained by analyzing the residuals between the measured and modeled FRF:

$$\varepsilon_F(k) = \frac{G(j\omega_k) - G(j\omega_k, \hat{\theta})}{\sigma_G(k)}. \tag{4-43}$$

If no model errors are left, ε_F should be white (Gaussian) noise. This is no longer so if some dynamics are missed (under modeling). Since these model errors have a smooth behavior, a correlation becomes visible that can be detected in a correlation test. Notice that this test does not protect against overmodeling, only under-modelling is detected. In Figure 4-41, the correlation analysis is shown for the 2/2 model where no statistical significant correlation is visible. Also the 1/2 model is analyzed, and here it is obvious that the residuals are strongly correlated, which is a very strong indication for model errors. This is also confirmed by the much larger cost function of this model.

(iv) Pole-zero cancellation

What happens with the extra poles-zeros of the 5/5 model if a 2/2 model does fit the data well? Figure 4-42 shows the answer to that question: the 3 additional poles are canceled by the 3 additional zeros in pole-zero pairs that almost completely coincide. In order to be sure that within the uncertainty bounds the pole and zero coincide, a statistical test would be needed, keeping also in mind that often a strong correlation between these poles and zeros is present. However, if the pole-zero plot shows very close pole/zero pairs, then these are good candidates to be eliminated in an order reduction step without affecting the quality of the estimated model.

The presence of these pole/zero pairs affects also the behavior of the uncertainty of the model, which is visualized in the next section.

(v) Cloud of models

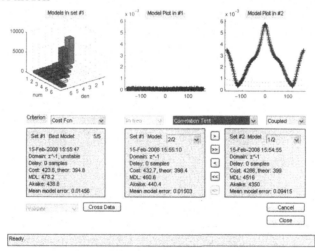

Figure 4-41 Evaluation and comparison of the estimated models: Evaluate or Compare Plant Models window; correlation analysis is selected.

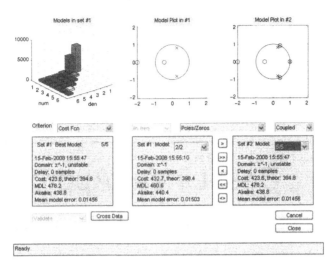

Figure 4-42 Evaluation and comparison of the estimated models:
`Evaluate or Compare Plant Models` window
`pole-zero cancellation` is selected.

To get an idea of the variability of the model, many possibilities exist. Before we have seen that uncertainty bounds can be generated (see Exercise 69). An alternative is to draw a cloud of models. From the estimated model parameters and covariance matrix, a series of model parameters are generated within the 95% uncertainty bounds, and the corresponding FRF is drawn for each of these. This results in a "cloud of models" that gives a very good visual impression about the noise sensitivity of the estimated FRF (or the poles and zeros). Because this approach relies less on intermediate realizations, it can give a more realistic impression of the system properties. In Figure 4-43, the cloud of models is shown for the 2/2 and the 6/6 model. It can be seen that the cloud of the 6/6 model is thicker than that of the 2/2 cloud. Also spikes can be seen in the 6/6 cloud, which is a very typical phenomenon that indicates the presence of coinciding pole/zero pairs, and hence a particular indication of over-modeling.

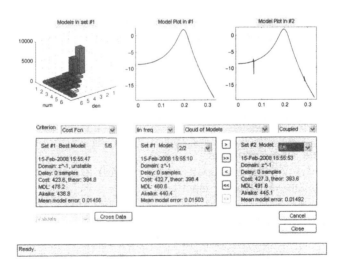

Figure 4-43 Evaluation and comparison of the estimated models: `Evaluate or Compare Plant Models` window; `Cloud of Models` is selected.

5

Best Linear Approximation
of Nonlinear Systems

What you will learn: This chapter gives you a full understanding of the impact of the nonlinear behavior of a system on classical frequency response function (FRF) measurements. This is important because all real life systems are, to some extent, nonlinear. Throughout the chapter you will get the answers to the following basic questions: Does the concept of FRF measurements still make sense for nonlinear systems? If so, under which conditions? What are the practical limitations of the linear framework? In the next chapter we will discuss what are the appropriate measurement procedures in the presence of nonlinear distortions. Before answering all these questions you will first get an in-depth understanding of the response of a nonlinear system to periodic inputs.

Chapter content:

5.1 RESPONSE OF A NONLINEAR SYSTEM TO A PERIODIC INPUT

5.1.1 Static nonlinear systems

A. Single Sine Response

The response $y(t)$ of a static nonlinear system $y = f(u)$ to a periodic input $u(t)$ is a periodic signal with the same period as the input:

$$\text{If } u(t + T_0) = u(t) \text{ for all } t, \text{ then } y(t + T_0) = f(u(t + T_0)) = f(u(t)) = y(t) \text{ for all } t. \quad (5\text{-}1)$$

The fundamental difference between a static nonlinear system and a linear time invariant (LTI) system is that a nonlinear system transfers energy from one frequency to other frequencies, which is impossible for an LTI system. This is illustrated by the following exercise.

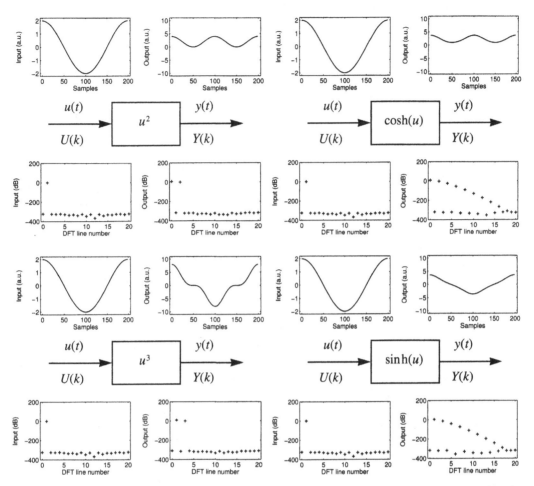

Figure 5-1   Response of even (top) and odd (bottom) static nonlinear systems to a single sine excitation: time signals (1 period) and DFT spectra (first 20 lines only).

Exercise 77.a (Single sine response of a static nonlinear system)   Consider the following static nonlinear systems

$$f(u) = u^2, \cosh(u), u^3, \text{ and } \sinh(u) \qquad (5\text{-}2)$$

Calculate the response $y = f(u)$ of these systems to the input $u(t) = A\cos(2\pi f_0 t)$, where $A = 2$ and $f_0 = 1 \text{ Hz}$. Choose $f_s = 200 \text{ Hz}$ and plot the input/output DFT spectra for one period $T_0 = 1/f_0$. Explain the results. Why has the sampling frequency been chosen that large?   □

Observations   From Figure 5-1 it can be seen that the response of $y = u^2$ to $u(t) = A\cos(2\pi f_0 t)$ contains energy at DC (line zero) and $2f_0$ (line two), while the response of $y = u^3$ contains energy at f_0 (line one) and $3f_0$ (line three). It nicely illustrates that the input energy at frequency f_0 is distributed over different frequencies at the output.

Discussion   It can be explained by calculating explicitly $y = u^2$, and u^3.

$$(A\cos(\omega_0 t))^2 = \left(\frac{A}{2}(e^{j\omega_0 t} + e^{-j\omega_0 t})\right)^2 = \frac{A^2}{4}(e^{j2\omega_0 t} + 2 + e^{-j2\omega_0 t}) = \frac{A^2}{2}(1 + \cos(2\omega_0 t)),$$

$$\tag{5-3}$$

$$(A\cos(\omega_0 t))^3 = \frac{A^3}{8}(e^{j3\omega_0 t} + 3e^{j\omega_0 t} + 3e^{-j\omega_0 t} + e^{-j3\omega_0 t}) = \frac{A^3}{4}(3\cos(\omega_0 t) + \cos(3\omega_0 t)),$$

where $\omega_0 = 2\pi f_0$. Both $y = u^2$, and u^3 are special cases of the following result

$$\cos^n(x) = \left(\frac{e^{jx} + e^{-jx}}{2}\right)^n = \begin{cases} \dfrac{1}{2^n}C_n^{n/2} + \dfrac{1}{2^{n-1}}\sum_{r=0}^{n/2-1} C_n^r \cos((n-2r)x) & n \text{ even,} \\[2mm] \dfrac{1}{2^{n-1}}\sum_{r=0}^{(n-1)/2} C_n^r \cos((n-2r)x) & n \text{ odd,} \end{cases}$$

$$\tag{5-4}$$

where $C_n^r = n!/(r!(n-r)!)$ and with $x = \omega_0 t$ (proof: use $(a+b)^n = \sum_{r=0}^{n} C_n^r a^r b^{n-r}$). From (5-4) it can be concluded that for all $n = 0, 1, 2, \ldots$ the response of $y = u^{2n}$ to $u(t) = A\cos(2\pi f_0 t)$ only contains energy at even multiples of f_0, while the response of $y = u^{2n+1}$ only contains energy at odd multiples of f_0. This result is also valid in general for even $(f(-u) = f(u))$ and odd $(f(-u) = -f(u))$ nonlinear functions; the only difference being that more harmonics are generated. Compare, for example, in Figure 5-1 the DFT spectrum of $\cosh(u)$ with that of u^2, and the DFT spectrum of $\sinh(u)$ with that of u^3. It can be seen that $\cosh(u)$ and $\sinh(u)$ contain significant energy till $16f_0$ (line sixteen) and $17f_0$ (line seventeen) respectively (-300 dB is the arithmetic precision of MATLAB®), while the highest harmonics of u^2 and u^3 are $2f_0$ (line two) and $3f_0$ (line three) respectively. These results also explain the choice of the sampling frequency f_s: to avoid aliasing in the output DFT spectrum the choice should be such that $f_s/2$ is larger than the largest significant output harmonic.

Note that u^2, $\cosh(u)$, u^3, and $\sinh(u)$ all satisfy (5-1), while u^2 and $\cosh(u)$ also satisfy $y(t + T_0/2) = y(t)$, and u^3 and $\sinh(u)$ also validate $y(t + T_0/2) = -y(t)$ (see the time signals in Figure 5-1). This is a consequence of the fact that an even/odd function only generates even/odd harmonics of f_0, respectively.

B. Multisine Response

With some care, the basic properties of the single sine response can be generalized to periodic signals consisting of the sum of harmonically related sinewaves (= multisine). This is illustrated in the following exercise.

Exercise 77.b (Multisine response of a static nonlinear system)   Calculate the response of $y = u^2$ and u^3 to

$$u(t) = \sum_{k=1}^{F} A_k \cos(k\omega_0 t + \phi_k) \quad \text{with} \quad \phi_k = -\pi k(k-1)/F \tag{5-5}$$

for the following choices of the multisine parameters, with $f_0 = 1$ Hz and $f_s = 200$ Hz:

- Full multisine: $F = 5$, and $A_k = \sqrt{2/5}$, $k = 1, 2, \ldots, 5$.
- Odd multisine: $F = 5$, $A_1 = A_3 = A_5 = \sqrt{2/3}$, and $A_2 = A_4 = 0$.

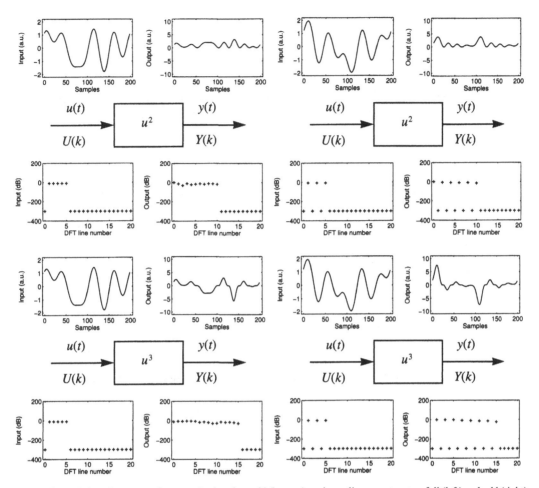

Figure 5-2 Response of an even (top) and an odd (bottom) static nonlinear system to a full (left) and odd (right) multisine excitation.

What is the rms value of the full and odd multisines? Plot one period of the input/output signals and the corresponding input/output DFT spectra. Compare the results with those of the single sine response in Figure 5-1. What do you conclude? Repeat the exercise for $y = \cosh(u)$ and $\sinh(u)$, for other choices of the phases ϕ_k, and for larger values of F (choose the amplitudes A_k such that the rms value of $u(t)$ equals one). Generalize your conclusions.   □

Observations From Figure 5-2 it can be seen that the responses of u^2 and u^3 to the full multisine (all harmonics of f_0 are excited) contain all even and odd harmonics of f_0, while the responses of u^2 and u^3 to the odd multisine (only the odd harmonics are excited) contain, respectively, even and odd harmonics only.

Discussion This can be explained as follows. Using $\cos(x) = (e^{jx} + e^{-jx})/2$ it can easily be seen that

$$\left(\sum_{k=1}^{F} A_k \cos(k\omega_0 t + \phi_k)\right)^n = \sum_{k_1, k_2, \ldots, k_n = 1}^{F} \prod_{i=1}^{n} A_{k_i} \cos(k_i \omega_0 t + \phi_{k_i}), \qquad (5\text{-}6)$$

which can be written as a sum of cosines with frequencies given by

$$(\textstyle\sum_{i=1}^{n} k_i)f_0 \quad \text{with} \quad k_i \in \{\pm k | k = 0, 1, ..., F\}, \tag{5-7}$$

which explains the responses to the full multisine. For example, for $n = 3$ the frequencies $(k_1 + k_2 + k_3)f_0$ will be excited. For the odd multisine (5-7) is replaced by

$$(\textstyle\sum_{i=1}^{n} k_i)f_0 \quad \text{with} \quad k_i \in \{\pm(2k+1) | k = 0, 1, ..., (F-1)/2\} \tag{5-8}$$

For example, for $n = 2$ the frequencies $(k_1 + k_2)f_0$ will be excited. Combining (5-8) with the property that the sum of an even/odd number of odd integers is always even/odd, shows that the responses of u^{2n} and u^{2n+1} to an odd multisine contain, respectively, even and odd harmonics only.

These properties of the odd multisine are valid for general even/odd nonlinear functions. Indeed, since the odd multisine satisfies $u(t + T_0/2) = -u(t)$, it follows that

$$y(t + T_0/2) = f(u(t + T_0/2)) = f(-u(t)) = \begin{cases} y(t) & \text{for } f(u) \text{ even,} \\ -y(t) & \text{for } f(u) \text{ odd,} \end{cases} \tag{5-9}$$

showing that the response of an even and odd nonlinear function to an odd multisine contains, respectively, even and odd harmonics only (Selby, 1973). Note that any nonlinear function $f(u)$ can be written as the sum of an even $f_E(u)$ and an odd $f_O(u)$ part

$$f(u) = f_E(u) + f_O(u) \text{ with } f_E(u) = (f(u) + f(-u))/2 \text{ and } f_O(u) = (f(u) - f(-u))/2 \tag{5-10}$$

Therefore, analyzing the harmonic content of the response to an odd multisine where some of the odd harmonics are not excited (e.g., 1 out of 3 consecutive odd harmonics) allows us to decide whether the nonlinear function contains only even, or only odd, or even and odd contributions. Indeed, it is sufficient to verify whether only even, or only odd, or even and odd harmonics are present in the response spectrum. This is impossible with the full multisine. See, for example, the response of u^3 to the full multisine in Figure 5-2: from the output spectrum it can be seen that a third-degree nonlinearity should be present (the highest output frequency equals three times the highest input frequency), but it is impossible to decide whether a second degree nonlinearity is present or not (in each case all output harmonics from DC to $2f_0$ are excited).

Conclusion Odd multisines where some of the odd harmonics are not excited are best suited for classifying the nonlinearities into even and odd contributions. The drawback w.r.t. the full multisine is the loss of a factor two in frequency resolution.

5.1.2 Dynamic nonlinear systems

A. Class of Nonlinear Dynamic Systems

The class of nonlinear dynamic systems considered in this book is restricted to those nonlinear dynamic systems which can be approximated arbitrarily well in least squares sense by a Volterra series [= generalization of a Taylor series for functions with memory, see Schetzen (1980)] on a given input domain. This class allows for an emphatic description of nonlinear

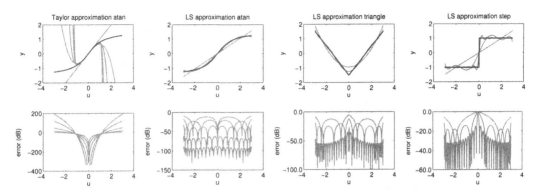

Figure 5-3 Uniform versus nonuniform polynomial approximation of a function. First column: Taylor series approximation `atan` function of degree 1, 3, 7, 15, and 23; second column: least squares approximation `atan` function of degree 1, 3, 7, 15, and 23; third column: least squares approximation triangle function of degree 2, 6, and 40; and fourth column: least squares approximation step function of degree 1, 7, and 41.

phenomena like saturation (e.g., amplifiers) and discontinuities (e.g., relays, quantizers). This is not in contradiction with the well-known fact that a Volterra series is only suitable for describing weakly nonlinear systems. Indeed, in a classical Volterra series expansion, the approximation error (difference between the true output and output of the Volterra series) converges everywhere to zero at the same rate as the number of terms in the series tends to infinity (= uniform convergence), while here it is only required that the power of the approximation error tends to zero (= pointwise convergence). The pointwise convergence does not exclude that the approximation error remains large at a discrete set of isolated points (similar to a Fourier series approximation of a discontinuous function), which is not the case for uniform convergence. These issues are illustrated in the following exercise.

Exercise 78 (Uniform versus Pointwise Convergence) Calculate the Taylor series expansion of degrees 1, 3, 7, 15, and 23 of the function `atan(u)` at $u = 0$, and make a polynomial least squares approximation of degrees 1, 3, 7, 15, and 23 of the `atan` function in the interval [-3, 3] (*Hint*: use the MATLAB$^®$ function `polyfit` with $N = 400$ linearly spaced data points, and scale the input by its standard deviation to improve the numerical conditioning.) Compare the approximation errors of the least squares fits and the Taylor series expansions in the interval [-3, 3]. Explain your results. Why do we select only odd degrees? Consider further the triangle and the step functions

$$\text{triangle}(u) = |u| - 1.5 \text{ and step}(u) = \begin{cases} -1 & \text{for } u < 0, \\ 0 & \text{for } u = 0, \\ 1 & \text{for } u > 0. \end{cases} \quad (5\text{-}11)$$

Do the Taylor series expansions of triangle(u) and step(u) exist at $u = 0$? Calculate for $u \in [-3, 3]$ a polynomial least squares approximation of degrees 2, 6, and 40 for triangle(u), and of degrees 1, 7, and 41 for step(u). Compare the approximation errors of the least squares fits of the triangle, step and `atan` functions. What do you conclude? Why do we select odd degrees for the step and even degrees for the triangle? □

Observations From Figure 5-3 it can be seen that the Taylor series expansion of the atan function is much better than the corresponding polynomial least squares approximation for $|u| \ll 1$, while it is much worse for $|u| \sim 1$, and even completely useless for $|u| > 1$. The least squares fits performs well over the whole interval [-3, 3], and the approximation error converges uniformly to zero as the degree of the polynomial increases to infinity.

Discussion The reason for the peculiar behavior of the Taylor series expansion of the atan(u) function is that the convergence radius of the Taylor series expansion equals one at $u = 0$ (Henrici, 1974). The uniform convergence of the least squares fit of atan(u) is a consequence of the fact that the atan function has continuous derivative over [-3, 3] (Kreider *et al.*, 1966).

Observations Note that although the Taylor series expansions of the triangle and step functions (5-11) do not exist at $u = 0$, it is possible to approximate these functions very well in least squares sense by polynomials (see Figure 5-3). The behavior of the least squares approximation error for increasing polynomial degree of the triangle and step functions (5-11) is different from that of the atan function. While the convergence to zero of the triangle approximation error is still uniform, but at a much slower rate than that of the atan function, the convergence to zero of the step approximation error is no longer uniform but pointwise. Indeed, irrespective of the polynomial degree, the step approximation error at $u = 0$ remains equal to one.

It can also be seen that for the odd functions (atan and step) only the coefficients of the odd powers are different from zero, while for the even functions only the even powers contribute to the solution.

Discussion These properties are a consequence of the fact that the atan function is infinitely differentiable over [-3, 3], while the triangle function has only a piecewise continuous derivative, and the step function is discontinuous [the more derivatives exist, the higher the convergence rate: see Zygmund (1979)]. Note, however, that the (uniform) convergence of the least squares approximation to a function does not imply the convergence of the derivatives.

When a function is approximated by a sum of basis functions, then we have in general that the odd behavior of the function is captured by the odd basis functions, and the even behavior is captured by the even basis functions.

Summarizing, the class of nonlinear dynamic systems considered in this book can be described roughly as those systems for which (i) the influence of the initial conditions vanishes asymptotically and (ii) the steady-state response to a periodic input is a periodic signal with the same period as the input. This class will be called PISPO (period in same period out) and it excludes nonlinear phenomena such as subharmonics, bifurcations, and chaos.

B. Single Sine Response
Simple nonlinear dynamic systems excited by a single sine can exhibit a very complex behavior. This is illustrated in the following exercise

Exercise 79.a (Normal operation, subharmonics, and chaos)   Consider the Duffing oscillator

$$\frac{d^2y(t)}{dt^2} + \frac{dy(t)}{dt} - 10y(t) + 100y^3(t) = A\cos(3.5t), \qquad (5\text{-}12)$$

where the input amplitude A is an arbitrary parameter. Note that (5-12) can be rewritten as an

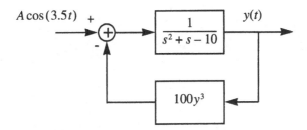

Figure 5-4   The Duffing oscillator (5-12) written as an unstable linear dynamic system
$1/(s^2 + s - 10)$ operating in closed loop with a static nonlinear feedback
branch $100y^3$.

unstable linear dynamic system operating in closed loop with a static nonlinear feedback
branch (see Figure 5-4). Calculate the response $y(t)$ of (5-12) over a hundred periods of the
input signal for the following four values of the amplitude: A = 0.67, 0.81, 0.82, and 0.9.
Take N = 1024 data points per period, with initial conditions $y(0)$ = 0.2 and
$y^{(1)}(0)$ = (−0.2), and integrate the differential equation using the MATLAB® function
ode45. Set the relative and absolute tolerance of the numerical integration method to
1×10^{-10} and 1×10^{-16} respectively, and plot the following:

- The input $A\cos(3.5t)$ and the response $y(t)$ over the hundred input periods.
- The input and the response over the last eight input periods.
- The input and response DFT spectra of the last eight input periods.
- The phase plane representation of the response: $y^{(1)}(t)$ as a function of $y(t)$ (the
  MATLAB® function ode45 calculates the solution and its derivative).

Interpret the results.   □

The solution of Exercise 79.a is shown in Figure 5-5. For A = 0.67 (first column) the
steady state output has the same period as the input (= normal operation), for A = 0.81 (second
column) the output period is twice the input period (= period doubling), for A = 0.82
(third column) the output period is four times the input period (= period quadrupling), while
for A = 0.9 (last column) the output is irregular and looks like colored noise (= chaos).

The first two rows of Figure 5-5 show the input and response over one hundred input
periods (note the transient response from t = 0 till t = 20 s), while rows three and four
show the input and response of the last eight input periods. Rows five and six show the first
twenty DFT lines of the input and response DFT spectra of the time signals in rows three and
four. Since the DFT of eight input periods is taken, DFT line number eight is the first nonzero line in the input DFT spectrum (-300 dB is the MATLAB® precision of the calculations). From the DFT spectra of the steady state responses (row six, columns one to three) it
can be seen that the numerical integration errors are about -200 dB below the DC level (line
zero). This corresponds to an accuracy of about ten significant digits.

For an input amplitude of A = 0.67 (first column), the response (output) DFT spectrum (column one, row six) contains DC (line zero), the fundamental (line eight), and the second harmonic (line sixteen). Harmonics three, four, … are also present but not shown in the
figure. For this input amplitude the steady state output has the same period as the input (=
normal operation).

For A = 0.81 (second column) period doubling occurs: the period of the response is
twice that of the input (compare the figures in column two, rows three and four). This can
also be seen in the output DFT spectrum (column two, row six) where, besides lines zero

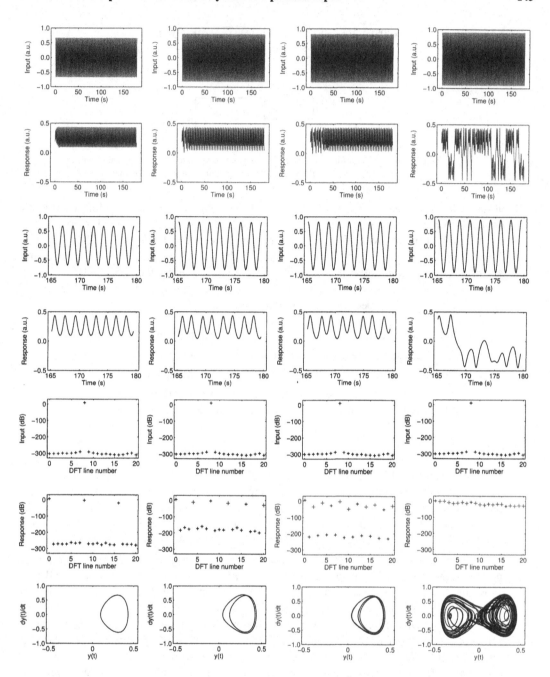

Figure 5-5   Response of the Duffing oscillator (5-12) for different values of the amplitude A. First column ($A = 0.67$): normal operation; second column ($A = 0.81$): period doubling; third column ($A = 0.82$): period quadrupling; fourth column ($A = 0.9$): chaos. First two rows: Simulated input and response signals. Rows three and four: Input and response over the last eight input periods. Rows five and six: First twenty DFT lines of the input and response DFT spectra (black "+": excited frequency and harmonics; gray "+": subharmonics). Last row: Phase plane representation.

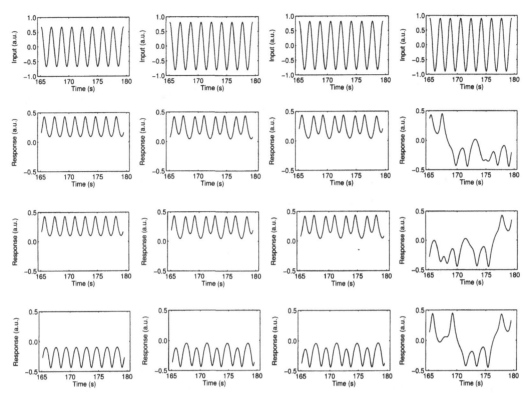

Figure 5-6   Steady state response of the Duffing oscillator (5-12) for different amplitudes A and different initial conditions $y(0)$ and $y^{(1)}(0)$. First column ($A = 0.67$): Normal operation. Second column ($A = 0.81$): Period doubling. Third column ($A = 0.82$): Period quadrupling; fourth column ($A = 0.9$): Chaos. First row: Input signal. Second row: Response for $y(0) = 0.2$ and $y^{(1)}(0) = (-0.2)$;. Third row: Response for $y(0) = 0.2$ and $y^{(1)}(0) = -0.201$. Last row: Response for $y(0) = -0.2$ and $y^{(1)}(0) = -0.2$.

(DC), eight (input fundamental), and sixteen (input second harmonic), also lines four (output fundamental), twelve (output third harmonic), and twenty (output fifth harmonic) are significantly different from zero.

Increasing the input amplitude to $A = 0.82$ (third column) gives period quadrupling: compare the figures in column three, rows three and four. It is more apparent in the output DFT spectrum (column three, row six) which contains now energy at all even DFT lines (two, four, six, ...).

Increasing the input amplitude further to $A = 0.9$ (last column) drives the system into chaotic regime, and all the output DFT lines are significantly different from zero (see row six, last column). The response of a chaotic system is very sensitive to small changes in the initial conditions. This is illustrated in the following exercise.

Exercise 79.b (Influence initial conditions) Change the initial conditions in Exercise 79.a to $y(0) = 0.2$ and $y^{(1)}(0) = -0.201$ and repeat the calculations. What do you observe? Change the initial conditions to $y(0) = -0.2$ and $y^{(1)}(0) = -0.2$ and repeat the calculations. What do you observe now? Verify that the phase plane figures are those of Exercise 79.a mirrored w.r.t. the vertical axis. Explain your results.   □

Observations From Figure 5-6 it can be seen that the steady state responses (A = 0.67, 0.81, and 0.82) are insensitive to small changes in the initial conditions, while the chaotic response (A = 0.9) is very sensitive to these small changes (see Figure 5-6, second and third rows).

Discussion The latter is typical for chaotic systems. Large changes in the initial conditions, however, result in large changes of the steady state responses (see Figure 5-6). This is due to the fact that (5-12) has two stable solutions (attractors) and, depending on the initial conditions, the steady-state response converges to one of both solutions. Note that these two solutions are mirrored ($y \rightarrow -y$), time-shifted versions of each other.

C. Multisine Response

The properties of the odd multisine for static nonlinear systems (see Section 5.1.1) are also valid for nonlinear dynamic PISPO systems. The following exercise illustrates this.

Exercise 80 (Multisine response of a dynamic nonlinear system)   Consider the following nonlinear dynamic PISPO systems

$$\frac{d^2y(t)}{dt^2} + 0.4\frac{dy(t)}{dt} + y(t) = g(u(t), u^{(1)}(t)) \text{ with } g(u(t), u^{(1)}(t)) = \begin{cases} u(t) + u^2(t)u^{(1)}(t), \\ u(t)u^{(1)}(t). \end{cases} \quad (5\text{-}13)$$

Calculate in MATLAB® the exact steady state response of these systems to the full and odd multisines $u(t)$ defined in equation (5-5) of Exercise 77.b (do not use numerical integration). What is the DC value of the steady-state responses? Explain. Compare the results with Figure 5-2 on page 140. Conclude.   □

Since the derivative of the input $u^{(1)}(t)$ is periodic with the same harmonic content as $u(t)$, exactly the same reasoning of equations (5-6) to (5-8) can be applied, showing that $u(t)u^{(1)}(t)$ and $u^2(t)u^{(1)}(t)$ excite, respectively, $(k_1 + k_2)f_0$ and $(k_1 + k_2 + k_3)f_0$ where

$$\begin{cases} k_i \in \{\pm k | k = 0, 1, ..., F\} & \text{full multisine,} \\ k_i \in \{\pm(2k+1) | k = 0, 1, ..., (F-1)/2\} & \text{odd multisine.} \end{cases} \quad (5\text{-}14)$$

Hence, exactly the same conclusions can be drawn as in Section 5.1.1: The response of the even/odd dynamic nonlinear system (5-13) to an odd multisine contains, respectively, even and odd harmonics only, while the response to the full multisine contains all harmonics. This result can be generalized to the class of PISPO systems. Indeed, the output of a PISPO system can be written as $y(t) = F(u(t))$, where $F(u(t))$ is an operator which contains, among other things, derivatives and/or integrals. Similar to (5-10), $F(u)$ can be split into an even $F_E(u)$ and an odd $F_O(u)$ part, thus showing that the response to an odd multisine can be classified/quantified as even and odd contributions. This is impossible for the response to a full multisine.

5.1.3 Detection, quantification, and classification of nonlinear distortions

From Exercises 77.b and 80 it follows that the steady-state response of a PISPO system to an odd multisine excitation can be split into even and odd nonlinear contributions (see

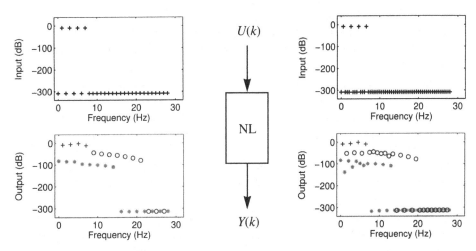

Figure 5-7   Steady-state response of the dynamic nonlinear system (5-15) to an odd (left column) and an odd–odd (right column) multisine. The output spectra (bottom row) are split in excited odd harmonics (black +), nonexcited even harmonics (gray *), and nonexcited odd harmonics (black o).

Figure 5-2). More specifically, one can distinguish even in-band (in the frequency band of the excitation), even out-band (outside the frequency band of the excitation), and odd out-band distortions. If every in-band odd frequency is excited, then it is impossible to visualize the in-band odd nonlinear contributions in the output spectrum. To reveal the in-band odd distortions it is necessary to exclude some of the in-band odd harmonics in the excitation signal. This is illustrated in the following exercise.

Exercise 81 (Detection, quantification, and classification of nonlinearities)   Consider the following nonlinear dynamic PISPO system

$$\frac{1}{\omega_1^2}\frac{d^2y(t)}{dt^2} + \frac{2\zeta}{\omega_1}\frac{dy(t)}{dt} + y(t) = u(t) + \alpha\left(\frac{du(t)}{dt}\right)^2 + \beta u^2(t)\frac{du(t)}{dt} \qquad (5\text{-}15)$$

with $\omega_1 = 30\,\text{Hz}$, $\zeta = 0.2$, $\alpha = 1\times10^{-7}\,\text{s}^2$, and $\beta = 1\times10^{-3}\,\text{s}$. Calculate the exact steady-state response of (5-15) to the following excitations

- Odd multisine: (5-5) with $f_0 = 1\,\text{Hz}$, $F = 7$, $A_{2k+1} = 1/\sqrt{2}$ for $k = 0, 1, 2$, and 3, and $A_{2k} = 0$.

- Odd–odd multisine: (5-5) with $f_0 = 0.5\,\text{Hz}$, $F = 13$, $A_{2k} = 0$, $A_{4k+1} = 1/\sqrt{2}$ for $k = 0, 1, 2$, and 3, and $A_{4k-1} = 0$ for $k = 1, 2$, and 3.

Choose $f_s = 200\,\text{Hz}$ and plot the input/output DFT spectra as a function of the frequency. Classify the output DFT spectrum in odd excited, odd nonexcited, and even nonexcited contributions, and make conclusions. Repeat the exercise for the odd–odd multisine but now with phases ϕ_k uniformly distributed in $(-\pi, \pi]$ (= random phase odd–odd multisine). Try out different phase realizations. How do the nonlinear distortions behave?   □

Observations   The output spectra in Figure 5-7 can be split into two parts: an in-band part [0 Hz, 7 Hz] being the frequency band covered by the input signal, and the out-band part (7 Hz, 100 Hz] being the remaining nonexcited frequency band. Due to the in-band nonexcited odd harmonics in the odd–odd multisine excitation it is possible to detect the in-band odd harmonic distortions, while this is impossible for the odd multisine (see Figure 5-7, bottom row).

Discussion The price to be paid for obtaining the in-band odd nonlinear distortion information, while maintaining the frequency resolution (1 Hz) of the original odd multisine, is the doubling of measurement time (2 s for the odd–odd multisine and 1 s for the odd multisine). If the multisine contains a lot of frequencies, then a few frequencies can be sacrificed for detecting the odd nonlinear distortions (e.g., one out of three, or one out of four), giving a better compromise between the frequency resolution of the excited odd harmonics, the frequency resolution of the nonexcited odd harmonics, and the measurement time (see Section 6.1.2).

Observations From the response to the odd–odd multisine (see Figure 5-7) it can be seen that the odd (o) and even (*) in-band distortions are about 40 dB and 60 dB below the contributions at the excited frequencies (+). Note also the amplification of the output spectra at the resonance frequency $\omega_1/(2\pi) \approx 4.8$ Hz (see Figure 5-7, bottom row, excited frequencies "+")

Discussion Due to the particular choice of α and β in (5-15), the nonlinear system has a dominant linear behavior. As a consequence, the input/output behavior at the excited frequencies (+) is mainly given by the linear transfer function

$$\frac{1}{(s/\omega_1)^2 + 2\zeta s/\omega_1 + 1} \tag{5-16}$$

This explains the amplification of the output spectra at the resonance frequency 4.8 Hz. Since the even nonlinearities have no contribution at the odd output harmonics, the deviation from the ideal linear behavior (5-16) at the excited frequencies (+) can only be explained by the odd nonlinearities. Comparing the levels of the nonexcited (o) and excited (+) odd output harmonics, one could predict a deviation of roughly 1% (-40 dB). Although it is tempting to make this extrapolation, there is no known reason as to why the levels of the nonlinear distortions at the excited and nonexcited odd harmonics should be the same. This important issue will be discussed in more details in Sections 6.1.2 and 6.1.6.

Repeating the simulation for different realizations of the random phase odd–odd multisine gives similar results as in Figure 5-7; the main difference being that the levels of the nonlinear distortions vary from realization to realization. This phenomenon is explained in detail in Sections 5.2 and 6.1.

5.1.4 What have we learned in Section 5.1?

- Unlike linear time invariant systems, nonlinear systems transfer energy from one frequency at the input to other frequencies at the output (Exercises 77.a and 79.a).

- For odd multisines the even and odd contributions of a nonlinear system are uniquely linked to, respectively, the even and odd harmonics in the output spectrum (Exercises 77.b and 80).

- Simple nonlinear dynamic systems can exhibit a very complex behavior, for example, period doubling and chaos (Exercise 79.a). Typical of the chaotic behavior is the extreme sensitivity of the response to small changes in the initial conditions (Exercise 79.b).

■ The class of nonlinear systems considered in this book is restricted to the so-called PISPO (period in, same period out) systems. These are systems whose steady-state response to a periodic input is a periodic signal with the same period as the input. It excludes phenomena such as subharmonics, bifurcations and chaos, but allows for strong nonlinearities such as relays and quantizers. The theory of the PISPO systems is based on the polynomial/Volterra mean square approximation of a static/dynamic nonlinear system (Exercise 78).

■ An odd multisine is well suited to detect the even in-band, the even out-band, and the odd out-band nonlinear distortions. To reveal the odd in-band nonlinear distortions, one has to sacrifice some of the odd in-band harmonics in the excitation. Hence, odd multisines with missing in-band odd harmonics allow for a full detection, classification, and quantification of the nonlinearities (Exercise 81).

5.2 BEST LINEAR APPROXIMATION OF A NONLINEAR SYSTEM

The standard procedure for measuring the impulse response $g(t)$ of a linear system is based on a correlation analysis

$$R_{yu}(t) = g(t) * R_{uu}(t) \tag{5-17}$$

with $u(t)$ the input signal, $y(t)$ the output signal, $*$ the convolution product, and $R_{yu}(t)$ and $R_{uu}(t)$ the cross- and autocorrelations, respectively

$$R_{yu}(\tau) = \mathbb{E}\{y(t)u(t-\tau)\} \text{ and } R_{uu}(\tau) = \mathbb{E}\{u(t)u(t-\tau)\} \tag{5-18}$$

Equation (5-17) is known as the Wiener–Hopf equation. For random excitations the solution to (5-17) minimizes

$$\mathbb{E}\{\|y(t) - g(t) * u(t)\|^2\} \tag{5-19}$$

w.r.t. $g(t)$ (Eykhoff, 1974; Bendat and Piersol, 1980). Taking the Fourier transform of (5-17) gives

$$G(j\omega) = \frac{S_{YU}(j\omega)}{S_{UU}(j\omega)}, \tag{5-20}$$

where the cross-power spectrum $S_{YU}(j\omega)$, the auto-power spectrum $S_{UU}(j\omega)$, and the frequency response function $G(j\omega)$ are the Fourier transforms of, respectively, $R_{yu}(t)$, $R_{uu}(t)$, and $g(t)$ (Eykhoff, 1974; Bendat and Piersol, 1980).

If the system has a weak nonlinear behavior, then one could use equations (5-17) to (5-20), where the mean values of the input/output signals have been removed, for determining some linear approximations of the nonlinear system (Pintelon and Schoukens, 2001; Enqvist and Ljung, 2005). For example, (5-19) becomes

$$\mathbb{E}\{\|\tilde{y}(t) - g(t) * \tilde{u}(t)\|^2\} \text{ with } \begin{cases} \tilde{y}(t) = y(t) - \mathbb{E}\{y(t)\}, \\ \tilde{u}(t) = u(t) - \mathbb{E}\{u(t)\} \end{cases} \tag{5-21}$$

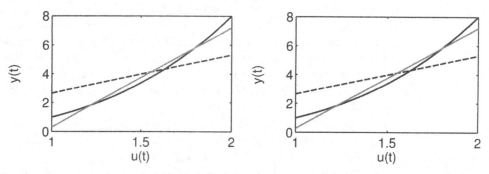

Figure 5-8   Linear approximation of $y(t) = u^3(t)$ (black solid line) with (gray solid line) and without (black dashed line) removal of the input/output DC values. Left: $u(t)$ uniformly distributed in $[1, 2]$ (mean $= 1.5$, and variance $= 1/12$). Right: $u(t)$ normally distributed with mean 1.5 and variance 1/12.

(see also Exercise 82). Since (5-21) minimizes the mean square error between the true (shifted) output of the nonlinear system $y(t) - \mathbb{E}\{y(t)\}$ and the output of the linear model $g(t)*(u(t) - \mathbb{E}\{u(t)\})$, this linear approximation will be called the *best linear approximation* $g_{BLA}(t)$ or $G_{BLA}(j\omega)$ of the nonlinear system for the given class of input signals $u(t)$. It is clear that the best linear approximation will depend on the class of excitation signals used, for example, changing the amplitude of the excitation will result in a different frequency response function. Hence, one could wonder whether the concept of a linear approximation of a nonlinear system makes sense at all. Other relevant questions that arise are "Which properties of the excitation influence $G_{BLA}(j\omega)$?" "How does $G_{BLA}(j\omega)$ depend on the nonlinearities?" "Can $G_{BLA}(j\omega)$ be measured using periodic excitation signals?" "What is the predictive power of $G_{BLA}(j\omega)$?" All these issues will be addressed in detail in the sequel of this chapter.

5.2.1 Static nonlinear systems

In the definition (5-21) of the best linear approximation (BLA) the mean values of the input and output signals are removed. The reason for this is that the best (in mean square sense) linearization around the operating point is sought. What happens if the DC signal values are not removed? This is illustrated by the following exercise.

Exercise 82 (Influence DC values signals on the linear approximation)   Consider the static nonlinear system $y(t) = u^3(t)$. Calculate the linear approximation $y(t) = gu(t)$ via (5-19) and (5-21) for the following input signal $u(t)$:

■ White uniformly distributed noise in $[1, 2]$ (mean $= 1.5$, and variance $= 1/12$).

■ White normally distributed noise with mean 1.5 and variance 1/12.

Use $N = 4000$ data points and calculate numerically the minimizers of (5-19) and (5-21) (*Hint*: replace the expected values by the sample means). Plot the nonlinear function and the linear approximations in the interval $[1, 2]$ of the input. Repeat the calculations for different random realizations of the input and compare the estimates. What do you conclude? Repeat the calculations for $N = 40,000$ data points. What do you conclude now? Explain! Compare the results with the analytic solutions of (5-19) and (5-21).   □

Observations From Figure 5-8 it can be seen that the estimates obtained without removal of the input/output DC values (black dashed lines) have much larger approximation errors than those obtained with removal of the DC values (gray solid lines).

Discussion The observation shows the importance of linearizing the nonlinear system around its operating point. Although the input/output signals are known exactly, one gets different (best) linear approximations for different random realizations of the input. The reason for this is that the expected values in (5-19) and (5-21) are approximated by their sample means, which depend on the actual random realization of the input. For a finite number of N input and output samples, the minimizers of (5-19) and (5-21) are respectively given by

$$\hat{g}_{LA} = \frac{\frac{1}{N}\sum_{t=0}^{N-1} y(t)u(t)}{\frac{1}{N}\sum_{t=0}^{N-1} u^2(t)} \text{ and } \hat{g}_{BLA} = \frac{\frac{1}{N}\sum_{t=0}^{N-1} (y(t)-\hat{y})(u(t)-\hat{u})}{\frac{1}{N}\sum_{t=0}^{N-1} (u(t)-\hat{u})^2}, \qquad (5\text{-}22)$$

where $\hat{x}$ is the sample mean of x

$$\hat{x} = \frac{1}{N}\sum_{t=0}^{N-1} x(t). \qquad (5\text{-}23)$$

From (5-22) it can be seen that the random nature of the input $u(t)$, together with the nonlinear relationship between the output $y(t)$ and the input $u(t)$, cause the variability of the estimates $\hat{g}_{LA}$ and $\hat{g}_{BLA}$. As N increases to infinity, the variability of the estimates decreases to zero, and the numerators and denominators in (5-22) converge to their expected values (law of large numbers, see Pintelon and Schoukens, 2001; Lukacs, 1975; and Chow and Teicher, 1988). For independent and identically distributed Gaussian and uniform inputs $u(t)$ with mean μ and variance σ^2, the asymptotic values of (5-22) are, respectively

$$g_{LA} = \frac{3\sigma^4 + \mu^4 + 6\mu^2\sigma^2}{\sigma^2 + \mu^2}, \quad g_{BLA} = 3\sigma^2 + 3\mu^2 \qquad \text{for Gaussian } u(t),$$

$$g_{LA} = \frac{(9/5)\sigma^4 + \mu^4 + 6\mu^2\sigma^2}{\sigma^2 + \mu^2}, \quad g_{BLA} = (9/5)\sigma^2 + 3\mu^2 \qquad \text{for uniform } u(t).$$

$$(5\text{-}24)$$

These values correspond to the analytic solutions of (5-19) and (5-21).

For small values of N, the variability of the estimates (5-22) is larger than the difference between the results for Gaussian and uniformly distributed inputs, and one would wrongly conclude that the probability density function (pdf) of the input does not influence the (best) linear approximations. As N increases, the variability of the estimates decreases, and the difference between the results of the Gaussian and uniform inputs becomes visible. Due to the particular choice of μ and σ^2 in Exercise 82, these differences are marginal in Figure 5-8. The influence of the input pdf on the best linear approximation is made more apparent in the following exercise.

Exercise 83.a (Influence of rms value and pdf on the BLA) Calculate numerically via (5-22) the best linear approximation of the `atan` and dead zone functions, respectively:

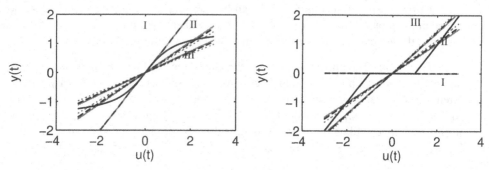

Figure 5-9   Best linear approximation of the atan (left) and dead zone (right) functions for different rms values (I: 0.1; II: 1.8; and III: 3), and inputs with different probability density function (gray dots: Gaussian; black dashes: uniform; and gray line: sine).

$$y(t) = \mathrm{atan}(u(t)) \text{ and } y(t) = \begin{cases} 0, & |u(t)| < 1, \\ u(t) - 1, & u(t) \geq 1, \\ u(t) + 1, & u(t) \leq -1 \end{cases} \tag{5-25}$$

for the following three classes of random input signals:

■ Zero mean, white normally distributed noise with standard deviation σ.

■ Zero mean, white uniformly distributed noise with standard deviation σ.

■ Zero mean, white sine distributed noise with standard deviation σ ($u(t) = \sqrt{2}\sigma \sin(\phi(t))$ with $\phi(t)$ uniformly distributed in $[-\pi, \pi]$).

where σ has the values 0.1, 1.8, and 3. (*Hint*: Use (5-22) with, for example, $N = 4000$.) Plot the functions $y(t)$ and their best linear approximations over the input interval $[-3, 3]$. What do you conclude? Explain.   □

Observations From Figure 5-9 it can be seen that the best linear approximations depend strongly on the rms value of the input. Except for $\sigma = 0.1$, they also depend on the probability density function of the input signal. For $\sigma = 0.1$, the best linear approximations (BLA) of the atan function almost coincide with the tangent line at $u = 0$ and, hence, are almost indistinguishable. Note that the BLAs of the dead zone function are exactly zero for $\sigma = 0.1$.

Discussion These observations are a consequence of the fact that $|u(t)|$ never exceeds one for the uniformly and sine distributed inputs, and that the probability of having a value that exceeds ten times the standard deviation is negligibly small for $N = 4000$ normally distributed samples. Indeed, the latter equals

$$\begin{aligned} \mathrm{Prob}(\max_t |u(t)| \geq 10\sigma) &= 1 - (\mathrm{Prob}(|u| \leq 10\sigma))^N \\ &= 1 - (1 - \mathrm{Prob}(|u| \geq 10\sigma))^N \\ &\approx N\mathrm{Prob}(|u| \geq 10\sigma) \\ &\approx 6.1 \times 10^{-20}, \end{aligned} \tag{5-26}$$

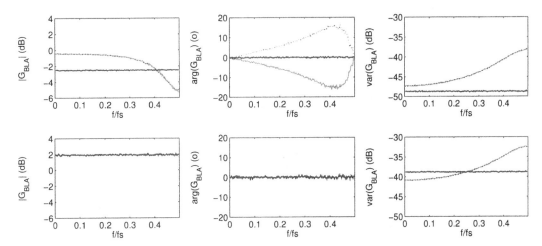

Figure 5-10   Best linear approximations $\hat{G}_{BLA}(j\omega)$ (left and middle column) of $y(t) = u^3(t)$ and their variance
var($\hat{G}_{BLA}(j\omega)$) (right column) for (filtered) white uniform noise (top row) and (filtered) white
Gaussian noise (bottom row) inputs $u(t)$, all with the same variance. Black lines: white noise
$u(t) = 0.5\sqrt{5}e(t)$. Gray lines: Minimum phase filtered white noise $u(t) = e(t) + 0.5e(t-1)$. Black
dotted lines: Nonminimum phase filtered white noise $u(t) = 0.5e(t) + e(t-1)$. Remark:
$dB(\sigma^2) \equiv 10\log_{10}(\sigma^2) = 20\log_{10}(\sigma) \equiv dB(\sigma)$.

where $Prob(|u| \geq 10\sigma)$ is calculated in MATLAB® as $erfc(10/\sqrt{2})$ with $erfc$ the complementary error function.

As one could expect, the best linear approximations (BLA) of the static nonlinear functions in Exercises 82 and 83.a are static functions (straight lines). In the following exercise the examples due to Enqvist (2005) show that this is not always true: the BLA of a static nonlinear function can be dynamic.

Exercise 83.b (Influence of power spectrum coloring and pdf on the BLA)   Consider the following three classes of excitation signals

$$u(t) = H(q)e(t) \text{ with } H(z^{-1}) = (\sqrt{5}/2), \ (1 + 0.5z^{-1}), \text{ and } 0.5 + z^{-1}, \qquad (5\text{-}27)$$

where $e(t)$ is zero mean, independent and identically distributed noise with variance $\sigma^2 = (1/3)$, and $H(z^{-1})$ are filters. Show that these three classes have the same rms value, but a different coloring of the power spectrum (the first class is white, while the second and the third classes have the same coloring). Calculate numerically the best linear approximation of the static nonlinear function $y(t) = u^3(t)$ via equation (5-20) for the three respective classes of input signals (5-27), where $e(t)$ is Gaussian and uniformly distributed noise (hint: first replace the expected values in (5-18) by their sample means over $M = 8000$ subrecords of $N = 512$ samples each; next replace the Fourier transforms in (5-20) by the DFTs over N samples; and finally, in order to suppress the leakage errors, difference the input/output DFT spectra of each subrecord). Repeat the calculations for $M = 800$ and $M = 80$. What do you conclude? Calculate analytically the uncertainty of the estimated best linear approximations. Compare the results obtained with the sample variance obtained from a hundred runs. Are the variations of the BLAs over the frequency significant? Why? Use the same approach to verify that the best linear approximations in Exercises 82 and 83.a are static functions. Why is this so?   □

Using the $M \times N$ input/output samples $u(t)$ and $y(t)$, $t = 0, 1, ..., (MN - 1)$, and following the lines of the linear case, the best linear approximation (5-20) and its variance are estimated as

$$\hat{G}_{\text{BLA}}(j\omega_{k+1/2}) = \frac{\hat{S}_{YU}(j\omega_{k+1/2})}{\hat{S}_{UU}(j\omega_{k+1/2})} = \frac{\frac{1}{M}\sum_{m=1}^{M} Y_{\text{diff}}^{[m]}(k)\,\overline{U_{\text{diff}}^{[m]}}(k)}{\frac{1}{M}\sum_{m=1}^{M} |U_{\text{diff}}^{[m]}(k)|^2},$$

$$\text{var}(\hat{G}_{\text{BLA}}(j\omega_{k+1/2})) = \frac{1}{M-1}\frac{\hat{S}_{YY}(j\omega_{k+1/2}) - |\hat{S}_{YU}(j\omega_{k+1/2})|^2/\hat{S}_{UU}(j\omega_{k+1/2})}{\hat{S}_{UU}(j\omega_{k+1/2})},$$

$$(5\text{-}28)$$

where $X_{\text{diff}}(k) = X(k+1) - X(k)$, and $X^{[m]}(k)$ is the DFT spectrum of the mth subrecord

$$X^{[m]}(k) = \frac{1}{\sqrt{N}}\sum_{t=0}^{N-1} x((m-1)N + t)e^{-2\pi jkt/N} \qquad (5\text{-}29)$$

The leakage errors are attenuated by the difference operation and tend to zero as N tends to infinity. For finite M, the random nature of the input $u(t)$ causes the variability of the estimated cross- and auto-power spectra $\hat{S}_{YU}$ and $\hat{S}_{UU}$, and, hence, also that of the estimated best linear approximation $\hat{G}_{\text{BLA}}$. Compared to the linear case, the static nonlinearity introduces an extra variability in the cross-power spectrum estimate $\hat{S}_{YU}$ (see Section 5.3). Therefore, a large number of averages are needed to obtain an accurate estimate. As both M and N tend to infinity, the estimate (5-28) converges to (5-20), and its variability decreases to zero.

Observations From Figure 5-10 it can be seen that for the (filtered) white Gaussian noise excitations the BLAs of the static nonlinearity are static (within its uncertainty, $\hat{G}_{\text{BLA}}$ has a frequency independent amplitude and zero phase), and independent of the coloring of the power spectrum. For the (filtered) white uniform noise excitations the picture is totally different: (i) the BLAs are dynamic for nonwhite power spectra; and (ii) the BLAs not only depend on the coloring of the power spectrum, but also on the phase of the signal filter $H(z^{-1})$ in (5-27).

Conclusion For non-Gaussian excitations the complete dependency over time matters and not only the correlation.

Observations The behavior of the uncertainty of the estimated BLAs (see Figure 5-10, right column) is somewhat different from that of the BLAs. Indeed, irrespective of the pdf of the input, the variance of the BLA estimate is frequency dependent for colored input power spectra, and frequency independent for white input power spectra.

Conclusion In the Gaussian case, the variance of a static BLA can be frequency dependent.

The reason why the BLAs in Exercise 82 are static is explained as follows. Consider the static nonlinear system $y(t) = f(u(t))$ where $u(t)$ is discrete-time, zero mean, independent and identically distributed non-Gaussian noise. Assume furthermore that $y(t)$ has zero mean. The best linear approximation is then calculated via the cross- and autocorrelations (5-18), to give

$$R_{yu}(\tau) = \mathbb{E}\{f(u(t))u(t-\tau)\} = \mathbb{E}\{y(t)u(t)\}\delta(\tau) = \sigma_{yu}^2\delta(\tau),$$
$$R_{uu}(\tau) = \mathbb{E}\{u(t)u(t-\tau)\} = \mathbb{E}\{u^2(t)\}\delta(\tau) = \sigma_u^2\delta(\tau),$$

$$(5\text{-}30)$$

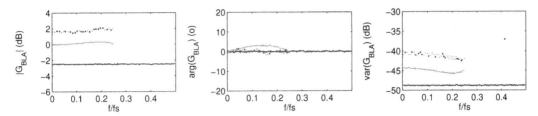

Figure 5-11 Best linear approximation (left and middle) of $y(t) = u^3(t)$ and its variance (right) for (filtered) white uniform noise inputs $u(t)$ with the same variance. Black solid lines: No filter (white noise). Gray solid lines: second-order filter. Black dotted lines: Thirty-fifth-order filter. Remark: $dB(\sigma^2) \equiv 10\log_{10}(\sigma^2) = 20\log_{10}(\sigma) \equiv dB(\sigma)$.

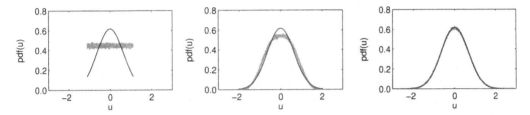

Figure 5-12 Probability density function of the input. Gray solid line: Distribution estimated from the time series $u(t)$ Black solid line: Normal distribution with same mean and variance as $u(t)$. Left: No filter. Middle: Second order filter. Right: Thirty-fifth-order filter.

where in (5-30) the second equality uses the independence over time of $u(t)$, and with $\delta(\tau)$ the Kronecker delta ($\delta(\tau) = 0$ if $\tau \neq 0$ and $\delta(0) = 1$). Since the corresponding cross- and auto-power spectra

$$S_{YU}(j\omega) = \sum_{\tau = -\infty}^{+\infty} R_{yu}(\tau)e^{-j\omega\tau} = \sigma_{yu}^2,$$

$$S_{UU}(j\omega) = \sum_{\tau = -\infty}^{+\infty} R_{uu}(\tau)e^{-j\omega\tau} = \sigma_u^2$$

(5-31)

are frequency independent, it follows that the BLA is static

$$G_{BLA}(j\omega) = S_{YU}(j\omega)/S_{UU}(j\omega) = \sigma_{yu}^2/\sigma_u^2.$$

Note that this result has been obtained without using the distribution function of $u(t)$.

For filtered white noise excitations $u(t) = H(q)e(t)$ where $e(t)$ is non-Gaussian, one may wonder what the effect of the length of the impulse response of the signal filter $H(z^{-1})$ is on the BLA. This is studied in the following exercise.

Exercise 83.c (Influence of length of impulse response of signal filter on the BLA)   Calculate numerically the best linear approximation of the static nonlinear function $y(t) = u^3(t)$ via (5-28), with $M = 8000$ and $N = 512$, for the following three classes of filtered input signals $u(t) = H(q)e(t)$:

■ No filtering: $H(z^{-1}) = \sqrt{5}/2$.

■ Second-order Chebyshev filter (cutoff frequency $f_s/4$; passband ripple 1 dB):
   ```
   [c, d] = cheby1(2, 1, 2*0.25);
   ```

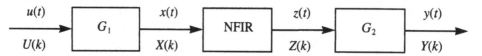

Figure 5-13   Generalized Wiener–Hammerstein (GWH) system: Cascade of a first linear dynamic block G_1, a nonlinear finite impulse response system (NFIR), and a second linear dynamic block G_2.

```
u = filter(c, d, e);
u = sqrt(5/12)*u/std(u);
```

■ Thirty-fifth order chebyshev filter (cutoff frequency $f_s/4$; passband ripple 1 dB):

```
[c, d] = cheby1(35, 1, 2*0.25);
u = filter(c, d, e);
u = sqrt(5/12)*u/std(u);
```

where $e(t)$ is uniformly distributed in [-1, 1] (Note: all inputs $u(t)$ have the same variance as in Exercise 83.b). Plot the best linear approximations and their uncertainty. Compare your results with those of Figure 5-10. What do you conclude? Explain! (*Hint*: Calculate numerically the probability density function of each input via the MATLAB® function `hist(u, sqrt(length(u)))`, and compare it to a normal distribution with the same mean value and variance).   □

Observations Comparing Figure 5-11 to Figure 5-10 on page 154 it can be seen that, within the excitation bandwidth, the best linear approximations of the filtered white uniform inputs converge for increasing order to the BLAs of the Gaussian inputs ($|G_{BLA}| = 1.9$ dB, and $\angle G_{BLA} = 0$).

Discussion The explanation for this can be found in Figure 5-12: as the order of the filter increases the input tends to a normally distributed random variable. This is a consequence of the central limit theorem (Pintelon and Schoukens, 2001; Billingsley, 1995; Chow and Teicher, 1988; Brillinger, 1981). Indeed, as the order of the filter increases, its impulse response $h(t)$ becomes longer, and the number of random variables that significantly contribute to the convolution product $u(t) = H(q)e(t)$

$$u(t) = \sum_{n=0}^{\infty} h(t-n)e(n) \qquad (5\text{-}32)$$

increases. The more significant random variables are added in (5-32), the more the sum $u(t)$ behaves as a Gaussian random variable (central limit theorem).

5.2.2 Dynamic nonlinear systems

For dynamic nonlinear systems the best linear approximation (BLA) is calculated theoretically via equation (5-20). Since in practice the input/output signals are always observed during a finite time, (5-20) is approximated by (5-28), which is prone to leakage errors. In the following exercise we verify whether the BLA can be measured using periodic excitation signals, which have the advantage that they are not subject to leakage errors if an integer number of periods is measured. As a test example we use a generalized Wiener–Hammerstein system (see Figure 5-13), which consists of the cascade of a first linear dynamic system G_1, a non-linear finite impulse response (NFIR) system, and a second linear dynamic system G_2.

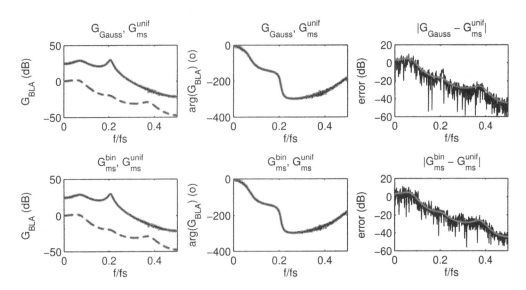

Figure 5-14   Best linear approximations (solid lines) and their standard deviation (dashed lines) of a generalized Wiener–Hammerstein system, for the classes of Gaussian noise and random phase multisines. Top row: Comparison of the BLAs obtained with Gaussian noise (G_{Gauss}, black lines) and random phase multisines with phases uniformly distributed in $[0, 2\pi)$ ($G_{\text{ms}}^{\text{unif}}$, gray lines). Bottom row: Comparison of the BLAs obtained with random phase multisines with phases uniformly distributed in $[0, 2\pi)$ ($G_{\text{ms}}^{\text{unif}}$, gray lines) and phases equal to zero or π with probability 1/2 ($G_{\text{ms}}^{\text{bin}}$, black lines). Right column: Magnitude of the complex difference between the 2 BLAs (black line), and standard deviation of the complex error (gray line).

Exercise 84.a (Comparison of Gaussian noise and random phase multisine)

Consider the generalized Wiener–Hammerstein (GWH) system shown in Figure 5-13 with

$$G_1(z^{-1}) = \frac{1}{1 - 0.5z^{-1} + 0.9z^{-2}}, \ z(t) = x^2(t-1)\,\text{atan}(x(t)), \ G_2(z^{-1}) = \frac{z^{-1} + 0.5z^{-2}}{1 - 1.5z^{-1} + 0.7z^{-2}}. \quad (5\text{-}33)$$

Calculate via (5-28), with $N = 1024$ and $M = 10^4$, the best linear approximation of the GWH system for the class of filtered, zero mean, white Gaussian noise excitations

$$u(t) = L(q)e(t) \text{ with } L(z^{-1}) = \frac{1 - 0.8z^{-1} + 0.1z^{-2}}{1 - 0.2z^{-1}} \text{ and } \text{var}(e(t)) = 1. \quad (5\text{-}34)$$

Do the same for the following class of random phase multisines

$$u(t) = \frac{1}{\sqrt{N}}\sum_{k=1}^{N/2-1} 2|L(z_k^{-1})|\cos(k\omega_0 t + \phi_k) \text{ with } z_k = \exp(j2\pi k/N), \quad (5\text{-}35)$$

where ϕ_k is independent and uniformly distributed in $[0, 2\pi)$. Use $N = 1024$ data points per period, and $M = 10^4$ independent realizations of the phases ϕ_k. Show that (5-35) has exactly the same power in each DFT bin of width f_s/N as the filtered white Gaussian noise excitation (5-34). Simplify (5-20) and (5-28) for periodic excitation signals. Is the `diff` operation in equation (5-28) still necessary for the periodic signal (5-35)? Explain! Compare the BLAs obtained with (5-34) and (5-35). Explain. Repeat the calculations for the signal class (5-35), where ϕ_k can only take the values 0 and π with equal probability. Is there any difference with the previously obtained BLAs (*Hint*: Compare the complex difference between the BLAs to the square root of the sum of the variances of the BLAs)? Explain.   □

For periodic excitation signals, (5-20) reduces to

$$G_{\mathrm{BLA}}(j\omega) = \frac{S_{YU}(j\omega)}{S_{UU}(j\omega)} = \frac{\mathbb{E}\{Y(k)\overline{U}(k)\}}{\mathbb{E}\{|U(k)|^2\}} = \frac{\mathbb{E}\{Y(k)\overline{U}(k)\}}{|U(k)|^2} = \mathbb{E}\left\{\frac{Y(k)}{U(k)}\right\}. \tag{5-36}$$

Using M realizations of the random phase multisine, the best linear approximation (5-36) and its uncertainty are estimated as

$$\hat{G}_{\mathrm{BLA}}(j\omega_k) = \frac{1}{M}\sum_{m=1}^{M} \frac{Y^{[m]}(k)}{U^{[m]}(k)},$$

$$\mathrm{var}(\hat{G}_{\mathrm{BLA}}(j\omega_k)) = \frac{1}{M(M-1)}\sum_{m=1}^{M}\left|\frac{Y^{[m]}(k)}{U^{[m]}(k)} - \frac{1}{M}\sum_{m=1}^{M}\frac{Y^{[m]}(k)}{U^{[m]}(k)}\right|^2, \tag{5-37}$$

where $X^{[m]}(k)$ is the DFT spectrum of the mth multisine realization. Note that (5-37) is a special case of (5-28), where the `diff` operation has been removed. Show this [*Hint*: follow the lines of (5-36)]. Compared with the linear case, the random nature of the multisine $u(t)$ increases the variability of the estimated BLA (5-37). Therefore, a larger number of averages M is needed to obtain an accurate estimate. As M tends to infinity, the estimate (5-37) converges to (5-36), and its variability decreases to zero.

Observations Although a very large number of averages has been taken ($M = 10^4$), the estimated BLAs (5-28) and (5-37) for, respectively, the class of Gaussian and random phase multisines are still noisy (see Figure 5-14). Note that for one realization ($M = 1$) the standard deviations of the BLAs are 40 dB ($\sqrt{10^4}$) larger than those given in Figure 5-14. Hence, for one realization, the uncertainty of $\hat{G}_{\mathrm{BLA}}$ is larger than $\hat{G}_{\mathrm{BLA}}$ over almost the whole frequency band.

Discussion The random nature of the BLA estimate shows that the frequency response function estimates can be written as

$$\hat{G}_{\mathrm{BLA}}(j\omega_k) = G_{\mathrm{BLA}}(j\omega_k) + G_S(j\omega_k), \tag{5-38}$$

where G_{BLA} are the BLAs (5-28) and (5-37), and with G_S the "nonlinear noise source." The large variability of the BLA estimate for one realization of the input motivates the many averages for calculating the best linear approximations.

From Figure 5-14 one can conclude that the BLAs of the three classes of excitation signals coincide within their uncertainty (the errors are of the same order of magnitude as their standard deviation). Since the best linear approximation of a nonlinear system depends on both the power spectrum (rms value and coloring) and the probability density function (pdf) of the excitation (see Section 5.2.1), and since the three classes of excitation signals have, by construction, the same power spectrum, it can be concluded that the three classes should have statistical properties (moments up to a certain degree dictated by the nonlinearity) similar to that of a normal distribution. This statement is verified in the following exercise, where it is shown that random phase multisines behave as Gaussian noise.

Exercise 84.b (Amplitude distribution of a random phase multisine)   Consider the random phase multisine (5-35), where $N = 1024$, and where the F excited frequencies are uniformly distributed over $(0, 0.5)f_s$. Calculate the probability density function of the multisine $u(t)$ for each of the following phase distributions ϕ_k and number of frequencies F,

- Independent and uniformly distributed in $[0, 2\pi)$, and $F = 1, 3$, and 15.
- Independent and equal to zero or π with equal probability, and $F = 1, 15$, and 255.

and for each of the following signal filters $L(z^{-1})$

- No signal filter: $(L(z^{-1}) = 1)$.
- Second order highpass filter: See (5-34).
- Sixth order Butterworth bandpass filter: $L(z^{-1}) = P(z^{-1})/Q(z^{-1})$ where the coefficients of P and Q are given by [p, q] = butter(3, 2*[0.2, 0.3]).

(hint: generate $M = 10^4$ random phase multisines $u(t)$ and calculate numerically the probability density function of the $M \times N$ data points via the MATLAB® function hist(u, sqrt(length(u))). Compare the pdf's with that of a normal distribution with the same mean value and variance. Explain the results.   □

Observations From Figure 5-15 it can be concluded that both classes of random phase multisines are normally distributed if the number of significantly contributing frequencies in the sum (5-35) is sufficiently large.

Discussion This is a consequence of the central limit theorem which roughly states that the sum of independently distributed random variables is asymptotically (as the number of significantly contributing random variables tends to infinity) normally distributed (Pintelon and Schoukens, 2001; Billingsley, 1995; Chow and Teicher, 1988; Brillinger, 1981). In fact, the only phase conditions for (5-35) to be asymptotically normally distributed are that the ϕ_k's are independent (over k) random variables, and that

$$\mathbb{E}\{e^{j\phi_k}\} = 0 \tag{5-39}$$

(Pintelon and Schoukens, 2001; Pintelon and Schoukens, 2002). The minimum number of frequencies required for an almost Gaussian behavior strongly depends on the coloring of the signal filter and the phase distribution. Indeed, fewer sinewaves are needed for the uniformly $[0, 2\pi)$ distributed phases than for the binary zero/π distributed phases, and many more sinewaves are needed for the bandpass signal filter than for the two other filters. The latter can be explained by the fact that, from the fifteen uniformly distributed frequencies, only five lie in the passband $[0.18, 0.32]f_s$ of the filter. For the 255 sines this number is 71. The slower convergence to normality of the binary zero/π random phase multisines compared with the uniform random phase multisines, is due to the discrete pdf of the former compared with the continuous pdf of the latter.

Observations The pdfs for 1 sinewave in Figure 5-15.b contain 3 dirac impulses, while those in Figure 5-15.a are continuous functions.

Discussion For 1 sinewave at DFT line $k = N/4$, the signal $u(t)$ (5-35) reduces to

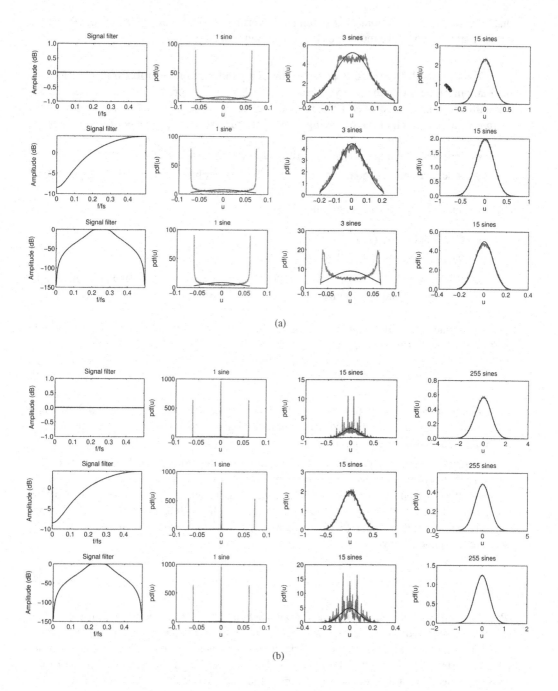

Figure 5-15 Probability density function (pdf) of random phase multisines $u(t)$ with a different number of frequencies uniformly distributed in $(0, 0.5)f_s$ (columns 2, 3, and 4), and different amplitudes (column 1). Gray solid lines: pdf estimated from the time series $u(t)$. Black solid lines: Normal pdf with the same mean and variance as $u(t)$. (a) Phases uniformly distributed in $[[0, 2\pi]]$. (b) Phases equal to zero or π with probability 1/2. Remark: $dB(\sigma^2) \equiv 10\log_{10}(\sigma^2) = 20\log_{10}(\sigma) \equiv dB(\sigma)$.

$$u(t) = \frac{2}{\sqrt{N}}|L(z_{N/4}^{-1})|\cos\left(\frac{\pi}{2}t + \phi_{N/4}\right), \tag{5-40}$$

where $t = 0, 1, ..., N-1$ and $\phi_{N/4} = 0$ or π with equal probability. Hence, the cosine function in (5-40) can only take the values -1, 0, and 1, which clarifies the 3 peaks in the second column of Figure 5-15.b. For a uniformly distributed phase $\phi_{N/4}$, the pdf of $u(t)$ (5-40) is continuous.

In Exercise 84.a, the equivalence between the class of Gaussian noise excitations and the class of random phase multisines for measuring the best linear approximation (BLA) has been illustrated for multisines which excite all harmonics (= full multisines). Provided that the power spectra (rms value and coloring) of both signal classes are the same, it followed that the BLA and its variance are exactly the same for the two classes. Although the sum of a moderate number of random phase sines is approximately normally distributed (see Exercise 84.b), one can wonder whether the full equivalence still holds when not all harmonics of the multisine are excited (e.g., odd multisines). This is analyzed in the following exercise.

Exercise 84.c (Influence of harmonic content multisine on BLA) Consider   the generalized Wiener Hammerstein (GWH) system shown in Figure 5-13 with NFIR part

$$z(t) = x^2(t-1)\text{atan}(x(t)) + 10|x(t-2)| \tag{5-41}$$

and where G_1 and G_2 defined in (5-33). Calculate via (5-37), with $N = 1024$ and $M = 10^4$, the best linear approximation of the GWH system for the class of random phase multisines (5-35) where all harmonics are excited (= full multisine). Repeat the same for the following classes of odd random phase multisines (even harmonics are not excited in (5-35)) with the same rms value as the full multisines:

- Odd multisines: harmonics $2k-1$, with $k = 1, 2, ..., N/4$.
- Odd–odd multisines: harmonics $4k-3$, with $k = 1, 2, ..., N/8$.
- Odd–random (harmonic grid) multisines: odd multisines where out of each group of two consecutive odd harmonics one randomly selected harmonic is not excited.
- Odd–sparse (harmonic grid) multisines: odd multisines where only a few odd harmonics are excited $mk-(m-1)$, with $k = 1, 2, ..., N/(2m)$, and $m = 8$, 16, and 32 (respectively 64, 32, and 16 excited frequencies).

Show that the amplitudes of the excited harmonics of the odd multisines are about $\sqrt{2}$ as large as those of the full multisine. For the odd–odd and odd–random multisines this factor is about two. Explain. Compare the BLAs and their variances. Are they equal (hint: compare the complex difference between the BLAs to the square root of the sum of the variances of the BLAs)? Why (not)? Compare the results of the oddsparse multisines with the odd multisines. What do you conclude?   □

Observations From Figure 5-16 it can be concluded that the BLAs are the same for the classes of full, odd, odd–odd, and odd–random multisines: The complex error is everywhere of the same order of magnitude as its standard deviation, except in the neighborhood of $0.2f/f_s$ (see below). However, their variances (strongly) differ (see Figure 5-17). If the number of frequencies in the multisine is not sufficiently large, then the BLA depends on the

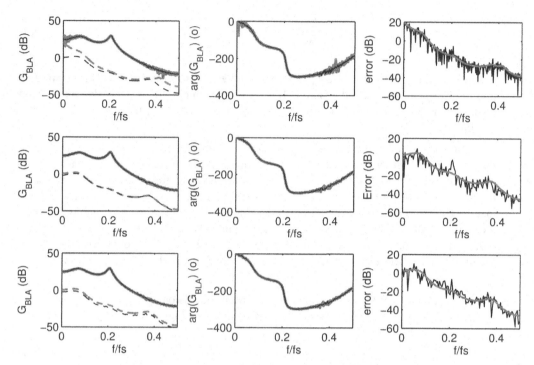

Figure 5-16 Best linear approximations (solid lines) and their standard deviation (dashed lines) of the generalized Wiener–Hammerstein system for the classes of full, odd, odd–odd, and odd–random (harmonic grid), random phase multisine excitations with the same power spectra (rms value and colouring). Top row: Comparison full (gray lines) and odd multisines (black lines). Middle row: Comparison odd (gray lines) and odd–odd multisines (black lines). Bottom row: Comparison odd (gray lines) and odd–random multisines (black lines). Right column: Magnitude of the complex difference between the BLAs (black line), and standard deviation of the complex error (gray line).

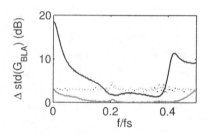

Figure 5-17 Difference between the standard deviations in decibels of the BLAs obtained by the classes of full, odd, odd–odd, and odd–random multisines (see Figure 5-16, left column). Black solid line: Difference between full and odd multisines. Gray solid line: Difference between odd and odd–odd multisines. Black dotted line: Difference between odd and odd–random multisines.

number of frequencies F (see Figure 5-18). From the right column of Figure 5-18 it can be seen that the difference between the BLAs decreases with about 6 dB per frequency doubling. The same is valid when comparing the peak around $0.2f/f_s$ in the bottom right error plot of Figure 5-18 (64 excited frequencies) with those in the middle right and bottom right error plots of Figure 5-16 (128 excited frequencies).

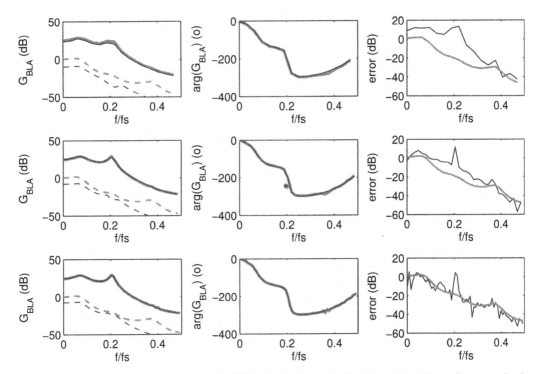

Figure 5-18   Best linear approximations (solid lines) and their standard deviation (dashed lines) of the generalized Wiener–Hammerstein system for the classes of odd, and odd–sparse (harmonic grid), random phase multisine excitations with the same power spectra (rms value and colouring). The BLAs and their uncertainty are only shown at the excited frequencies of the odd–sparse multisine. Top row: Comparison odd (gray lines) and odd–sparse with 16 frequencies (black lines). Middle row: Comparison odd (gray lines) and odd–sparse with 32 frequencies (black lines) Bottom row: Comparison odd (gray lines) and odd–sparse with 64 frequencies (black lines). Right column: Magnitude of the complex difference between the BLAs (black line), and standard deviation of the complex error (gray line).

Discussion   This can be explained as follows. In Schoukens *et al.* (1998) and Pintelon and Schoukens (2001) it has been shown that the dependence on the number of frequencies is given by

$$G_{\text{BLA}, F}(j\omega_k) = G_{\text{BLA}}(j\omega_k) + O(F^{-1}) \tag{5-42}$$

The $O(F^{-1})$ term in (5-42) is consistent with the observation from Figure 5-18 that the difference between the BLAs decreases with about 6 dB per frequency doubling. It also shows that the peaks around $0.2f/f_s$ in the error plots of Figure 5-16 are due to the $O(F^{-1})$ term in equation (5-42). This peak is no longer visible in the top right error plot of Figure 5-16 (256 excited frequencies), showing that for $F = 256$ the $O(F^{-1})$ term is smaller than or equal to the uncertainty of the BLA estimate.

Conclusion   As long as the effective power per frequency band is the same, and the number of frequencies is sufficiently large, the BLA does not depend on the harmonic content of the multisine (Schoukens *et al.*, 2009a).

Observations   The variability of the full multisine BLA strongly differs from those of the odd multisine BLAs (see Figure 5-17). From Figure 5-17 it can also be seen that the standard deviation of the odd–odd multisine BLA only differs from that of the odd multisine BLA for frequencies below $0.1f_s$ and above $0.4f_s$ with a maximal difference of 3 dB (factor $\sqrt{2}$). The difference between the standard deviations in decibels of the odd and odd–random multisines is about 3 dB over the whole frequency band.

Discussion Since the response of a nonlinear system to an odd multisine has no even nonlinear contributions at the odd harmonics (see Section 5.1), the best linear approximation - which is calculated at the excited frequencies - only contains odd nonlinear contributions. The response of the GWH to the full multisine, however, contains both the even and odd nonlinear contributions at the excited harmonics (see Section 5.1). It explains the larger variability of the full multisine BLA compared with all the odd multisine BLAs. The difference between the standard deviations in decibels of the odd and odd–random multisines of about 3 dB corresponds exactly to the square root of the ratio of the number of excited harmonics in the odd and odd–random multisines (see Section 6.1.4 for a detailed discussion).

Conclusion All these observations show that the "nonlinear noise source" G_S in the estimated best linear approximation (5-38) depends on the input power spectrum (rms value, coloring, and harmonic content) and the even and/or odd nonlinear distortions (Schoukens *et al.*, 1998; Pintelon and Schoukens, 2001).

Comparing Figure 5-16 to Figure 5-14 it can be seen that, although the NFIR parts (5-41) and (5-33) are different, the BLAs of the two GWH systems are remarkably similar (in fact they are exactly the same). The explanation for this can be found in the following exercise.

Exercise 85 (Influence of even and odd nonlinearities on BLA)   Consider the generalized Wiener–Hammerstein (GWH) systems shown in Figure 5-13 on page 157 with NFIR parts $z(t) = x(t) + z_{NL}(t)$ where $z_{NL}(t)$ is given by

- Even nonlinearity: $z_{NL}(t) = |x^2(t-1)\text{atan}(x(t))|$.
- Odd nonlinearity: $z_{NL}(t) = x^2(t-1)\text{atan}(x(t))$.

and where G_1 and G_2 are defined in (5-33). Calculate via (5-37), with $N = 1024$ and $M = 10^4$, the best linear approximation of the GWH systems for the class of odd random phase multisine inputs

$$u(t) = \alpha\sum\nolimits_{k=1}^{N/4} \cos((2k-1)\omega_0 t + \phi_{2k-1}) \qquad (5-43)$$

with ϕ_{2k-1} uniformly in $[0, 2\pi]$ and where α is chosen such that the rms value of $u(t)$ equals $1/\sqrt{2}$. Show that α equals $\sqrt{2/N}$. Compare the best linear approximations with the product $G_1 G_2$ of the transfer functions of the linear dynamic blocks. What do you conclude? Explain. Repeat all calculations for the following class of chi-squared distributed random excitations $u(t) = (e^2(t) - 1)/2$, where $e(t)$ is normally distributed with zero mean and unit variance [*Hint*: Use (5-28)]. Show that the standard deviation of $u(t)$ equals $1/\sqrt{2}$. Compare the BLAs with the underlying linear system. Are the conclusions different from those of the odd random phase multisines? Explain why. (*Hint*: Verify the skewness - third-order central moment divided by the cube of its standard deviation - of the signals.)   □

Observations It follows from Figure 5-19 that the best linear approximation of the GWH system with the even nonlinearity coincides with the underlying linear system $G_1 G_2$ (the difference is at the level of the arithmetic precision of MATLAB®), while the BLA of the

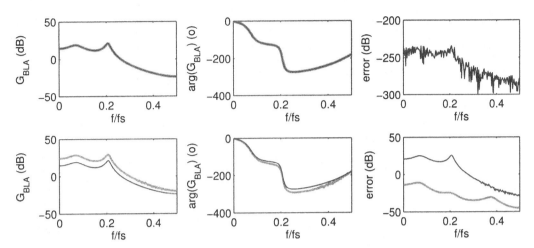

Figure 5-19   Comparison of the best linear approximation (BLA) of a generalised Wiener–Hammerstein (GWH) system (gray solid lines) with the true underlying linear system (black solid lines), for the class of odd random phase multisines. Top row: GWH with even nonlinear part. Bottom row: GWH with odd nonlinear part. Right column: Magnitude of the complex difference between the BLA and the true underlying linear system (black line), and standard deviation of the complex error (gray line).

GWH system with the odd nonlinearity clearly differs from $G_1 G_2$ (the difference is significantly larger than its uncertainty).

Discussion   This can be explained as follows. Since the response of an even nonlinearity to an odd multisine will only contain contributions at the even harmonics (see Section 5.1.3), the best linear approximation (5-36) - which is calculated at the excited frequencies only - is not affected by even nonlinearities. Odd nonlinearities, however, create contributions at the odd harmonics that are phase-coherent with the input and, hence, push the BLA away from the underlying linear system (Schoukens *et al.*, 1998; Pintelon and Schoukens, 2001). Both observations show that the best linear approximation (5-36) can be written as

$$G_{\text{BLA}}(j\omega) = G_0(j\omega) + G_{\text{B}}(j\omega) \tag{5-44}$$

where G_0 is the true underlying linear system (if it exists), and where G_{B} is the bias contribution that depends on the power spectrum of the input (rms value and coloring) and the odd nonlinearities only. Note that since the random phase multisines behave as Gaussian noise (Exercises 84.a and 84.b), these conclusions are also valid for the class of Gaussian noise excitations.

Observations   For the class of chi-squared distributed excitations the conclusions are different from those of the class of random phase and Gaussian noise excitations (compare Figure 5-20 with Figure 5-19): The BLA depends on the odd as well as the even nonlinearities.

Discussion   This can be explained by the (a)symmetry (skewness) of the pdf of the signals. Indeed, the BLA of an even nonlinearity is exactly zero for symmetrically distributed signals, while it is nonzero for asymmetric distributions. For random phase multisines and Gaussian noise excitations the pdf's of the signals $u(t)$ and $x(t)$ in Figure 5-13 are perfectly symmetric (zero skewness), while for the chi-squared distributed excitations they are nonsymmetric (nonzero skewness). This can be verified by calculating the sample skewness of

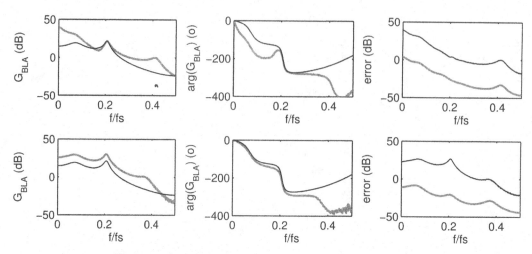

Figure 5-20 Comparison of the best linear approximation (BLA) of a generalised Wiener–Hammerstein (GWH) system (gray solid lines) with the true underlying linear system (black solid lines), for the class of chi-squared distributed noise. Top row: GWH with even nonlinear part. Bottom row: GWH with odd nonlinear part. Right column: Magnitude of the complex difference between the BLA and the true underlying linear system (black line), and standard deviation of the complex error (gray line).

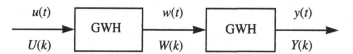

Figure 5-21 Cascade of two generalised Wiener–Hammerstein (GWH) systems. Each block consists a nonlinear FIR system sandwiched between two linear dynamic systems (see Figure 5-13).

$u(t)$ and $x(t)$. For the chi-squared distributed excitations we find, respectively, 2.84 and 0.147, which should be compared with the value 1.13×10^{-4} for Gaussian random variables. Although the second-order linear dynamic system G_1 (see Figure 5-13) reduces the skewness of the signal (from 2.84 to 0.147), it is still large enough to generate significant even nonlinear contributions to the BLA.

Observations Comparing Figure 5-20 with Figure 5-19, it can be seen that the best linear approximations (gray lines) and the corresponding standard deviations are different for both classes of excitations signals. It illustrates once more the influence of the pdf on the estimated BLA.

A final issue that merits attention is the question of whether the best linear approximation of the cascade of two nonlinear systems equals the cascade of the BLAs of each system separately. This is analyzed in the following exercise.

Exercise 86 (BLA of a cascade) Consider the cascade of two identical generalized Wiener–Hammerstein (see Figure 5-21) systems, where each system has the NFIR part

$$z(t) = 1 + 0.5x^2(t-1)\mathrm{atan}(x(t)) \qquad (5\text{-}45)$$

and linear dynamic blocks G_1 and G_2 defined in (5-33). Calculate via (5-37), with $N = 1024$ and $M = 4\times10^4$, the best linear approximation of the cascade of the GWH sys-

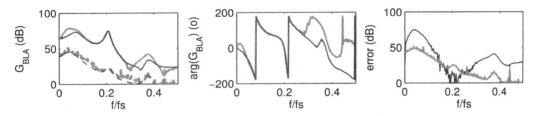

Figure 5-22   Comparison of the best linear approximation (BLA) of the cascade of two dynamic nonlinear systems (black solid lines) with the cascade of the BLAs of each nonlinear system separately (gray solid lines). Black dashed line: Standard deviation of the BLA of the cascade. Gray dashed line: Standard deviation of the cascade of the BLAs. Right column: Complex difference between the BLA of the cascade and the cascade of the BLAs (black solid line), and standard deviation of the difference (gray solid line).

tems (see Figure 5-21: From $u(t)$ to $y(t)$) for the class of odd random phase multisines

$$u(t) = \frac{1}{\sqrt{N}}\sum_{k=1}^{N/4} 2|L(z_{2k-1}^{-1})|\cos((2k-1)\omega_0 t + \phi_{2k-1}), \qquad (5\text{-}46)$$

where $L(z^{-1})$ is defined in (5-34). Calculate for the same input $u(t)$ the BLA of the first (see Figure 5-21: from $u(t)$ to $w(t)$) and the second (see Figure 5-21: from $w(t)$ to $y(t)$) GWH system. Compare the BLA of the cascade with the cascade of the BLAs. Are they the same? Why (not)? Calculate the coherence function $\gamma(j\omega) = |S_{YU}(j\omega)|/\sqrt{S_{YY}(j\omega)S_{UU}(j\omega)}$ of the BLA of the cascade and compare it with the previous figure. What do you conclude? Replace one of the two GWH systems by a linear dynamic system, and repeat the calculations. What do you conclude now? Explain.   □

Observations As could be expected, the BLA of the cascade of the two nonlinear systems is not equal to the cascade of the BLAs of each nonlinear system separately (see Figure 5-22). Indeed, the difference between both is everywhere larger than the standard deviation of the difference, except in the band $[0.17, 0.23]f_s$ (see Figure 5-22, right plot). In the band $[0.17, 0.23]f_s$, however, the BLA of the cascade equals the cascade of the BLAs (see Figure 5-22, left and middle plots). The method for detecting the frequency bands where the cascade rule applies is discussed in the next paragraphs.

Using M realizations of the random phase multisines the coherence function of the estimated BLA of the cascade is calculated as

$$\hat{\gamma}(k) = \frac{|\hat{S}_{YU}(j\omega_k)|}{\sqrt{\hat{S}_{YY}(j\omega_k)\hat{S}_{UU}(j\omega_k)}} = \frac{1}{\sqrt{1 + M\dfrac{\text{var}(\hat{G}_{\text{BLA}}(j\omega_k))}{|\hat{G}_{\text{BLA}}(j\omega_k)|^2}}}, \qquad (5\text{-}47)$$

where $\hat{G}_{\text{BLA}}$ and $\text{var}(\hat{G}_{\text{BLA}})$ are defined in (5-37).

Observations Comparing the BLAs and the coherence function (see Figure 5-23), it can be seen that the BLA of the cascade equals the cascade of the BLAs in those frequency bands where the coherence function of the estimated BLA of the cascade is larger than or equal to -6 dB ($0.5 \le \gamma \le 1$). This simple rule of thumb allows us to verify a posteriori in which frequency bands the cascade rule applies to the sub-blocks of a complex nonlinear system.

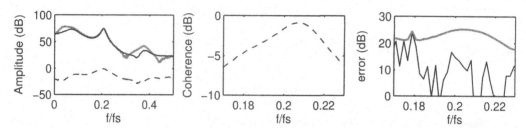

Figure 5-23   Comparison between the best linear approximations (BLAs) and the coherence function. Left: BLA of the cascade (black solid lines); cascade of the BLAs (gray solid lines); and coherence function of the estimated BLA of the cascade (black dashed lines). Middle: Zoom of the coherence function in the band $[0.17, 0.23]f_s$. Right: Zoom of the difference between the BLA of the cascade and the cascade of the BLAs (black solid line), and of the standard deviation of the difference (gray solid line) in the band $[0.17, 0.23]f_s$.

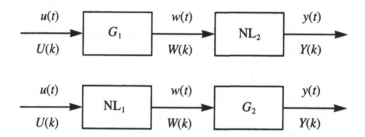

Figure 5-24   Cascade of a linear and a nonlinear dynamic system (top), or cascade of a nonlinear and a linear dynamic system (bottom).

Discussion Note that the cascade rule only applies if the BLAs of the first and second subsystem are measured using respectively the input $u(t)$ and the intermediate signal $w(t)$ of the cascade in Figure 5-21. It is also remarkable that the cascade rule remains valid for coherence values as low as 0.5. Indeed, the standard deviation of the FRF measurement using one random phase multisine ($\sqrt{M}\text{std}(\hat{G}_{\text{BLA}})$ in (5-47)) equals then $\sqrt{3}|\hat{G}_{\text{BLA}}|$! If one of the two nonlinear systems in the cascade is linear (see Figure 5-24), then the cascade rule is always valid. Indeed, using the following property of the cross-power spectrum $S_{(W_1 Y)(W_2 U)} = W_1 S_{YU}\overline{W}_2$ (Bendat and Piersol, 1980) it can easily be verified that

$$G_{\text{BLA}} = \frac{S_{YU}}{S_{UU}} = \frac{S_{Y(G_1^{-1}W)}}{S_{(G_1^{-1}W)(G_1^{-1}W)}} = \frac{S_{YW}\overline{G_1}^{-1}}{S_{WW}|G_1^{-1}|^2} = G_{\text{BLA2}}G_1 \tag{5-48}$$

for the cascade of a linear dynamic system G_1 and a nonlinear dynamic system with BLA G_{BLA2} (see Figure 5-24, top row), and

$$G_{\text{BLA}} = \frac{S_{YU}}{S_{UU}} = \frac{S_{(G_2 W)U}}{S_{UU}} = \frac{G_2 S_{WU}}{S_{UU}} = G_2 G_{\text{BLA1}} \tag{5-49}$$

for the cascade of a nonlinear dynamic system with BLA G_{BLA1} and a linear dynamic system G_2 (see Figure 5-24, bottom row). We refer the reader to the paper of Dobrowiecki and Schoukens (2002) for more information.

5.2.3 What Have we Learned in Section 5.2?

∎ The best linear approximation (BLA) of a nonlinear system minimizes the mean square error between the true (shifted) output of the nonlinear system $y(t) - \mathbb{E}\{y(t)\}$ and the output of the linear model $g_{\text{BLA}}(t) * (u(t) - \mathbb{E}\{u(t)\})$ [see equation (5-21)], and is defined for a *class* of excitation signals. Before calculating the best linear approximation, the DC values of the input/output signals should be removed (Exercise 82). In the frequency domain, the solution of minimization problem (5-21) boils down to the division of the cross-power by the auto-power spectrum: $G_{\text{BLA}}(j\omega) = S_{YU}(j\omega) / S_{UU}(j\omega)$. In practice this formula is approximated by (5-28) and (5-37) for, respectively, random and periodic excitations $u(t)$ (Exercises 83.b and 84.a).

∎ Assuming that the input/output signals are observed without errors, the BLA measured via (5-28) and (5-37) can be written as

$$\hat{G}_{\text{BLA}}(j\omega_k) = G_{\text{BLA}}(j\omega_k) + G_S(j\omega_k), \tag{5-50}$$

where G_S denotes the "nonlinear noise source" ($\mathbb{E}\{G_S\} = 0$). The stochastic nonlinear contributions G_S depend on the power spectrum (rms value and coloring) and the probability density function (pdf) of the input (Exercises 83.b, 83.c, and 85), and on the even and odd nonlinearities (Exercises 84.c and 85). They decrease to zero for increasing number of averages over random phase realizations of the multisine.

∎ The best linear approximation of a nonlinear system can be written as

$$G_{\text{BLA}}(j\omega_k) = G_0(j\omega_k) + G_B(j\omega_k), \tag{5-51}$$

with G_0 the true underlying linear system (if it exists), and where G_B denotes the "nonlinear bias contribution." The systematic nonlinear contributions G_B depend on the power spectrum (rms value and coloring) and the pdf of the input (Exercises 83.a, 83.b, and 85), and, in general, on the even and odd nonlinearities (Exercise 85). For non-Gaussian inputs not only does the autocorrelation (power spectrum) affect G_B, but also all higher order correlations over time (the complete dependency over time). Consequently, for filtered white non-Gaussian noise the BLA also depends on the phase of the filter (Exercise 83.b). For Gaussian noise and random phase multisines the systematic nonlinear contributions G_B do *not* depend on the *even* nonlinearities (Exercises 84.c and 85).

∎ For Gaussian noise excitations, the BLA of a static nonlinearity is static and independent of the coloring of the power spectrum (Exercise 83.b). However, the BLA of a static nonlinearity is dynamic for nonwhite, non-normally distributed inputs (Exercise 83.b). Although this is, at first glance, a completely counter intuitive result, one should in fact not be surprised. Indeed, since $S_{UU}(j\omega)$ is frequency-dependent for nonwhite noise, and since $S_{YU}(j\omega)$ is in general not proportional to $S_{UU}(j\omega)$ for a nonlinear system, the best linear approximation $G_{\text{BLA}}(j\omega) = S_{YU}(j\omega) / S_{UU}(j\omega)$ will in general be frequency-dependent. In fact, it is surprising that the BLA of a static nonlinear system is still static for colored Gaussian noise. In both (Gaussian and non-Gaussian) cases the variance of the BLA measurement depends on the coloring of the power spectrum (Exercise 83.b).

∎ Filtered white non-Gaussian noise behaves asymptotically (as the length of the impulse response of the filter tends to infinity) as Gaussian noise (Exercise 83.c).

■ A random phase multisine

$$u(t) = \frac{1}{\sqrt{N}}\sum_{k=1}^{N/2-1}A_k\cos(k\omega_0 t + \phi_k) \tag{5-52}$$

A with A_k the user defined amplitudes, and ϕ_k the independent (over k) distributed phases such that $\mathbb{E}\{e^{j\phi_k}\} = 0$, behaves asymptotically (for $N \to \infty$) as Gaussian noise (Exercises 84.a and 84.b). Hence, the BLA for the class of the random phase multisines is asymptotically (for $N \to \infty$) the same as the BLA for the class of Gaussian noise with the same power spectrum (rms value and coloring). For finite values of N, the random phase multisine BLA $G_{\text{BLA}N}$ is related to the Gaussian noise BLA G_{BLA} as

$$G_{\text{BLA}N}(j\omega_k) = G_{\text{BLA}}(j\omega_k) + O(N^{-1}) \tag{5-53}$$

(Exercise 84.c). If N is sufficiently large, then the systematic nonlinear contributions G_{B} in (5-51) do not depend on the harmonic content (full, odd, odd–odd, odd random harmonic grid, ...) of the multisine, as long as the (equivalent) power per frequency band remains the same (Exercise 84.c). The stochastic nonlinear contributions G_{S}, however, strongly depend on the harmonic content (Exercise 84.c).

■ Summarizing, in the absence of input/output measurement errors the best linear approximations estimated via (5-28) and (5-37) can be written as

$$\hat{G}_{\text{BLA}}(j\omega_k) = (G_{\text{BLA}}(j\omega_k) + O_{\text{B}}(N^{-1})) + (G_{\text{S}}(j\omega_k) + O_{\text{S}}(N^{-1/2})), \tag{5-54}$$

with $G_{\text{BLA}}(j\omega_k) = G_0(j\omega_k) + G_{\text{B}}(j\omega_k)$, $O_{\text{B}}(N^{-1})$ the bias term on the BLA measurement due to the finite number of frequencies (random phase multisine excitations) or the leakage errors (noise excitations), and $O_{\text{S}}(N^{-1/2})$ the stochastic term ($\mathbb{E}\{O_{\text{S}}\} = 0$) due to the leakage errors (noise excitations only) which is zero for the random phase multisines. G_{B} and G_{S} are, respectively, the bias and stochastic ($\mathbb{E}\{G_{\text{S}}\} = 0$) contributions of the nonlinearities to the BLA. These contributions cannot be eliminated by increasing N (G_{B} is independent of N, and $\text{var}(G_{\text{S}}) = O(N^0)$). G_{S} is suppressed by averaging the measurements over a large number of realizations (multisines) or blocks (noise excitations) M [see (5-28) and (5-37)]. To reduce O_{B}, the period length (number of frequencies in the multisines) or the block length (noise excitations) N is increased. Finally, O_{S} (noise excitations only) is diminished by increasing M and/or N.

■ The best linear approximation of the cascade of two nonlinear systems is not equal to the cascade of the best linear approximation of each nonlinear system separately (Exercise 86). From a practical point of view the cascade rule is valid in those frequency bands where the coherence function of the BLA measurement of the cascade is larger than -6 dB. The cascade rule is always valid if one of the two nonlinear systems is linear.

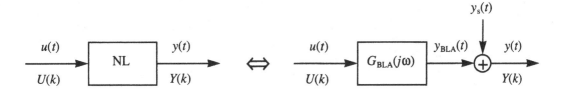

Figure 5-25   The output of the nonlinear system (left diagram) can exactly be written as the sum of the output of the best linear approximation and the stochastic nonlinear distortions (right diagram).

5.3 PREDICTIVE POWER OF THE BEST LINEAR APPROXIMATION

Although no noise is added in the simulations, it follows from the previous section that the best linear approximation calculated via (5-28) and (5-37) is noisy. It strongly suggests that a part of the nonlinear distortions acts as "noise" and that the actual output $y(t)$ of the nonlinear system can be written as the sum of the output of the BLA $y_{BLA}(t)$ and the stochastic nonlinear distortions $y_s(t)$

$$y(t) = y_{BLA}(t) + y_s(t) \quad \text{with} \quad y_{BLA}(t) = g_{BLA}(t) * u(t) \tag{5-55}$$

(see Figure 5-25). Note that an equation of the form (5-55) is EXACTLY true for ANY linear approximation (LA) $g_{LA}(t)$, viz:

$$y(t) = y_{LA}(t) + v(t) \quad \text{with} \quad y_{LA}(t) = g_{LA}(t) * u(t) \tag{5-56}$$

as long as the properties of the residuals $v(t)$ are not specified. What makes (5-55) peculiar is that $y_s(t)$ has similar stochastic properties as measurement noise. This will be shown in the sequel of this section. It explains why $y_s(t)$ is indistinguishable from the measurement noise in classical frequency response function measurements using random excitations (5-28).

Summarizing, in this section we analyze the stochastic properties of the residuals (= stochastic nonlinear distortions) $y_s(t)$, and establish the link between the output residuals $y_s(t)$ (5-55) and the stochastic nonlinear distortions $G_S(j\omega_k)$ on the BLA measurement (5-54). Further, the predictive power of the best linear approximation is verified.

5.3.1 Static nonlinear systems

In the following exercise we analyze some time domain properties of the residuals $y_s(t)$ (probability density function, and cross-correlation with the input $u(t)$) of the (best) linear approximations of a static nonlinear system excited by uniformly distributed white noise. The predictive power of the (best) linear approximations is verified within the class of uniformly distributed white noise.

Exercise 87.a (Predictive power BLA — static NL system)   Calculate via (5-22), with $N = 10^4$, the best linear approximation g_{BLA} of $y(t) = u^3(t)$ for the class of uniformly $[-1, 1]$ distributed white noise $u(t)$. Define further the linear approximations $g_{LA} = 0.9g_{BLA}$ and $g_{LA} = 1.1g_{BLA}$, and calculate the cross-correlation

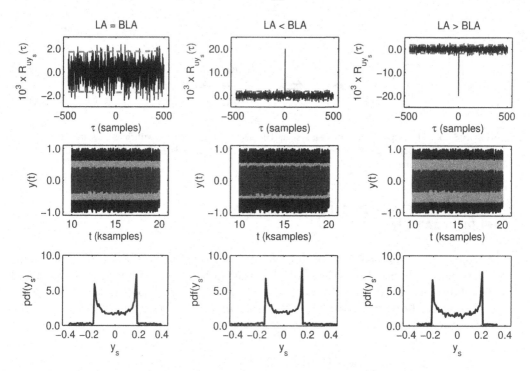

Figure 5-26   Properties of the linear approximation (LA) of the static nonlinear system $y(t) = u^3(t)$
excited by white uniformly $[-1, 1]$ distributed noise: best linear approximation (BLA)
g_{BLA} (left column), LA with gain $0.9g_{BLA}$ (middle column), and LA with gain $1.1g_{BLA}$
(right column). Top row: Cross-correlation between the input $u(t)$ and the stochastic nonlinear distortions $y_s(t)$ (black), and its 95% uncertainty bounds (gray dashed lines). Middle
row: True output (black), output of the (B)LA (light gray), and output error of the (B)LA
(dark gray). Bottom row: Probability density function (pdf) of the output error of the
(B)LA.

$$R_{uy_s}(\tau) = \mathbb{E}\{u(t)y_s(t-\tau)\} \quad \text{with} \quad y_s(t) = y(t) - g_{(B)LA}u(t) \qquad (5\text{-}57)$$

between the input $u(t)$ and the residuals $y_s(t)$ of the (best) linear approximations $g_{(B)LA}$
(*Hint*: Approximate (5-57) by the sample cross-correlation, and calculate its uncertainty
assuming that $y_s(t)$ and $u(t)$ are independent). What do you conclude? Next generate a
validation set of 10^4 data points and compare the output predicted by the (B)LA's with the
output of the static nonlinearity. At a glance the difference between the true output and the
output of the (B)LA (residual $y_s(t)$) appears to be the smallest for the linear approximation
$g_{LA} = 1.1g_{BLA}$. Explain the apparent contradiction [*Hint*: Analyze the variance and the
probability density function of the residuals $y_s(t)$.]   □

The sample cross-correlation $\hat{R}_{uy_s}(\tau)$ between the input $u(t)$ and output residuals $y_s(t)$
($t = 0, 1, ..., N-1$) is given by

$$\hat{R}_{uy_s}(\tau) = \frac{1}{N-\tau}\sum_{t=\tau}^{N-1}u(t)y_s(t-\tau). \qquad (5\text{-}58)$$

Assuming that $u(t)$ and $y_s(t)$ are independent, it can easily be verified that the variance of

$\hat{R}_{uy_s}(\tau)$ is given by

$$\text{var}(\hat{R}_{uy_s}(\tau)) = \sigma_u^2 \sigma_{y_s}^2 / (N - \tau) \qquad (5\text{-}59)$$

($u(t)$ and $y_s(t)$ are zero mean stationary stochastic processes). Since $\hat{R}_{uy_s}(\tau)$ (5-58) is asymptotically normally distributed (central limit theorem), calculating $\pm 2\text{std}(\hat{R}_{uy_s}(\tau))$ via (5-58) gives the 95% confidence bound of the sample cross-correlation under the hypothesis that $u(t)$ and $y_s(t)$ are independently distributed.

Observations The results of the cross-correlation analysis are shown in the top row of Figure 5-26. It can be seen that the output residuals of the BLA are uncorrelated with the input (the fraction inside the 95% confidence bounds equals 94.8%), while those of the LAs are correlated with the input at $\tau = 0$ (the cross-correlation at $\tau = 0$ lies significantly outside the 95% confidence interval).

Discussion The observation that the output residuals of the BLA are uncorrelated with the input is in fact not surprising. Indeed, it is a direct consequence of the general property of least squares estimators that the residuals are uncorrelated with the regressors (Ljung, 1999), applied to the linear least squares problem (5-21) (Enqvist, 2005).

Observations Comparing the output predicted by the (best) linear approximations with the true output of the static nonlinear system (see Figure 5-26, middle row), one would wrongly conclude that the linear approximation $g_{LA} = 1.1 g_{BLA}$ has the smallest output error. Indeed, the standard deviation of the output errors $y_s(t)$ equals 0.151, 0.155, and 0.155 for, respectively, g_{BLA}, $g_{LA} = 0.9 g_{BLA}$, and $g_{LA} = 1.1 g_{BLA}$.

Discussion The apparent contradiction can be explained by the probability density function of $y_s(t)$ (see Figure 5-26, bottom row): (i) although the extreme values of $y_s(t)$ are larger for g_{BLA} and $g_{LA} = 0.9 g_{BLA}$ than for $g_{LA} = 1.1 g_{BLA}$, they have a low probability; and (ii) for the linear approximation $g_{LA} = 1.1 g_{BLA}$ the pdf is maximal for $|y_s(t)| > 0.2$, while for the two other approximations the pdf is maximal for $|y_s(t)| < 0.2$.

Observations Note that the pdf of the output residuals (stochastic nonlinear distortions) $y_s(t)$ strongly deviates from the uniform pdf of the input (see Figure 5-26, bottom row).

Discussion It emphasizes that the best linear approximation of a static nonlinear function is a second order equivalent, and shows that the higher order moments of $y_s(t)$ contain information about the nonlinear behavior of the system.

5.3.2 Dynamic nonlinear systems

In this section we analyze the time and frequency domain properties of the output residuals of the best linear approximation (BLA) of a generalized Wiener–Hammerstein system excited by odd random phase multisines. The number of odd excited frequencies is chosen sufficiently large such that the odd random phase multisines behave as Gaussian noise (see Section 5.2.2). Next, the predictive power of the BLA is verified within the class of odd random phase multisines.

Exercise 87.b (Properties of output residuals — dynamic NL system) Consider the generalized Wiener Hammerstein (GWH) system shown in Figure 5-13 on page 157 with NFIR part

$$z(t) = x(t) + 0.01x^2(t-1)\tanh(x(t)), \qquad (5\text{-}60)$$

where G_1 and G_2 are defined in (5-33). Calculate via (5-37), with $N = 1024$ and $M = 10^4$, the best linear approximation (BLA) of the GWH system for the class of odd random phase multisines

$$u(t) = \sqrt{\frac{2}{N}} \sum_{k=1}^{N/4} \cos((2k-1)\omega_0 t + \phi_{2k-1}) \tag{5-61}$$

with ϕ_{2k-1} uniformly distributed in $[0, 2\pi]$. Using the BLA, calculate the output residuals

$$y_s^{[m]}(t) = y^{[m]}(t) - g_{\mathrm{BLA}}(t) * u^{[m]}(t) \tag{5-62}$$

for $m = 1, 2, ..., M$ and $t = 0, 1, ..., N-1$ (*Hint*: make the calculations in the frequency domain and convert to the time domain), and draw the following figures:

- $\mathrm{var}(y_s(t))$ as a function of time (*Hint*: approximate the variance by the sample variance of $y_s^{[m]}(t)$ over the realizations m as a function of time t; next, to verify the variability of the sample variance over time, calculate the sample mean and sample standard deviation over t of the sample variance over the realizations). Is $\mathrm{var}(y_s(t))$ constant?

- The probability density function (pdf) of y_s (hint: use $y_s^{[m]}(t)$ for $m = 1, 2, ..., M$ and $t = 0, 1, ..., N-1$). Compare with a Gaussian pdf with the same mean and variance. Is y_s normally distributed? Why (not)?

- The cross-correlation between the input $u(t)$ and the output residual $y_s(t)$

$$R_{uy_s}(\tau) = \mathbb{E}\{u(t)y_s(t-\tau)\} \text{ with } y_s(t) = y(t) - g_{\mathrm{BLA}}(t) * u(t) \tag{5-63}$$

(*Hint*: approximate (5-63) by the sample cross-correlation over time via the MATLAB® instruction `xcorr(x,y,'unbiased')`; next, to verify its variability, calculate the sample mean and sample standard deviation over the last thousand realizations). Are $u(t)$ and $y_s(t)$ uncorrelated?

- The probability density function of $|Y_S(k)|$ and $\angle Y_S(k)$ at one excited frequency (hint: use $Y_S^{[m]}(k)$, $m = 1, 2, ..., M$). Compare the pdf's to respectively a Rayleigh distribution with second order moment $\mathrm{var}(Y_S(k))$ (*Hint*: use the MATLAB® instruction `raylpdf(x,b)` with $2b^2 = \mathrm{var}(Y_S(k))$), and a uniform $[0, 2\pi]$ distribution. What do you conclude?

- The autocorrelation of frequency domain output residuals $Y_S(k)$ at the excited frequencies

$$R_{Y_S}(k) = \mathbb{E}\{Y_S(l)\overline{Y_S}(l-k)\} \text{ with } Y_S(k) = Y(k) - G_{\mathrm{BLA}}(j\omega_k)U(k) \tag{5-64}$$

(*Hint*: approximate (5-64) by the sample cross-correlation over frequency via the MATLAB® instruction `xcorr(x,x,'unbiased')`; next, to verify its variability, calculate the sample mean and sample standard deviation over the last thousand realizations). Is $Y_S(k)$ correlated over the frequency?

- The variance of $Y_S(k)$ at the excited frequencies (*Hint*: calculate the sample variance of $Y_S^{[m]}(k)$ over m). Is the variance a smooth function of the frequency? Repeat the calculations for $N = 512$, and 256. What do you conclude?

Discuss the results.   □

First of all note that the output residual $y_s(t)$ of a nonlinear PISPO systems (see Section 5.1.2.A) excited by a periodic input, is a periodic signal with the same period as the input. Therefore, the sample cross-correlation between the input $u(t)$ and the output residuals (stochastic nonlinear distortions) $y_s(t)$ is calculated as

$$\hat{R}_{uy_s}(\tau) = \frac{1}{N}\sum_{t=0}^{N-1} u(t)y_s(t-\tau) \tag{5-65}$$

[a time-efficient implementation of (5-65) uses the FFT: see Rabiner and Gold (1975)]. By definition of the best linear approximation (5-21), and by construction of the random phase multisines, $y_s(t)$ has zero mean value over, respectively, time (one period) and realizations, viz.,

$$\frac{1}{N}\sum_{t=0}^{N-1} y_s(t) = 0 \quad \text{and} \quad \mathbb{E}\{y_s(t)\} = 0 \tag{5-66}$$

Observations Although $y_s(t)$ is periodic, it follows from (5-66) and the top row of Figure 5-27 that, within one period, $y_s(t)$ behaves as second order stationary, non-Gaussian distributed noise: $\mathbb{E}\{y_s(t)\} = 0$, $\text{var}(y_s(t))$ is independent of t, and $y_s(t)$ is uncorrelated with the input $u(t)$ (fractions inside the 95% confidence bounds are, respectively, 94.7% and 98.4%).

Conclusion $y_s(t)$ is a non-Gaussian, zero mean, second order cyclo-stationary random process, whose higher-order moments contain information about the nonlinear system. It confirms the close connection between Gaussian noise and random phase multisines with a sufficiently large number of excited frequencies (see Section 5.2.2).

A zero mean complex random variable $z = x + jy$ with variance $\sigma_z^2 = \mathbb{E}\{|z|^2\}$ is circular complex distributed if $\mathbb{E}\{z^2\} = 0$:

$$\mathbb{E}\{z^2\} = 0 \Leftrightarrow \begin{cases} \mathbb{E}\{x^2\} = \mathbb{E}\{y^2\}, \\ \mathbb{E}\{xy\} = 0 \end{cases} \tag{5-67}$$

(x and y are uncorrelated random variables with zero mean and variance $\sigma_z^2/2$). The complex random variable z is circular complex normally distributed if in addition to (5-67) x and y are normally distributed. For circular complex normally we have that $\mathbb{E}\{z^n\} = 0$ for each finite $n \in \mathbb{N}$. It can be shown that z is circular complex normally distributed if and only if the amplitude $|z|$ and phase $\angle z$ are, respectively, Rayleigh and uniformly distributed

$$f_{|z|}(|z|) = \frac{2|z|}{\sigma_z^2} e^{-\frac{|z|^2}{\sigma_z^2}} \quad \text{and} \quad f_{\angle z}(\angle z) = \begin{cases} 1/(2\pi), & \angle z \in [-\pi, \pi] \\ 0, & \text{elsewhere} \end{cases} \tag{5-68}$$

(Papoulis, 1981).

Observations Using result (5-68), it can be concluded from the middle row of Figure 5-27 (left and middle plots) that $Y_S(k)$ is circular complex normally distributed. This is consis-

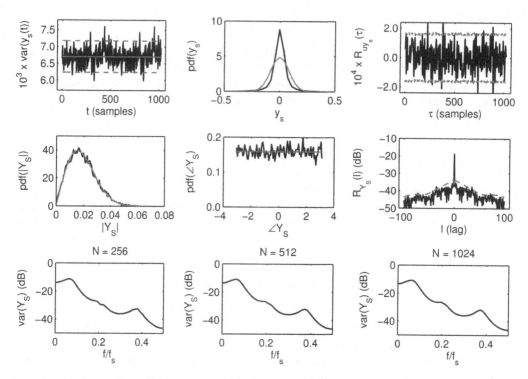

Figure 5-27   Properties of the stochastic nonlinear distortion $y_s(t)$ of a generalized Wiener–Hammerstein system excited by an odd random phase multisine $u(t)$. Top row left: Variance over the realizations of $y_s(t)$ (black solid line), mean value of the variance over time (gray solid line), and 95% confidence interval of the variance (gray dashed lines). Top row middle: Probability density function (pdf) of y_s (black), and gaussian pdf with the same variance (gray). Top row right: Cross-correlation between $u(t)$ and $y_s(t)$ (black solid line), and its 95% confidence interval (gray dashed lines). Middle row left: pdf of $|Y_s|$ (black), and Rayleigh pdf with the same second order moment (gray). Middle row middle: pdf of $\angle Y_s$ (black), and uniform pdf in $[-\pi, \pi]$ (gray). Middle row right: Auto-correlation of $Y_s(k)$ at the excited odd harmonics (black), and its 95% confidence interval (gray dashed lines). Bottom row: Variance Y_s at the excited frequencies for three different values of N (total number of odd excited frequencies equals $N/4$). Remark: $dB(\sigma^2) \equiv 10\log_{10}(\sigma^2) = 20\log_{10}(\sigma) \equiv dB(\sigma)$.

tent with the theoretical result that $Y_s(k)$ is asymptotically (as the number of excited frequencies tends to infinity) circular complex normally distributed (Pintelon and Schoukens, 2001).

The sample autocorrelation of the frequency domain output residuals $Y_s(2k-1)$ at the excited odd harmonics is calculated as

$$\hat{R}_{Y_s}(l) = \frac{1}{N/4-l-1}\sum_{k=l+1}^{N/4} Y_s(2k-1)\overline{Y}_s(2k-1-2l) \qquad (5\text{-}69)$$

with l the lag between the odd harmonics. Since $\hat{R}_{Y_s}(l)$ (5-69) is asymptotically circular complex normally distributed (central limit theorem applied to a sum of circular complex distributed random variables), $\sqrt{3}\,\text{std}(\hat{R}_{Y_s}(l))$ gives the 95% confidence bound of $\left|\hat{R}_{Y_s}(l)\right|$.

Observations From the middle right plot of Figure 5-27 it can be concluded that $Y_s(2k-1)$ is uncorrelated over the frequency (only $\hat{R}_{Y_s}(0)$ lies significantly outside the 95% confidence bound).

Discussion This is consistent with the theoretical result that $Y_S(k)$ is mixing over k of order infinity (see Pintelon and Schoukens, 2001), which means that the dependency between $Y_S(k)$ and $Y_S(l)$ vanishes sufficiently fast as the difference $k - l$ tends to infinity.

Observations From the bottom row of Figure 5-27 it can be concluded that $\text{var}(Y_S(k))$ is a smooth function of the frequency. It can also be seen that the variance does not decrease as N increases ($\text{var}(Y_S(k)) = O(N^0)$).

Discussion The smooth behavior of $\text{var}(Y_S(k))$ explains why the stochastic nonlinear distortions can be approximated very well by filtered white noise (see Chapter 5 on parametric modeling). Since ($\text{var}(Y_S(k)) = O(N^0)$), the influence of the stochastic nonlinear distortions will not decrease by increasing the measurement time (period length N or the number of excited frequencies). Although this seems to be counter-intuitive, the observation is normal because the nonlinear distortion at the output is an intrinsic property of the time-invariant nonlinear system that cannot vanish by measuring the output over longer times.

In the following exercise we verify the predictive power of the BLA obtained in Exercise 87.b for well chosen signals within the class of odd random phase excitations.

Exercise 87.c (Predictive power of BLA — dynamic NL system) Use the best linear approximation obtained in Exercise 87.b to predict the output of the GWH system defined in Exercise 87.b, for the following three excitations:

- Odd random phase multisine: (5-61) with ϕ_{2k-1} uniformly distributed in $[0, 2\pi]$.
- Odd Schroeder multisine: (5-43) with ($\phi_{2k-1} = 2\pi(2k-1)(2k-2)/N$).
- Zero phase multisine: (5-43) with ($\phi_{2k-1} = 0$).

Compare the output of the BLAs with the output of the GWH system. What do you observe? Explain.   □

Observations The best linear approximation obtained from $M = 10^4$ realizations of the odd random phase multisine (5-61) gives output residuals (stochastic nonlinear distortions) y_s with an average standard deviation of 0.082 (see the top left plot of Figure 5-27). It means that on the average, the rms value of the output error $y(t) - y_{\text{BLA}}(t)$ of the BLA will not be smaller than 0.082. This is verified for the following three multisines specially selected from among a class of odd random phase multisines: a random phase, a Schroeder phase, and a zero phase multisine. From Figure 5-28 it can be seen that the output error of the BLA is very good for the random phase, but not good enough for the Schroeder phase, and bad for the zero phase. The corresponding rms values of the output error $y(t) - y_{\text{BLA}}(t)$ equal 0.082, 0.17, and 0.46. Note that the prediction via the BLA fails at the extreme values of $y(t)$.

Discussion The observed rms values of the output error of the Schroeder phase (0.17) and zero phase (0.46) multisines are not in contradiction with that of the class of random phase multisines (0.082), because the probability of selecting the Schroeder phase or zero phase multisines by random phase generation is exactly zero for a continuous phase distribution, and is negligibly small for a discrete phase distribution if the number of excited frequencies is sufficiently large. Hence, one can always find signals from the signal class for which the output error is much worse than the average behavior, however, the probability of randomly selecting such a signal is very small or zero. Note that the output error can be much larger if the BLA is used for signals which do not belong to the class of excitations the BLA was calculated for, for example, other rms value, and/or other coloring power spectrum, and/

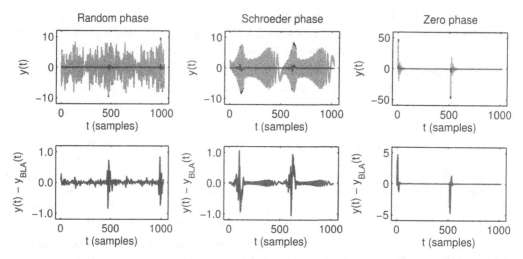

Figure 5-28 Predictive power of the best linear approximation (BLA) of a generalised Wiener–Hammerstein system. Top row: Output nonlinear system $y(t)$ (black), output $y_{BLA}(t)$ of the BLA (light gray), and output error $y(t) - y_{BLA}(t)$ for respectively a random phase (left), Schroeder phase (middle), and zero phase (right) odd multisine. Bottom row: Zoom of the output errors in the top row.

or other probability density function. This is of course an extrapolation which is known to be potentially dangerous.

5.3.3 Relationship between the stochastic nonlinear distortions at the output and those on the best linear approximation

In the previous section the stochastic properties of the output residuals $y_s(t)$ have been studied in detail. In this section we establish the relationship between the output residuals $Y_S(k)$ and the stochastic nonlinear distortions $G_S(j\omega_k)$ on the BLA measurement (5-54).

For the mth realization of a random phase multisine, (5-55) can be written in the frequency domain as

$$Y^{[m]}(k) = G_{BLA}(j\omega_k)U^{[m]}(k) + Y_S^{[m]}(k) \tag{5-70}$$

Dividing (5-70) by $U^{[m]}(k)$

$$Y^{[m]}(k)/U^{[m]}(k) = G_{BLA}(j\omega_k) + Y_S^{[m]}(k)/U^{[m]}(k) \tag{5-71}$$

and comparing the result (5-71) to (5-37) and (5-50) gives

$$G_S(j\omega_k) = \frac{1}{M}\sum_{m=1}^{M}\frac{Y_S^{[m]}(k)}{U^{[m]}(k)} \quad \text{with} \quad \text{var}(G_S(j\omega_k)) = \frac{\text{var}(Y_S(k))}{M|U(k)|^2}. \tag{5-72}$$

It shows that G_S and Y_S have exactly the same stochastic properties. Hence, $G_S(j\omega_k)$ is asymptotically (for the number of excited frequencies going to infinity) circular complex normally distributed, and mixing of order infinity over k (the dependency over the frequency index k tends quickly to zero as the index difference increases). It explains why the stochastic nonlinear distortions behave as measurement noise in FRF measurements.

For random excitations (5-70) is replaced by

$$Y^{[m]}(k) = G_{\text{BLA}}(j\omega_k) U^{[m]}(k) + Y_S^{[m]}(k) + T_{\text{BLA}}(j\omega_k), \tag{5-73}$$

where $T_{\text{BLA}} = O(N^{-1/2})$ accounts for the leakage errors (see Chapter 3). Applying the diff operator $X_{\text{diff}}(k) = X(k+1) - X(k)$ to the input/output DFT spectra suppresses the leakage error

$$Y_{\text{diff}}^{[m]}(k) = G_{\text{BLA}}(j\omega_{k+1/2}) U_{\text{diff}}^{[m]}(k) + Y_{S\text{diff}}^{[m]}(k). \tag{5-74}$$

Multiplying (5-74) by $\overline{U_{\text{diff}}^{[m]}}(k)$ and summing over m gives

$$\hat{S}_{YU}(j\omega_{k+1/2}) = G_{\text{BLA}}(j\omega_{k+1/2})\hat{S}_{UU}(j\omega_{k+1/2}) + \hat{S}_{Y_SU}(j\omega_{k+1/2}), \tag{5-75}$$

where the (cross-)power spectra are calculated as in (5-28). Comparing (5-75) to (5-28) and (5-50) shows that

$$G_S(j\omega_{k+1/2}) = \frac{\hat{S}_{Y_SU}(j\omega_{k+1/2})}{\hat{S}_{UU}(j\omega_{k+1/2})} \text{ with } \text{var}(G_S(j\omega_{k+1/2})) = \frac{\text{var}(Y_S(k+1/2))}{MS_{UU}(j\omega_{k+1/2})}. \tag{5-76}$$

Hence, the same conclusions hold as for the random phase multisines.

5.3.4 What have we learned in Section 5.3?

■ The output $y(t)$ of a PISPO (period in, same period out) system excited by a stationary random process or a random phase multisine $u(t)$ can be written as

$$y(t) = g_{\text{BLA}}(t)*u(t) + y_s(t) \tag{5-77}$$

with $g_{\text{BLA}}(t)$ the impulse response of the best linear approximation (BLA), and $y_s(t)$ the output residuals (stochastic nonlinear distortions). For stationary random processes $y_s(t)$ is a zero mean, non-Gaussian distributed, stationary process, that is uncorrelated with the input $u(t)$ (Exercise 87.a). For random phase multisines with a sufficiently large number of excited frequencies, $y_s(t)$ is a zero mean, non-Gaussian distributed, second order cyclo-stationary process, that is uncorrelated with (but not independent of) $u(t)$ (Exercise 87.b). In both cases the higher order moments of the residuals $y_s(t)$ contain information about the nonlinear behavior of the system.

■ In the frequency domain (5-77) is written as

$$Y(k) = G_{\text{BLA}}(j\omega_k)U(k) + T_{\text{BLA}}(j\omega_k) + Y_S(k), \tag{5-78}$$

where $T_{\text{BLA}} = O(N^{-1/2})$ contains the initial and final conditions of the experiment (leakage errors) for random excitations, and is exactly zero for random phase multisines. The input DFT spectrum $U(k)$ and the output residuals $Y_S(k)$ are both an

$O(N^0)$. It means that $Y_S(k)$ does not decrease as the measurement time (proportional to N) increases (Exercise 87.b). For the class of Gaussian excitations and random phase multisines with a sufficiently large number of frequencies, the output residual has the following properties (Pintelon and Schoukens, 2001; Exercise 87.b): (i) $Y_S(k)$ is a zero mean, circular complex ($\mathbb{E}\{Y_S^2(k)\} = 0$) normally distributed random variable; (ii) var($Y_S(k)$) is a smooth function of the frequency; and (iii) $Y_S(k)$ is mixing of order infinity, which means that $Y_S(k)$ and $Y_S(l)$ rapidly become independent as the difference $k - l$ increases. All these properties allow us to calculate uncertainty bounds with a given confidence level.

■ The best linear approximation (BLA) of a nonlinear system excited by a certain class of excitation signals can be used to predict the output of that nonlinear system within that particular class of excitation signals. Within this particular class of excitation signals, the rms value of the output error $y(t) - y_{BLA}(t)$ (= difference between the true and the output of the BLA) equals std(y_s). The latter gives the lower bound on the predictive power of the BLA of the nonlinear system. It is always possible to find signals within the considered class (e.g., Schroeder phase, or zero phase) for which the rms value of the output error is much larger than std(y_s) (Exercise 87.c). However, the probability of randomly selecting such a signal is negligibly small or zero. Using the BLA for signals which do not belong to the class of excitations that the BLA was calculated for is an extrapolation which is potentially dangerous (nothing can be guaranteed about the rms value of the output error).

■ If the linear approximation used for predicting the output of a nonlinear system is different from the BLA of that system, then the variance of the output error $y(t) - y_{LA}(t)$ is always larger than var(y_s) (Exercise 87.a). Hence, the BLA is the best second-order linear time invariant equivalent (Enqvist, 2005).

■ The stochastic nonlinear distortions $G_S(j\omega_k)$ on the BLA measurements (5-28) and (5-37) have exactly the same stochastic properties as the output residual $Y_S(k)$. This explains why it is difficult to distinguish $G_S(j\omega_k)$ from the measurement noise. The variances of $G_S(j\omega_k)$ and $Y_S(k)$ are related by

$$\text{var}(G_S(j\omega_k)) = \frac{\text{var}(Y_S(k))}{|U(k)|^2} \text{ and var}(G_S(j\omega_k)) = \frac{\text{var}(Y_S(k))}{S_{UU}(j\omega_k)} \qquad (5\text{-}79)$$

for, respectively, random phase multisines and random excitations.

□

6

Measuring the Best Linear Approximation of a Nonlinear System

What you will learn: In the previous chapter we learned to understand the behavior of a non-linear systems. Its output was split in the best linear approximation (BLA) and a nonlinear noise source. In this chapter we learn how to measure the FRF of the BLA. We also show how to measure the nature and the level of the nonlinear distortions. Efficient algorithms are explained to obtain all this information in one single measurement procedure.

Chapter content:

6.1 MEASURING THE BEST LINEAR APPROXIMATION

This section describes nonparametric methods for identifying the best linear approximation (BLA) $G_{BLA}(j\omega_k)$ and the variance of the stochastic nonlinear contributions $G_S(j\omega_k)$ (or $Y_S(k)$) from noisy input–output observations of a nonlinear system excited by random phase multisines (see Figure 6-1). Two measurement procedures are discussed and compared: (i) a robust (slow) method that measures the stochastic nonlinear contributions via different random phase multisine experiments directly and (ii) a fast (nonrobust) method that estimates $G_S(j\omega_k)$ via one single experiment with a specially designed random phase multisine. Finally, an attempt is made to estimate the order of magnitude of the bias contribution G_B on the BLA (deviation from the true underlying linear system) starting from the variance of the stochastic nonlinear distortions, $\text{var}(G_S)$.

6.1.1 Robust method

We distinguish three cases in the measurement setup of Figure 6-1: (i) known input $u_0(t)$ and noisy output measurements $y(t)$, (ii) noisy input/output measurements $u(t), y(t)$, and (iii) noisy input/output measurements $u(t), y(t)$ with known reference signal $r(t)$ (typically the

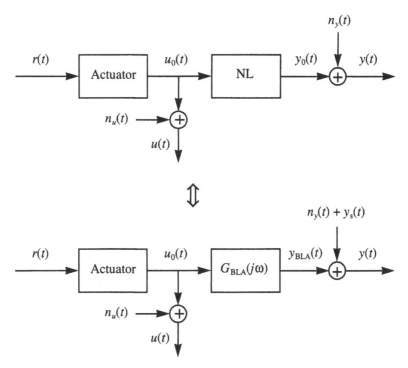

Figure 6-1   Noisy input/output measurements $u(t), y(t)$ of a nonlinear system driven by an actuator with Gaussian input $r(t)$ (top diagram), and the corresponding best linear approximation representation (bottom diagram).

signal stored in the arbitrary waveform generator). Since each case results in a different processing of the measured input/output data, we handle them separately.

A. Measurement in the Presence of Output Noise

Assuming that the input is known exactly ($n_u = 0$ in Figure 6-1), the nonparametric estimation of the BLA ($G_{BLA}(j\omega_k)$), the variance of the stochastic nonlinear distortions ($var(G_S(j\omega_k))$ or $var(Y_s(k))$), and the noise variance ($var(N_G(k))$ or $var(N_Y(k))$), is based on the analysis of the sample mean and sample variance of the frequency response function over different multisine periods and multisine realizations (see Figure 6-2):

1. Choose the frequency resolution f_0, the excited harmonics $k \in \{1, 2, ..., N/2 - 1\}$, the amplitudes A_k, and the rms value of the random phase multisine $r(t)$ (see eq. (5-52).

2. Make a random choice of the phases ϕ_k of the non-zero harmonics of the random phase multisine (5-52) such that $\mathbb{E}\{e^{j\phi_k}\} = 0$ (e.g. ϕ_k is uniformly distributed in $[0, 2\pi)$), and calculate the corresponding time signal $r(t)$.

3. Apply the excitation $r(t)$ to the actuator (see Figure 6-1) and measure P frequency response functions from $P \geq 2$ consecutive periods of the steady-state response $u_0(t)$, $y(t)$ (at least two periods are needed to calculate the noise variance).

4. Repeat steps 2 and 3 preferably $M \geq 7$ times (at least seven realizations are needed to preserve the properties of the maximum likelihood estimator in the parametric modelling step; see Chapter 4).

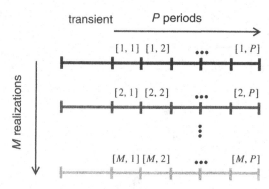

Figure 6-2   Robust measurement procedure: $P \geq 2$ periods of the steady-state response to a random phase multisine excitation are measured; and this experiment is repeated for $M \geq 7$ different random phase multisine realizations.

Since $n_y(t)$ is a stochastic process, and $y_s(t)$ is a periodic signal, depending on the phase realization of the reference signal $r(t)$, the FRF of the mth realization and pth period is related to the best linear approximation, the stochastic nonlinear distortions, and the output noise as

$$G^{[m,p]}(j\omega_k) = \frac{Y^{[m,p]}(k)}{U_0^{[m]}(k)} = G_{\mathrm{BLA}}(j\omega_k) + \frac{Y_S^{[m]}(k)}{U_0^{[m]}(k)} + \frac{N_Y^{[m,p]}(k)}{U_0^{[m]}(k)}. \qquad (6\text{-}1)$$

This shows that the sample variance over the P periods only depends on the output noise, while the sample variance over the M realizations depends on both the stochastic nonlinear distortions and the output noise. From the $M \times P$ noisy frequency response functions $G^{[m,p]}(j\omega_k)$ (6-1), $m = 1, 2, ..., M$ and $p = 1, 2, ..., P$, one can calculate for each experiment (random phase multisine realization) the average frequency response function (FRF) $\hat{G}^{[m]}(j\omega_k)$ and its sample variance $\hat{\sigma}^2_{\hat{G}^{[m]}}(k)$ over the P periods, viz.,

$$\hat{G}^{[m]}(j\omega_k) = \frac{1}{P}\sum\nolimits_{p=1}^{P} G^{[m,p]}(j\omega_k)\,,\ \hat{\sigma}^2_{\hat{G}^{[m]}}(k) = \sum\nolimits_{p=1}^{P}\frac{\left|G^{[m,p]}(j\omega_k) - \hat{G}^{[m]}(j\omega_k)\right|^2}{P(P-1)}. \qquad (6\text{-}2)$$

Additional averaging over the M experiments gives the final FRF $\hat{G}_{\mathrm{BLA}}(j\omega_k)$ and its sample variance $\hat{\sigma}^2_{\hat{G}_{\mathrm{BLA}}}(k)$ of the whole measurement procedure

$$\hat{G}_{\mathrm{BLA}}(j\omega_k) = \sum\nolimits_{m=1}^{M}\frac{\hat{G}^{[m]}(j\omega_k)}{M}\,,\ \hat{\sigma}^2_{\hat{G}_{\mathrm{BLA}}}(k) = \sum\nolimits_{m=1}^{M}\frac{\left|\hat{G}^{[m]}(j\omega_k) - \hat{G}_{\mathrm{BLA}}(j\omega_k)\right|^2}{M(M-1)}\,, \qquad (6\text{-}3)$$

and an improved estimate of the noise variance

$$\hat{\sigma}^2_{\hat{G}_{\mathrm{BLA}},\,\mathrm{n}}(k) = \frac{1}{M^2}\sum\nolimits_{m=1}^{M}\hat{\sigma}^2_{\hat{G}^{[m]}}(k) \qquad (6\text{-}4)$$

Using (6-1) to (6-4) and the fact that $\left|U_0^{[m]}\right|$ is independent of m, it can easily be verified that

$$\mathbb{E}\{\hat{\sigma}^2_{G_{BLA}, n}(k)\} = \frac{\text{var}(N_G(k))}{MP}$$

$$\mathbb{E}\{\hat{\sigma}^2_{G_{BLA}}(k)\} = \frac{\text{var}(G_S(j\omega_k))}{M} + \frac{\text{var}(N_G(k))}{MP} \qquad (6\text{-}5)$$

with

$$N_G(k) = \frac{N_Y(k)}{U_0(k)}, \quad G_S(j\omega_k) = \frac{Y_S(k)}{U_0(k)}, \qquad (6\text{-}6)$$

where $U_0(k)$, $N_Y(k)$ and $Y_S(k)$ are DFT spectra calculated over one signal period. Hence, the scaled difference between the total variance and the noise variance is an estimate of the variance of the stochastic nonlinear contributions, viz.,

$$\text{var}(G_S(j\omega_k)) \approx M(\hat{\sigma}^2_{G_{BLA}}(k) - \hat{\sigma}^2_{G_{BLA}, n}(k)). \qquad (6\text{-}7)$$

Two special cases are worth mentioning: (i) either $P \geq 7$ periods of $M = 1$ realization or (ii) $P = 1$ period of $M \geq 7$ realizations are available. In the first case the FRF and its noise variance are obtained via (6-2). No indication about the level of the nonlinear distortions can be given. In the second case the FRF and its total variance is calculated via (6-3). Since no noise variance information is available, the contribution of the nonlinear distortions in the total variance cannot be quantified.

Exercise 88.a (Robust method for noisy FRF measurements) Implement the robust method for noisy FRF measurements in MATLAB®.

- Apply this program to the setup of Figure 6-1 where the input is observed without errors ($n_u = 0$), the reference signal $r(t)$ is a full random phase multisine [see equation (5-52)] consisting of five hundred frequencies in the band [4 Hz, 2 kHz] and rms value equal to one; the actuator is a fourth order analog Chebyshev filter with a passband ripple of 6 dB and a cutoff frequency of 2 kHz, and the nonlinear system is a Wiener–Hammerstein system (see Figure 5-13 on page 157) with linear dynamic blocks

$$G_1(s) = \frac{1}{1 + s/(Q\omega_0) + s^2/\omega_0^2} \quad \text{and} \quad G_2(s) = \frac{1}{1 + \tau s} \qquad (6\text{-}8)$$

($Q = 10$, $\omega_0 = 2\pi f_0$, $f_0 = 1$ kHz, and $\tau = 1/(600\pi)$ s), and static nonlinear block

$$z(t) = 5\tanh(x(t)/5). \qquad (6\text{-}9)$$

At the sampling instances $n_y(t)$ is zero mean white noise with standard deviation $1/(10\sqrt{2})$. Choose $f_s = 50$ kHz (What is the corresponding number of points per period N? Why is f_s chosen that large?) and apply the robust method for noisy FRF measurements with $P = 2$ and $M = 60$.

- Using (6-3), (6-4), and (6-7), plot the frequency response function, its total variance

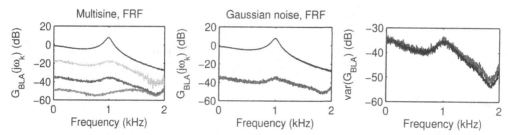

Figure 6-3 Measurement of the best linear approximation (BLA) of a Wiener–Hammertein system from noiseless input and noisy output measurements - robust method. Left: BLA of the multisine experiment (black), its total variance (dark gray), its noise variance (medium gray), and the variance of the stochastic nonlinear distortions w.r.t. one multisine realization (6-7) (light gray). Middle: BLA of the Gaussian noise experiment (black), and its total variance (dark gray). Right: Comparison of the total variances of the multisine BLA (black) and the Gaussian noise BLA (dark gray).

(noise and stochastic nonlinear distortions), its noise variance, and the variance of the stochastic nonlinear distortions w.r.t. one realization of the random phase multisine. Verify the expected values (6-5).

■ Compare the previous results with the FRF measurement obtained using a Gaussian noise excitation with the same bandwidth and rms value as the multisine excitation. Generate, for this purpose, $N \times M \times P$ data points of the input/output signals (*Hint*: neglect the transient terms and calculate the output of the linear blocks via the inverse DFT of $G_1(j\omega_k)U(k)$ and $G_2(j\omega_k)Z(k)$), split these signals into M blocks of $N \times P$ data points, and calculate the BLA and its uncertainty according to equation (5-28). Proceeding in this way the averaging of the stochastic nonlinear distortions is exactly the same for the random phase multisines and the Gaussian noise excitation. Is the averaging of the output noise the same in both cases? If not, what is the difference? Explain.

What would happen with the variances of the BLA estimates if the rms value of the excitation is doubled? Explain.   □

The ratio of the sampling frequency (50 kHz) to the largest excitation frequency (2 kHz) equals 25, resulting in an oversampling factor of 12.5. It is chosen that large to avoid aliasing in the output spectrum of the harmonics created by the static nonlinearity. The number of time domain samples $N \times P = 75,000$ per block, for calculating the BLA of the Gaussian noise experiment via (5-28), is large enough to allow for neglecting the transient (leakage) effects when transforming the random input/output samples to the frequency domain and vice versa.

Observations Figure 6-3 compares the estimated best linear approximation (BLA) and its variance obtained via the robust random phase multisine procedure, with that of the Gaussian noise excitation. It can be seen that the total variances (noise and stochastic nonlinear distortions) of both estimates (see Figure 6-3, right plot) almost coincide from 4 Hz till about 1.2 kHz. Above 1.2 kHz the total variance of the Gaussian noise BLA is larger than that of the random phase multisine BLA.

Discussion The reason for this is that the output noise is $P = 2$ times less averaged for the Gaussian noise BLA, giving an increase in noise variance of 3 dB w.r.t. the random phase multisine case. Since the stochastic nonlinear distortions are dominant, this increase in noise variance is not apparent below 1.2 kHz. The frequency resolution of the Gaussian BLA estimate is, however, twice that of the random phase multisine estimate.

Observations Doubling the rms value of the excitation reduces the noise variance of the BLA by 6 dB (explain!), and increases the variance of the stochastic nonlinear distortions by about 8 dB. Since the latter are dominant in the band 4 Hz to about 1.2 kHz (compare the dark gray line with the medium gray line in the left plot of Figure 6-3), the total variance of the BLA will increase in that frequency band.

Conclusion This counterintuitive phenomenon - increasing the excitation rms value results in a larger FRF variance - can easily be understood via the robust random phase multisine procedure, but would remain a mystery if only Gaussian noise experiments were available (see Figure 6-3, middle plot).

B. Measurement in the Presence of Input/Output Noise - Unknown Reference Signal

If the reference signal $r(t)$ in Figure 6-1 is unknown, then averaging of the input/output spectra over the different realizations in the measurement scheme of Figure 6-2 would be impossible. Indeed, since $\mathbb{E}\{e^{j\phi_k}\} = 0$ by definition of the random phase multisine $r(t)$ [see equation (5-52)], we have

$$
\begin{aligned}
\mathbb{E}\{U_0(k)\} &= \mathbb{E}\{A(j\omega_k)R(k)\} = A(j\omega_k)|R(k)|\mathbb{E}\{e^{j\phi_k}\} = 0, \\
\mathbb{E}\{Y_{\text{BLA}}(k)\} &= \mathbb{E}\{G_{\text{BLA}}(j\omega_k)U_0(k)\} = G_{\text{BLA}}(j\omega_k)\mathbb{E}\{U_0(k)\} = 0,
\end{aligned}
\tag{6-10}
$$

where $A(j\omega)$ and $G_{\text{BLA}}(j\omega)$ are, respectively, the actuator characteristic and the best linear approximation (see Figure 6-1). Hence, the sample means of the measured input/output spectra over different realizations of the random phase multisine excitation $r(t)$ converge to zero as the number of realizations M tends to infinity. This explains why averaging of the input/output spectra is only performed over the different periods in the following procedure (steps one, two, and four are the same as those in Section 6.1.1.A):

1. Choose the frequency resolution f_0, the excited harmonics $k \in \{1, 2, ..., N/2 - 1\}$, the amplitudes A_k, and the rms value of the random phase multisine $r(t)$ [see equation (5-52)].

2. Make a random choice of the phases ϕ_k of the nonzero harmonics of the random phase multisine (5-52), and calculate the corresponding time signal $r(t)$.

3. Apply the excitation $r(t)$ to the actuator (see Figure 6-1) and measure $P \geq 2$ consecutive periods of the steady-state response $u(t)$, $y(t)$ (at least two periods are needed to calculate the input/output noise (co-)variances).

4. Repeat steps 2 and 3 $M \geq 7$ times (at least seven realizations are needed to preserve the properties of the maximum likelihood estimator in the parametric modelling step; see Chapter 4).

Since $n_u(t)$ and $n_y(t)$ are stochastic processes, and $y_s(t)$ is a periodic signal depending on the phase realization of the reference signal $r(t)$, the input/output spectra of the mth realization and pth period are related to the input/output noise, the best linear approximation, and the stochastic nonlinear distortions as

$$U^{[m,p]}(k) = U_0^{[m]}(k) + N_U^{[m,p]}(k),$$

$$Y^{[m,p]}(k) = G_{\text{BLA}}(j\omega_k)U_0^{[m]}(k) + Y_S^{[m]}(k) + N_Y^{[m,p]}(k). \tag{6-11}$$

From the $M \times P$ noisy input/output spectra $U^{[m,p]}(k)$, $Y^{[m,p]}(k)$ (6-11), $m = 1, 2, ..., M$ and $p = 1, 2, ..., P$, one can calculate for each experiment (random phase multisine realization) the average input/output spectra $\hat{U}^{[m]}(k)$, $\hat{Y}^{[m]}(k)$ and the corresponding sample (co-)variances $\hat{\sigma}_{\hat{U}^{[m]}}^2(k)$, $\hat{\sigma}_{\hat{Y}^{[m]}}^2(k)$, $\hat{\sigma}_{\hat{Y}^{[m]}\hat{U}^{[m]}}^2(k)$ over the P periods

$$\hat{U}^{[m]}(k) = \frac{1}{P}\sum_{p=1}^{P} U^{[m,p]}(k), \ \hat{Y}^{[m]}(k) = \frac{1}{P}\sum_{p=1}^{P} Y^{[m,p]}(k),$$

$$\hat{\sigma}_{\hat{U}^{[m]}}^2(k) = \sum_{p=1}^{P} \frac{|U^{[m,p]}(k) - \hat{U}^{[m]}(k)|^2}{P(P-1)}, \ \hat{\sigma}_{\hat{Y}^{[m]}}^2(k) = \sum_{p=1}^{P} \frac{|Y^{[m,p]}(k) - \hat{Y}^{[m]}(k)|^2}{P(P-1)}, \tag{6-12}$$

$$\hat{\sigma}_{\hat{Y}^{[m]}\hat{U}^{[m]}}^2(k) = \sum_{p=1}^{P} \frac{(Y^{[m,p]}(k) - \hat{Y}^{[m]}(k))\overline{(U^{[m,p]}(k) - \hat{U}^{[m]}(k))}}{P(P-1)}$$

with $\bar{x}$ the complex conjugate of x. Using (6-12), the frequency response function of the mth experiment is calculated as

$$\hat{G}^{[m]}(j\omega_k) = \frac{\hat{Y}^{[m]}(k)}{\hat{U}^{[m]}(k)} \tag{6-13}$$

Additional averaging over the M experiments gives (i) the final FRF $\hat{G}_{\text{BLA}}(j\omega_k)$ and its sample variance $\hat{\sigma}_{\hat{G}_{\text{BLA}}}^2(k)$ of the whole measurement procedure as

$$\hat{G}_{\text{BLA}}(j\omega_k) = \sum_{m=1}^{M} \frac{\hat{G}^{[m]}(j\omega_k)}{M}, \ \hat{\sigma}_{\hat{G}_{\text{BLA}}}^2(k) = \sum_{m=1}^{M} \frac{|\hat{G}^{[m]}(j\omega_k) - \hat{G}_{\text{BLA}}(j\omega_k)|^2}{M(M-1)}, \tag{6-14}$$

(ii) an improved estimate of the input/output noise (co-)variances w.r.t. one multisine experiment

$$\hat{\sigma}_{\hat{U}^{[m]},\text{n}}^2(k) = \frac{1}{M}\sum_{m=1}^{M} \hat{\sigma}_{\hat{U}^{[m]}}^2(k), \ \hat{\sigma}_{\hat{Y}^{[m]},\text{n}}^2(k) = \frac{1}{M}\sum_{m=1}^{M} \hat{\sigma}_{\hat{Y}^{[m]}}^2(k),$$

$$\hat{\sigma}_{\hat{Y}^{[m]}\hat{U}^{[m]},\text{n}}^2(k) = \frac{1}{M}\sum_{m=1}^{M} \hat{\sigma}_{\hat{Y}^{[m]}\hat{U}^{[m]}}^2(k), \tag{6-15}$$

and (iii) an estimate of the noise variance of the BLA

$$\hat{\sigma}_{\hat{G}_{\text{BLA}},\text{n}}^2(k) = \frac{|\hat{G}_{\text{BLA}}(j\omega_k)|^2}{M}\left(\frac{\hat{\sigma}_{\hat{Y}^{[m]},\text{n}}^2(k)}{\hat{S}_{Y_0Y_0}(k)} + \frac{\hat{\sigma}_{\hat{U}^{[m]},\text{n}}^2(k)}{\hat{S}_{U_0U_0}(k)} - 2\text{Re}(\frac{\hat{\sigma}_{\hat{Y}^{[m]}\hat{U}^{[m]},\text{n}}^2(k)}{\hat{S}_{Y_0U_0}(k)})\right), \tag{6-16}$$

where $\hat{S}_{U_0U_0}(k)$, $\hat{S}_{Y_0Y_0}(k)$ and $\hat{S}_{Y_0U_0}(k)$ are calculated as

$$\hat{S}_{U_0 U_0}(k) = \frac{1}{M}\sum_{m=1}^{M} |\hat{U}^{[m]}(k)|^2 - \hat{\sigma}^2_{\hat{U}^{[m]}, \mathrm{n}}(k), \ \hat{S}_{Y_0 Y_0}(k) = \frac{1}{M}\sum_{m=1}^{M} |\hat{Y}^{[m]}(k)|^2 - \hat{\sigma}^2_{\hat{Y}^{[m]}, \mathrm{n}}(k),$$

$$\hat{S}_{Y_0 U_0}(k) = \frac{1}{M}\sum_{m=1}^{M} \hat{Y}^{[m]}(k)\overline{\hat{U}^{[m]}(k)} - \hat{\sigma}^2_{\hat{Y}^{[m]}\hat{U}^{[m]}, \mathrm{n}}(k). \tag{6-17}$$

Note that $\hat{S}_{U_0 U_0}(k)$ and $\hat{S}_{Y_0 U_0}(k)$ are unbiased estimates of, respectively, $|U_0(k)|^2$ and $Y_0(k)\overline{U_0(k)}$, while $\hat{S}_{Y_0 Y_0}(k)$ is biased by the stochastic nonlinear distortions. This can easily be verified via (6-11) and

$$\mathbb{E}\{\hat{\sigma}^2_{\hat{U}^{[m]}, \mathrm{n}}(k)\} = \frac{\mathrm{var}(N_U(k))}{P}, \ \mathbb{E}\{\hat{\sigma}^2_{\hat{Y}^{[m]}, \mathrm{n}}(k)\} = \frac{\mathrm{var}(N_Y(k))}{P},$$

$$\mathbb{E}\{\hat{\sigma}^2_{\hat{Y}^{[m]}\hat{U}^{[m]}, \mathrm{n}}(k)\} = \frac{\mathrm{covar}(N_Y(k), N_U(k))}{P}. \tag{6-18}$$

Using (6-11) to (6-18) and the fact that $|U_0^{[m]}|$ is independent of m, it can be verified that

$$\mathbb{E}\{\hat{\sigma}^2_{G_{\mathrm{BLA}}, \mathrm{n}}(k)\} \approx \frac{\mathrm{var}(N_G(k))}{MP},$$

$$\mathbb{E}\{\hat{\sigma}^2_{G_{\mathrm{BLA}}}(k)\} \approx \frac{\mathrm{var}(G_S(j\omega_k))}{M} + \frac{\mathrm{var}(N_G(k))}{MP}, \tag{6-19}$$

with

$$N_G(k) = \frac{N_Y(k) - G_{\mathrm{BLA}}(j\omega_k)N_U(k)}{U_0(k)}, \ G_S(j\omega_k) = \frac{Y_S(k)}{U_0(k)}, \tag{6-20}$$

where $U_0(k)$, $N_U(k)$, $N_Y(k)$, and $Y_S(k)$ are DFT spectra calculated over one signal period. Hence, the noise (co-)variances and the variance of the stochastic nonlinear distortions can be estimated as

$$\mathrm{var}(N_U(k)) \approx P\hat{\sigma}^2_{\hat{U}^{[m]}, \mathrm{n}}(k), \ \mathrm{var}(N_Y(k)) \approx P\hat{\sigma}^2_{\hat{Y}^{[m]}\hat{U}^{[m]}, \mathrm{n}}(k),$$

$$\mathrm{covar}(N_Y(k), N_U(k)) \approx P\hat{\sigma}^2_{\hat{Y}^{[m]}\hat{U}^{[m]}, \mathrm{n}}(k), \tag{6-21}$$

$$\mathrm{var}(G_S(j\omega_k)) \approx M(\hat{\sigma}^2_{G_{\mathrm{BLA}}}(k) - \hat{\sigma}^2_{G_{\mathrm{BLA}}, \mathrm{n}}(k)),$$

$$\mathrm{var}(Y_S(k)) = \hat{S}_{U_0 U_0}(k)\mathrm{var}(G_S(j\omega_k)), \tag{6-22}$$

where $\hat{S}_{U_0 U_0}(k)$ is defined in (6-17).

Exercise 88.b (Robust method for noisy input/output measurements without reference signal)　Write a MATLAB® program for the robust method using input/output data without a reference signal.

■　Apply this program to the setup of Figure 6-1 where the reference signal $r(t)$ is a full random phase multisine [see equation (5-52)] consisting of five hundred frequencies in the band [4 Hz, 2 kHz] and rms value equal to one. The actuator is a

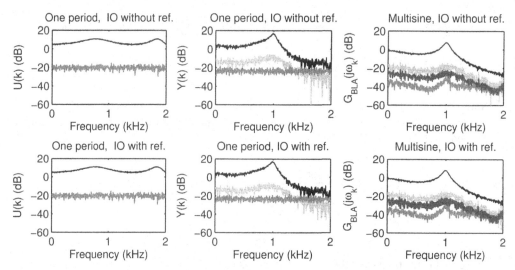

Figure 6-4   Measurement of the best linear approximation (BLA) of a Wiener–Hammerstein system from noisy input/output measurements without (top row) and with (bottom row) reference signal - robust method. Left and middle: Respectively, input and output DFT spectrum of one multisine period. Right: BLA of the multisine experiment. Black: Amplitude of the input/output DFT spectra (left and middle) or the BLA (right). Light gray: Variance of the stochastic nonlinear distortions w.r.t. one multisine realization (middle and right). Medium gray: The noise variance (left, middle, and right). Dark gray: The total variance (right).

fourth-order analog Chebyshev filter with a passband ripple of 6 dB and a cutoff frequency of 2 kHz, and the nonlinear system is a Wiener–Hammerstein system (see Figure 5-13 on page 157) with linear dynamic blocks and static nonlinear block defined in, respectively, (6-8) and (6-9). At the sampling instances ($f_s = 50 \text{ kHz}$) the input/output errors $n_u(t)$ and $n_y(t)$ are zero mean white noise sources with standard deviations 0.1 and $1/(10\sqrt{2})$, respectively.

■ Using (6-14) to (6-16), (6-21) and (6-22), plot the input/output DFT spectra of one signal period together with the noise variance and the variance of the output stochastic nonlinear distortions. Plot also the frequency response function, its total variance (noise and stochastic nonlinear distortions), its noise variance, and the variance of the stochastic nonlinear distortions w.r.t. one realization of the random phase multisine.

■ Compare the stochastic nonlinear distortions with those obtained in Exercise 88.a. What do you conclude? Verify the expected values in (6-18).

Using (6-18), show that $\hat{S}_{U_0 U_0}(k)$ and $\hat{S}_{Y_0 U_0}(k)$ in (6-17) are unbiased, and that the bias of $\hat{S}_{Y_0 Y_0}(k)$ is given by $\text{var}(Y_S(k))$. Justify expression (6-16) for the noise variance of the BLA and verify the expected values (6-19).   □

Observations   The simulation results are shown in Figure 6-4. Compared with the robust method for noisy FRF measurements (see Figure 6-3), the robust method for noisy input/output measurements can recover the noise variances of the input/output DFT spectra (respectively, -20 dB and -23 dB), and the variance of the stochastic nonlinear distortions at the output. From the output DFT spectrum (see Figure 6-4, top row, middle plot) it can be seen that the stochastic nonlinear distortions are about 20 dB below the signal level. Hence, the

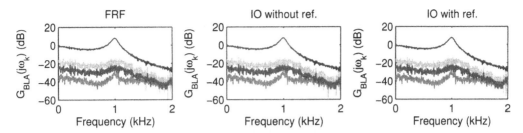

Figure 6-5   Measurement of the best linear approximation (BLA) of a Wiener–Hammertein system from noisy input/output measurements: Robust FRF method (left), Robust IO method without reference (middle), and robust IO method with reference (right).

difference between the actual output of the nonlinear system and the output of the BLA is about 10%.

Discussion   One can also apply the robust FRF method of Section 6.1.1.A to the noisy input/output measurements, and the systematic errors introduced by the input noise both on the estimated FRF and its confidence bounds are negligibly small if the input signal-to-noise ratio (SNR) $|U_0(k)|/\text{std}(N_U(k))$ of one signal period is larger than or equal to 20 dB (Pintelon *et al.*, 2003). Since this is the case for the simulation example of Exercise 88.b, the results obtained by the robust FRF method coincide with those obtained by the robust input/output (IO) method without a reference signal (see Figure 6-5). Due to the averaging of the input/output spectra over P periods, the condition on the input SNR of one signal period is relaxed to $20 \text{ dB} - 10\log_{10}(P)$ for the robust IO method.

The underlying reasoning leading to equation (6-16) is the linearization of the input/output noise contribution to the measured frequency response function. Rewriting the FRF as

$$G(j\omega_k) = \frac{Y(k)}{U(k)} = \frac{Y_0(k) + N_Y(k)}{U_0(k) + N_U(k)} = G_0(j\omega_k)\frac{1 + N_Y(k)/Y_0(k)}{1 + N_U(k)/U_0(k)} \qquad (6\text{-}23)$$

and using the approximation $1/(1 + x) \approx 1 - x$ for $|x| \ll 1$, it can be seen that

$$G(j\omega_k) \approx G_0(j\omega_k)\left(1 + \frac{N_Y(k)}{Y_0(k)}\right)\left(1 - \frac{N_U(k)}{U_0(k)}\right) \approx G_0(j\omega_k)\left(1 + \frac{N_Y(k)}{Y_0(k)} - \frac{N_U(k)}{U_0(k)}\right) \qquad (6\text{-}24)$$

if $|N_U(k)/U_0(k)| \ll 1$ and $|N_Y(k)/Y_0(k)| \ll 1$. From (6-24) it immediately follows that

$$\text{var}(G(j\omega_k)) \approx |G_0(j\omega_k)|^2\left(\frac{\text{var}(N_Y(k))}{|Y_0(k)|^2} + \frac{\text{var}(N_U(k))}{|U_0(k)|^2} - 2\text{Re}(\frac{\text{covar}(N_Y(k), N_U(k))}{Y_0(k)\overline{U_0(k)}})\right), \qquad (6\text{-}25)$$

which is conceptually similar to formula (6-16) for the noise variance of the BLA. It also explains why unbiased estimates of the (cross-)power spectra $|Y_0(k)|^2$, $|U_0(k)|^2$, and $Y_0(k)\overline{U_0(k)}$ are needed in (6-16).

C. Measurement in the Presence of Input/Output Noise — Known Reference Signal

If the reference signal $r(t)$ in Figure 6-1 is known, then the measured input/output spectra can be projected on this signal, giving

$$U_R(k) = \frac{U(k)}{R(k)} = U_{R0}(k) + \frac{N_U(k)}{R(k)},$$

$$Y_R(k) = \frac{Y(k)}{R(k)} = Y_{R0}(k) + \frac{Y_S(k)}{R(k)} + \frac{N_Y(k)}{R(k)} \tag{6-26}$$

with

$$U_{R0}(k) = A(j\omega_k),$$

$$Y_{R0}(k) = G_{\text{BLA}}(j\omega_k) U_{R0}(k), \tag{6-27}$$

where $A(j\omega)$ is the linear actuator characteristic. Since the phases of $U_{R0}(k)$ and $Y_{R0}(k)$ are only dependent on both the actuator transfer function and the best linear approximation, it follows from (6-26) that the projected input/output DFT spectra $U_R(k)$, $Y_R(k)$ can be averaged over both the periods and the realizations in the measurement scheme of Figure 6-2.

Note that in (6-26), one can as well divide the input/output DFT spectra by $e^{j\angle R(k)}$ instead of $R(k)$

$$U_R(k) = \frac{U(k)}{e^{j\angle R(k)}} = U_{R0}(k) + \frac{N_U(k)}{e^{j\angle R(k)}},$$

$$Y_R(k) = \frac{Y(k)}{e^{j\angle R(k)}} = Y_{R0}(k) + \frac{Y_S(k)}{e^{j\angle R(k)}} + \frac{N_Y(k)}{e^{j\angle R(k)}} \tag{6-28}$$

with

$$U_{R0}(k) = A(j\omega_k)|R(k)|,$$

$$Y_{R0}(k) = G_{\text{BLA}}(j\omega_k) U_{R0}(k). \tag{6-29}$$

The advantage of (6-28) over (6-26) is that the projected input/output spectra $U_R(k)$ and $Y_R(k)$ have the same physical interpretation as the original input/output DFT spectra.

Applying steps 1 to 4 of Section 6.1.1.B, the projected input/output spectra (6-26) or (6-28) of the mth realization and pth period are related to the input/output noise, the best linear approximation, and the stochastic nonlinear distortions as

$$U_R^{[m,\,p]}(k) = U_{R0}^{[m]}(k) + \frac{N_U^{[m,\,p]}(k)}{R^{[m]}(k)},$$

$$Y_R^{[m,\,p]}(k) = G_{\text{BLA}}(j\omega_k) U_{R0}^{[m]}(k) + \frac{Y_S^{[m]}(k)}{R^{[m]}(k)} + \frac{N_Y^{[m,\,p]}(k)}{R^{[m]}(k)}. \tag{6-30}$$

Calculating the sample means and sample (co-)variance of $U_R^{[m,\,p]}(k)$ and $Y_R^{[m,\,p]}(k)$ over the P periods gives

$$\hat{U}_R^{[m]}(k) = \frac{1}{P}\sum_{p=1}^{P} U_R^{[m,p]}(k), \ \hat{Y}_R^{[m]}(k) = \frac{1}{P}\sum_{p=1}^{P} Y_R^{[m,p]}(k),$$

$$\hat{\sigma}_{\hat{U}_R^{[m]}}^2(k) = \sum_{p=1}^{P} \frac{\left|U_R^{[m,p]}(k) - \hat{U}_R^{[m]}(k)\right|^2}{P(P-1)}, \ \hat{\sigma}_{\hat{Y}_R^{[m]}}^2(k) = \sum_{p=1}^{P} \frac{\left|Y_R^{[m,p]}(k) - \hat{Y}_R^{[m]}(k)\right|^2}{P(P-1)}, \quad (6\text{-}31)$$

$$\hat{\sigma}_{\hat{Y}_R^{[m]}\hat{U}_R^{[m]}}^2(k) = \sum_{p=1}^{P} \frac{(Y_R^{[m,p]}(k) - \hat{Y}_R^{[m]}(k))\overline{(U_R^{[m,p]}(k) - \hat{U}_R^{[m]}(k))}}{P(P-1)}.$$

Additional averaging over the M experiments gives

$$\hat{U}_R(k) = \frac{1}{M}\sum_{m=1}^{M} \hat{U}_R^{[m]}(k), \ \hat{Y}_R(k) = \frac{1}{M}\sum_{m=1}^{M} \hat{Y}_R^{[m]}(k),$$

$$\hat{\sigma}_{\hat{U}_R}^2(k) = \sum_{m=1}^{M} \frac{\left|\hat{U}_R^{[m]}(k) - \hat{U}_R(k)\right|^2}{M(M-1)}, \ \hat{\sigma}_{\hat{Y}_R}^2(k) = \sum_{m=1}^{M} \frac{\left|\hat{Y}_R^{[m]}(k) - \hat{Y}_R(k)\right|^2}{M(M-1)}, \quad (6\text{-}32)$$

$$\hat{\sigma}_{\hat{Y}_R\hat{U}_R}^2(k) = \sum_{m=1}^{M} \frac{(\hat{Y}_R^{[m]}(k) - \hat{Y}_R(k))\overline{(\hat{U}_R^{[m]}(k) - \hat{U}_R(k))}}{M(M-1)}$$

and an improved estimate of the input/output noise (co-)variances, viz.,

$$\hat{\sigma}_{\hat{U}_{R,\,n}}^2(k) = \frac{1}{M^2}\sum_{m=1}^{M} \hat{\sigma}_{\hat{U}_R^{[m]}}^2(k), \ \hat{\sigma}_{\hat{Y}_{R,\,n}}^2(k) = \frac{1}{M^2}\sum_{m=1}^{M} \hat{\sigma}_{\hat{Y}_R^{[m]}}^2(k),$$

$$\hat{\sigma}_{\hat{Y}_R\hat{U}_{R,\,n}}^2(k) = \frac{1}{M^2}\sum_{m=1}^{M} \hat{\sigma}_{\hat{Y}_R^{[m]}\hat{U}_R^{[m]}}^2(k). \quad (6\text{-}33)$$

Finally, the FRF $\hat{G}_{\text{BLA}}(j\omega_k)$, its total variance $\hat{\sigma}_{\hat{G}_{\text{BLA}}}^2(k)$, and its noise variance $\hat{\sigma}_{\hat{G}_{\text{BLA}},\,n}^2(k)$ are found as

$$\hat{G}_{\text{BLA}}(j\omega_k) = \frac{\hat{Y}_R(k)}{\hat{U}_R(k)},$$

$$\hat{\sigma}_{\hat{G}_{\text{BLA}}}^2(k) = \left|\hat{G}_{\text{BLA}}(j\omega_k)\right|^2\left(\frac{\hat{\sigma}_{\hat{Y}_R}^2(k)}{\left|\hat{Y}_R(k)\right|^2} + \frac{\hat{\sigma}_{\hat{U}_R}^2(k)}{\left|\hat{U}_R(k)\right|^2} - 2\text{Re}(\frac{\hat{\sigma}_{\hat{Y}_R\hat{U}_R}^2(k)}{\hat{Y}_R(k)\hat{U}_R(k)})\right), \quad (6\text{-}34)$$

$$\hat{\sigma}_{\hat{G}_{\text{BLA}},\,n}^2(k) = \left|\hat{G}_{\text{BLA}}(j\omega_k)\right|^2\left(\frac{\hat{\sigma}_{\hat{Y}_{R,\,n}}^2(k)}{\left|\hat{Y}_R(k)\right|^2} + \frac{\hat{\sigma}_{\hat{U}_{R,\,n}}^2(k)}{\left|\hat{U}_R(k)\right|^2} - 2\text{Re}(\frac{\hat{\sigma}_{\hat{Y}_R\hat{U}_{R,\,n}}^2(k)}{\hat{Y}_R(k)\hat{U}_R(k)})\right).$$

Using (6-30) to (6-34) and the fact that $|R^{[m]}|$ is independent of m, it can easily be verified that

$$\mathbb{E}\{\hat{\sigma}_{\hat{U}_{R,\,n}}^2(k)\} = \frac{\text{var}(N_U(k))}{MP|R(k)|^2}, \ \mathbb{E}\{\hat{\sigma}_{\hat{Y}_{R,\,n}}^2(k)\} = \frac{\text{var}(N_Y(k))}{MP|R(k)|^2},$$

$$\mathbb{E}\{\hat{\sigma}_{\hat{Y}_R\hat{U}_{R,\,n}}^2(k)\} = \frac{\text{covar}(N_Y(k), N_U(k))}{MP|R(k)|^2}, \quad (6\text{-}35)$$

$$\mathbb{E}\{\hat{\sigma}^2_{\hat{U}_R}(k)\} = \frac{\text{var}(N_U(k))}{MP|R(k)|^2}, \ \mathbb{E}\{\hat{\sigma}^2_{\hat{Y}_R}(k)\} = \frac{\text{var}(Y_S(k))}{M|R(k)|^2} + \frac{\text{var}(N_Y(k))}{MP|R(k)|^2},$$

$$\mathbb{E}\{\hat{\sigma}^2_{\hat{Y}_R\hat{U}_R}(k)\} = \frac{\text{covar}(N_Y(k), N_U(k))}{MP|R(k)|^2}, \tag{6-36}$$

$$\mathbb{E}\{\hat{\sigma}^2_{\hat{G}_{\text{BLA}}}(k)\} \approx \frac{\text{var}(G_S(j\omega_k))}{M} + \frac{\text{var}(N_G(k))}{MP},$$

$$\mathbb{E}\{\hat{\sigma}^2_{\hat{G}_{\text{BLA},n}}(k)\} \approx \frac{\text{var}(N_G(k))}{MP}, \tag{6-37}$$

where $N_G(k)$, $G_S(j\omega_k)$ are defined in (6-20) and where $R(k)$, $N_U(k)$, $N_Y(k)$ and $Y_S(k)$ are DFT spectra calculated over one signal period. Hence, the noise (co-)variances and the variance of the stochastic nonlinear distortions can be estimated as

$$\text{var}(N_U(k)) \approx MP|R(k)|^2\hat{\sigma}^2_{\hat{U}_R,n}(k), \ \text{var}(N_Y(k)) \approx MP|R(k)|^2\hat{\sigma}^2_{\hat{Y}_R,n}(k),$$

$$\text{covar}(N_Y(k), N_U(k)) \approx MP|R(k)|^2\hat{\sigma}^2_{\hat{Y}_R\hat{U}_R,n}(k), \ \text{var}(Y_S(k)) \approx M|R(k)|^2(\hat{\sigma}^2_{\hat{Y}_R}(k) - \hat{\sigma}^2_{\hat{Y}_R,n}(k)), \tag{6-38}$$

$$\text{var}(N_G(k)) \approx MP\hat{\sigma}^2_{\hat{G}_{\text{BLA}},n}(k), \ \text{var}(G_S(j\omega_k)) \approx M(\hat{\sigma}^2_{\hat{G}_{\text{BLA}}}(k) - \hat{\sigma}^2_{\hat{G}_{\text{BLA}},n}(k)). \tag{6-39}$$

Exercise 88.c (Robust method for noisy input/output measurements with reference signal) Program the robust method for input/output data with reference signal in MATLAB®.

- Apply this program to the simulation example of Exercise 88.b. Using (6-32) to (6-34), (6-38), and (6-39), plot the DFT spectrum of one noisy output period together with the noise variance and the variance of the stochastic nonlinear distortions.

- Plot the frequency response function, its total variance (noise and stochastic nonlinear distortions), its noise variance, and the variance of the stochastic nonlinear distortions for one multisine experiment. Compare the results with those of Exercises 88.a and 88.b. What do you conclude? Explain.

So far it has been assumed that the actuator in the setup of Figure 6-1 is linear. What happens if the actuator is nonlinear? How would you detect the nonlinear behavior of the actuator? □

Observations Figure 6-4 shows the simulation results. It can be seen that they coincide with those of Exercise 88.b (robust IO method without reference), and Exercise 88.a (robust FRF method, see Figure 6-5).

Discussion The reason for this is that the input signal-to-noise ratio of one period of the simulation ($25 \text{ dB} \leq 20\log_{10}(|U_0(k)/N_U(k)|) \leq 31 \text{ dB}$) is sufficiently large for each of the three methods: at least 20 dB for the robust FRF method (no averaging of the input/output spectra), $20 \text{ dB} - 10\log_{10}(P)$ for the robust IO method without reference (averaging of the input/output spectra over the P periods), and $20 \text{ dB} - 10\log_{10}(MP)$ for the robust IO method with reference (averaging of the input/output spectra over the P periods and the M realizations).

If the actuator behaves nonlinearly, then the measured input/output spectra are given by

$$U^{[m,p]}(k) = A_{\text{BLA}}(j\omega_k)R^{[m]}(k) + U_S^{[m]}(k) + N_U^{[m,p]}(k),$$

$$Y^{[m,p]}(k) = H_{\text{BLA}}(j\omega_k)R^{[m]}(k) + Y_S^{[m]}(k) + N_Y^{[m,p]}(k),$$

(6-40)

where A_{BLA} and H_{BLA} are the best linear approximations of, respectively, the actuator and the cascade of the actuator and the nonlinear plant, and where the input stochastic nonlinear distortions U_S are correlated with the output stochastic nonlinear distortions Y_S. Projection of the input/output spectra on the reference signal gives

$$U_R^{[m,p]}(k) = A_{\text{BLA}}(j\omega_k) + U_S^{[m]}(k)/R^{[m]}(k) + N_U^{[m,p]}(k)/(R^{[m]}(k),)$$

$$Y_R^{[m,p]}(k) = H_{\text{BLA}}(j\omega_k) + Y_S^{[m]}(k)/R^{[m]}(k) + N_Y^{[m,p]}(k)/R^{[m]}(k)$$

(6-41)

with $X_R = X/(R.)$ Applying the averaging procedure (6-31) to (6-34) to these projected input/output spectra shows that:

■ $\hat{G}_{\text{BLA}}(j\omega_k)$ (6-34) is an estimate of $H_{\text{BLA}}(j\omega_k)/A_{\text{BLA}}(j\omega_k)$ and, hence, depends on the actuator characteristics (see also Exercise 86),

■ the expected values of the sample (co-)variances (6-32) are given by

$$\mathbb{E}\{\hat{\sigma}_{\hat{U}_R}^2(k)\} = \frac{\text{var}(U_S(k))}{M|R(k)|^2} + \frac{\text{var}(N_U(k))}{MP|R(k)|^2},$$

$$\mathbb{E}\{\hat{\sigma}_{\hat{Y}_R}^2(k)\} = \frac{\text{var}(Y_S(k))}{M|R(k)|^2} + \frac{\text{var}(N_Y(k))}{MP|R(k)|^2},$$

(6-42)

$$\mathbb{E}\{\hat{\sigma}_{\hat{Y}_R\hat{U}_R}^2(k)\} = \frac{\text{covar}(Y_S(k), U_S(k))}{M|R(k)|^2} + \frac{\text{covar}(N_Y(k), N_U(k))}{MP|R(k)|^2},$$

■ the stochastic nonlinear distortions G_S on the BLA are related to the input U_S and output Y_S stochastic nonlinear contributions as

$$G_S(j\omega_k) = \frac{Y_S(k) - G_{\text{BLA}}(j\omega_k)U_S(k)}{U_0(k)}$$

(6-43)

with $U_0(k) = A_{\text{BLA}}(j\omega_k)R(k)$,

■ the expected values of the noise (co-)variances in (6-35) and (6-37) remain valid.

To reveal the nonlinear behavior of the actuator, it is sufficient to verify whether $\hat{\sigma}_{\hat{U}_R}^2(k)$ is significantly larger than $\hat{\sigma}_{\hat{U}_R,n}^2(k)$ [compare (6-35) with (6-42)]. Note that this test cannot be performed if the reference signal is not available.

6.1.2 Fast method

The basic idea of the fast method consists of measuring the system with one random phase multisine excitation where some of the in-band harmonics are omitted (= detection lines). If these detection lines are randomly chosen among the excited in-band harmonics (= random phase multisine with random harmonic grid), then the output level at the detection lines corresponds to the stochastic nonlinear distortions at the neighboring excited lines (Schoukens *et al.*, 2005). However, the interpretation of the level of the detection lines at the output can be jeopardized by spectral impurities at the input. These impurities can be due to the nonlinear behavior of the generator/actuator, and/or the nonlinear interaction between the nonlinear system under test and the generator/actuator. A method is proposed to deal with this problem. We first explain the design of the random phase multisines with random harmonic grid, and next describe the fast measurement procedure. Finally, it is shown that spurious signal energy at the even input detection lines may introduce a bias on the estimated level of the odd or even nonlinear distortions; even if the output has been corrected for the spectral impurity of the input.

A. Design of Random Phase Multisines with Random Harmonic Grid

A random phase multisine with random harmonic grid is a random phase multisine given by

$$u(t) = \sum_{k=1}^{N/2-1} A_k \cos(k\omega_0 t + \phi_k) \quad \text{with} \quad \mathbb{E}\{e^{j\phi_k}\} = 0, \qquad (6\text{-}44)$$

where in-band harmonics are randomly omitted (= detection lines). To guarantee a minimal resolution of the detection lines, the excited harmonics are split into groups of equal number of consecutive lines, and one harmonic randomly chosen from each group is eliminated (for example, a hundred excited harmonics are split into twenty five groups of four consecutive excited harmonics, and one out of the four harmonics is randomly eliminated in each group). The amplitudes A_k of the harmonics, the frequency distribution of the harmonics f_k (e.g. uniform or logarithmic), the type of the excited harmonics (odd or even and odd), and the number of consecutive harmonics in each group F_{group}, are set by the user. The design of such multisines is illustrated in the following exercise.

Exercise 89.a (Design of baseband odd and full random phase multisines with random harmonic grid)   Design baseband random phase multisines (6-44) with a linear (uniform) frequency distribution f_k in the band $[1 \text{ Hz}, 100 \text{ Hz}]$, equal amplitudes A_k, rms value equal to one, and phases ϕ_k uniformly distributed in $[0, 2\pi)$

- odd–random multisine: $k = 1, 3, 5, 7, ..., k_{\max}$ in (6-44), with a frequency resolution of 2 Hz between the odd harmonics, and one odd detection line randomly chosen in each group of three consecutive odd harmonics ($F_{\text{group}} = 3$). $k_{\max}$ is chosen such that $k_{\max}f_0 \approx 100 \text{ Hz}$.
- full-random multisine: $k = 1, 2, 3, 4, ..., k_{\max}$ in (6-44), with a frequency resolution of 1 Hz, and one detection line randomly chosen in each group of four consecutive harmonics ($F_{\text{group}} = 4$). $k_{\max}$ is chosen such that $k_{\max}f_0 \approx 100 \text{ Hz}$.

Compare the period length and the amplitudes of the harmonics of both multisines. Explain. What is the (average) frequency resolution of the even and odd detection lines for the odd–random and full-random multisines? What is the average frequency resolution of the BLA measurement? Choose a sampling frequency of 400 Hz and plot one period of the time signal and the corresponding DFT spectrum. □

Exercise 89.b (Design of bandpass odd and full random phase multisines with random harmonic grid)   Repeat Exercise 89.a for bandpass random phase multisines (6-44) with a logarithmic frequency distribution f_k in the band $[100 \text{ Hz}, 10 \text{ kHz}]$, equal harmonic amplitudes A_k, rms value equal to one, and phases ϕ_k uniformly distributed in $[0, 2\pi)$

- Odd–random multisine: $k = 2r + 1$, $r = r_{min}, r_{min} + 1, ..., r_{max}$, in (6-44), with a frequency resolution of 10 Hz between the odd harmonics, a frequency ratio between two consecutive odd excited harmonics of 1.1, and one odd detection line randomly chosen in each group of three consecutive odd harmonics ($F_{group} = 3$). r_{min} and r_{max} are chosen such that $(2r_{min} + 1)f_0 \approx 100$ Hz and $(2r_{max} + 1)f_0 \approx 10$ kHz.

- Full–random multisine: $k = k_{min}, k_{min} + 1, ..., k_{max}$ in (6-44), with a frequency resolution of 5 Hz, a frequency ratio between two consecutive excited harmonics of $\sqrt{1.1}$, and one detection line randomly chosen in each group of four consecutive harmonics ($F_{group} = 4$). k_{min} and k_{max} are chosen such that $k_{min}f_0 \approx 100$ Hz and $k_{max}f_0 \approx 10$ kHz.

Justify the choice of the frequency resolution, and the frequency ratio between the excited harmonics for the logarithmic full-random multisine. Choose a sampling frequency of 120 kHz and plot one period of the time signal and the corresponding DFT spectrum. Can the frequency distribution of the multisine be exactly logarithmic? Explain?   □

For each frequency distribution (linear or logarithmic), the frequency resolution ($f_{k+1} - f_k = f_0$) of the full-random multisine is twice the frequency resolution of the odd harmonics ($f_{2k+1} - f_{2k-1} = 2f_0$) of the odd–random multisine, giving the same period length for both multisines. Since the odd–random and full-random multisines have the same rms value, and since the respective one third and one fourth of their excited harmonics serve as detection lines, the ratio between the amplitudes of the harmonics of the odd–random and full-random multisines is given by

$$\frac{(A_{2k+1})_{\text{odd-random}}}{(A_{2k+1})_{\text{full-random}}} = \sqrt{\frac{\frac{3}{4} \times 2F}{\frac{2}{3} \times F}} = 1.5 = 3.5 \text{ dB} \tag{6-45}$$

with F the total number of in-band odd harmonics (Proof: Use the fact that the signal power is equally distributed over the excited harmonics.) If the probability $p = 1/F_{group}$ that an in-band (odd) harmonic is used as detection line is the same for the odd–random and full-random multisines, then (6-45) is exactly equal to $\sqrt{2}$, viz.,

$$\frac{(A_{2k+1})_{\text{odd-random}}}{(A_{2k+1})_{\text{full-random}}} = \sqrt{\frac{(1-p) \times 2F}{(1-p) \times F}} = \sqrt{2} = 3 \text{ dB}. \tag{6-46}$$

This is achieved by grouping the (odd) harmonics in the same number F_{group} of consecutive (odd) harmonics for, respectively, the odd–random and full-random multisines.

The frequency resolution of the detection lines of the odd–random multisine with uniform frequency distribution (linear tone) equals $2f_0$ for the even harmonics, and on the average $F_{group} \times 2f_0$ for the odd harmonics (minimal $(2F_{group} - 1) \times 2f_0$ and maximal $2f_0$), with F_{group} the number of consecutive odd harmonics in one group. For the full-random multisine (linear tone) the average frequency resolution equals $F_{group} \times f_0$ (minimal $(2F_{group} - 1) \times f_0$ and maximal f_0), with F_{group} the number of consecutive harmonics in one group.

The average frequency resolution of the BLA measurement is $2f_0 \times F_{group}/(F_{group} - 1)$ and $f_0 \times F_{group}/(F_{group} - 1)$ for, respectively, the odd–random and full-random multisines. To-

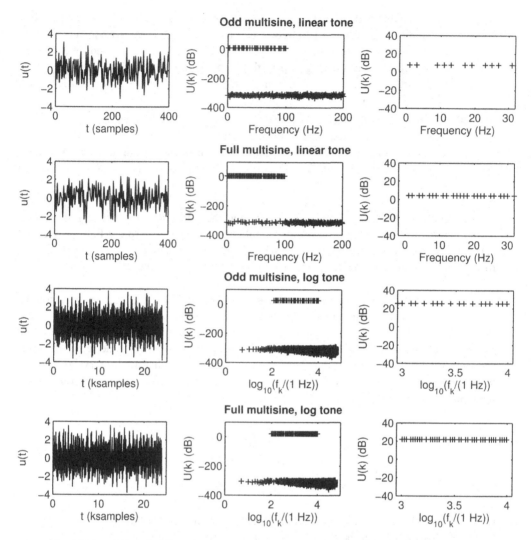

Figure 6-6   Odd and full random phase multisines with linear (uniform) and logarithmic frequency distribution (tone), and random distribution of the detection lines (odd: one out of three consecutive harmonics. Full: One out of four consecutive harmonics). Left column: One period of the time signals. Middle column: DFT spectrum. Right column: Zoom of the in-band harmonics.

gether with the expressions for the detection lines this allows us to make a motivated choice for F_{group}, which is a trade-off between the frequency resolution of the detection lines and the BLA measurement.

Observations Figure 6-6 shows the designed multisines. It can clearly be seen that the frequency resolution of the full-random multisines is twice that of the odd–random multisines. The frequency distribution of the log tones is quasi-logarithmic since the logarithmic distributed frequencies are rounded to the nearest DFT frequency; otherwise the signal would not be periodic.

Conclusion The log tones are very useful when the frequency band of interest covers several decades. This is the case in, for example, acoustics, electrochemical impedance spectroscopy, and electrical machines. In those applications a linear tone would have too much power in the highest decade.

B. Fast Measurement Procedure

Consider the setup of Figure 6-1 on page 184 where the reference signal $r(t)$ is a random phase multisine with a random harmonic grid. The nonparametric estimation of the BLA ($G_{\text{BLA}}(j\omega_k)$), the variance of the stochastic nonlinear distortions ($\text{var}(G_S(j\omega_k))$ or $\text{var}(Y_s(k))$), and the noise variance ($\text{var}(N_G(k))$ or $\text{var}(N_Y(k))$), is based on the analysis of the sample mean and sample variance of the input/output signals over the periods (measurement scheme of Figure 6-2 with $M = 1$):

1. Choose the frequency resolution f_0, the frequency distribution (e.g., linear or logarithmic), the excited harmonics $k \in \{1, 2, ..., N/2 - 1\}$ (odd–random multisine: odd only; full-random multisine: even and odd), the amplitudes A_k, and the rms value of the random phase multisine $r(t)$ [see equation (6-44) on page 197].

2. Create the random harmonic grid: choose F_{group} and split the harmonics (odd–random multisine: odd only; full-random multisine: even and odd) into groups of equal number of consecutive harmonics F_{group}; next, randomly eliminate one harmonic from each group.

3. Choose randomly the phases ϕ_k of the nonzero harmonics of the random phase multisine (5-52) such that $\mathbb{E}\{e^{j\phi_k}\} = 0$ (e.g., ϕ_k is uniformly distributed in $[0, 2\pi)$), and calculate the corresponding time signal $r(t)$.

4. Apply the excitation $r(t)$ to the actuator (see Figure 6-1 on page 184) and measure $P \geq 2$ consecutive periods of the steady-state response $u(t)$, $y(t)$ (at least two periods are needed to calculate the noise (co-)variances; however, seven periods are needed to preserve the properties of the maximum likelihood estimator in the parametric modeling step; see Chapter 4).

The input/output spectra of the pth period are related to the input/output noise, the best linear approximation, and the stochastic nonlinear distortions as

$$
\begin{aligned}
U^{[p]}(k) &= U_0(k) + N_U^{[p]}(k), \\
Y^{[p]}(k) &= G_{\text{BLA}}(j\omega_k) U_0(k) + Y_S(k) + N_Y^{[p]}(k),
\end{aligned}
\tag{6-47}
$$

where $U_0(k) \neq 0$ at the excited harmonics, and $U_0(k) = 0$ at the nonexcited in-band and out-band harmonics (detection lines). From the P noisy input/output spectra $U^{[p]}(k)$, $Y^{[p]}(k)$, $p = 1, 2, ..., P$, one can calculate the average input/output spectra $\hat{U}(k)$, $\hat{Y}(k)$ and the corresponding sample (co-)variances $\hat{\sigma}^2_{\hat{U}, \text{n}}(k)$, $\hat{\sigma}^2_{\hat{Y}, \text{n}}(k)$, $\hat{\sigma}^2_{\hat{Y}\hat{U}, \text{n}}(k)$

$$
\hat{U}(k) = \frac{1}{P}\sum_{p=1}^{P} U^{[p]}(k), \ \hat{Y}(k) = \frac{1}{P}\sum_{p=1}^{P} Y^{[p]}(k),
$$

$$
\hat{\sigma}^2_{\hat{U}, \text{n}}(k) = \sum_{p=1}^{P} \frac{\left|U^{[p]}(k) - \hat{U}(k)\right|^2}{P(P-1)}, \ \hat{\sigma}^2_{\hat{Y}, \text{n}}(k) = \sum_{p=1}^{P} \frac{\left|Y^{[p]}(k) - \hat{Y}(k)\right|^2}{P(P-1)},
\tag{6-48}
$$

$$
\hat{\sigma}^2_{\hat{Y}\hat{U}, \text{n}}(k) = \sum_{p=1}^{P} \frac{(Y^{[p]}(k) - \hat{Y}(k))\overline{(U^{[p]}(k) - \hat{U}(k))}}{P(P-1)}
$$

at all the DFT frequencies (excited and nonexcited harmonics). Using (6-48) at the excited harmonics gives the FRF estimate $\hat{G}_{\text{BLA}}(j\omega_k)$ and its noise sample variance $\hat{\sigma}^2_{\hat{G}_{\text{BLA}},\,\text{n}}(k)$

$$\hat{G}_{\text{BLA}}(j\omega_k) = \frac{\hat{Y}(k)}{\hat{U}(k)},$$

$$\hat{\sigma}^2_{\hat{G}_{\text{BLA}},\,\text{n}}(k) = \left|\hat{G}_{\text{BLA}}(j\omega_k)\right|^2 \left(\frac{\hat{\sigma}^2_{\hat{Y},\,\text{n}}(k)}{\left|\hat{Y}(k)\right|^2} + \frac{\hat{\sigma}^2_{\hat{U},\,\text{n}}(k)}{\left|\hat{U}(k)\right|^2} - 2\,\text{Re}\!\left(\frac{\hat{\sigma}^2_{\hat{Y}\hat{U},\,\text{n}}(k)}{\hat{Y}(k)\overline{\hat{U}(k)}}\right)\right).$$

$$(6\text{-}49)$$

Equation (6-48) also allows us to verify whether the detection lines (nonexcited harmonics) in the input/output spectra contain signal energy or not. This is done as follows: Assuming that no signal energy is present ($X_0(k) = 0$), the ratio $\left|\hat{X}(k)\right|^2 / \hat{\sigma}^2_{\hat{X},\,\text{n}}(k)$ is F-distributed with $2, 2P-2$ degrees of freedom (Stuart and Ord, 1987), and the following null hypothesis test can be constructed: If

$$\left|\hat{X}(k)\right|^2 / \hat{\sigma}^2_{\hat{X},\,\text{n}}(k) \leq F_{0.95,\,2,\,2P-2} \qquad (6\text{-}50)$$

with $F_{0.95,\,2,\,2P-2}$ the 95% percentile of a $F_{2,\,2P-2}$ distributed random variable, then the null hypothesis $X_0(k) = 0$ is accepted, otherwise it is rejected ($F_{0.95,\,2,\,2P-2}$ can be calculated via the MATLAB® function `finv(0.95,2,2*P-2)` of the Statistics Toolbox). If the detection lines at the input contain no signal energy (the null hypothesis $U_0(k) = 0$ is accepted), then the presence of signal energy at the output detection lines is directly linked to the nonlinear behavior of the system. However, if the input detection lines contain signal energy (the null hypothesis $U_0(k) = 0$ is rejected), then the signal energy at the output detection lines is (partially) due to input energy at those frequencies, and a direct link with the nonlinear behavior of the system cannot be established. Indeed, the signal energy at the input detection lines (= spectral impurity) can be due to the nonlinear behavior of the actuator and/or the nonlinear interaction between the nonlinear plant and the actuator (feedback or nonlinear loading). A first order correction to compensate for the influence of the impurities at the output can be made as explained below. Impurities due to actuator distortions should be small in order to get valid results (the level at the input detection lines is at least 10 dB below that of the excited harmonics), while there are no constraints to compensate for impurities due to a nonlinear feedback.

1. The BLA at the nonexcited in-band harmonic ω_l is calculated via linear interpolation of the BLA (6-49) at the closest excited harmonics $\omega_k < \omega_l$ and $\omega_m > \omega_l$

$$\hat{G}_{\text{BLA}}(j\omega_l) = \frac{(m-l)\hat{G}_{\text{BLA}}(j\omega_k) + (l-k)\hat{G}_{\text{BLA}}(j\omega_m)}{m-k}. \qquad (6\text{-}51)$$

This is automatically performed by the function `interp1` in MATLAB®.

2. The output spectrum at the nonexcited in-band harmonic ω_l is corrected as

$$\hat{Y}_c(l) = \hat{Y}(l) - \hat{G}_{\text{BLA}}(j\omega_l)\hat{U}(l) \qquad (6\text{-}52)$$

where $\hat{G}_{\text{BLA}}(j\omega_l)$ is defined in (6-51). The noise variance of (6-52) is calculated as

$$\hat{\sigma}^2_{\hat{Y}_c, n}(l) = \hat{\sigma}^2_{Y, n}(l) + |\hat{G}_{BLA}(j\omega_l)|^2 \hat{\sigma}^2_{U, n}(l) - 2\text{Re}(\hat{\sigma}^2_{YU, n}(l)\overline{\hat{G}_{BLA}(j\omega_l)}) \tag{6-53}$$

with $\hat{\sigma}^2_{U, n}(l)$, $\hat{\sigma}^2_{Y, n}(l)$, and $\hat{\sigma}^2_{YU, n}(l)$ the estimated input/output noise (co-)variances (6-48).

Formula (6-52) is motivated as follows. If the input signal amplitude at the nonexcited frequencies (detection lines) is small compared with that at the excited frequencies, then the nonlinear contributions of input detection lines at the output spectrum (interactions between the detection lines and the detection lines and the excited harmonics) can be neglected w.r.t. those of the excited input harmonics. Hence, the linear contribution of the input detection lines at the output is dominant and can be removed via (6-52).

The corrected output spectrum (6-52) can be analyzed without ambiguity: if the detection lines $\hat{Y}_c(l)$ contain significant signal energy (the null hypothesis $Y_{c0}(l) = 0$ (6-50) is rejected), then the plant behaves nonlinearly; otherwise the plant behaves linearly and the signal energy at the detection lines is solely due to the nonlinear behavior of the generator/actuator. For odd multisines the even and odd detection lines of the corrected output spectrum correspond to, respectively, the even and odd nonlinear contributions of the plant (see Section 5.2.2).

Finally, the level of the nonlinear distortion on the BLA measurement is estimated by extrapolating the level of the detection lines in the corrected output spectrum to the excited frequencies:

1. The level of the stochastic nonlinear distortions at the excited harmonics $|\hat{Y}_c(k)|$ is calculated via linear interpolation of the power of the closest in-band detection lines $l < k$ and $m > k$

$$|\hat{Y}_c(k)|^2 = \frac{(m - k)|\hat{Y}_c(l)|^2 + (k - l)|\hat{Y}_c(m)|^2}{m - l} \tag{6-54}$$

(use the MATLAB® function `interp1`). Here, all detection lines are used for the full-random multisines, and only the odd detection lines are used for the odd–random multisines (for odd–random multisines the odd output harmonics are disturbed by the odd nonlinear distortions only, see Section 5.2.2).

2. Since $\hat{Y}_c(l)$ in (6-52) also contains the contribution of the input and output noise $N_{\hat{U}}(l)$ and $N_{\hat{Y}}(l)$, the ratio

$$\hat{\sigma}^2_{\hat{G}_{BLA}}(k) = \frac{|\hat{Y}_c(k)|^2}{|\hat{U}(k)|^2}, \tag{6-55}$$

where k denotes an excited harmonic, is an estimate of the total variance (stochastic nonlinear distortion + input/output noise) of $\hat{G}_{BLA}(j\omega_k)$ (see Exercise 89.c).

3. Finally, the variance of the stochastic nonlinear distortions are calculated as

$$\text{var}(G_S(j\omega_k)) \approx (\hat{\sigma}^2_{\hat{G}_{BLA}}(k) - \hat{\sigma}^2_{\hat{G}_{BLA}, n}(k)),$$
$$\text{var}(Y_S(k)) \approx |\hat{U}(k)|^2 \text{var}(G_S(j\omega_k)), \tag{6-56}$$

where $\hat{U}(k)$, $\hat{\sigma}^2_{G_{BLA},n}(k)$, and $\hat{\sigma}^2_{G_{BLA}}(k)$ are defined in (6-48), (6-49), and (6-55), respectively.

Note that the variance (6-55) should always be calculated using the corrected output spectrum (6-52) and (6-54), even if it is known beforehand that the input is spectrally pure (the noiseless detection lines have zero magnitude). Indeed, if $\hat{Y}_c(k)$ is replaced by $\hat{Y}(k)$ in (6-55), then $\hat{\sigma}^2_{G_{BLA}}(k)$ does not contain the influence of the input noise and the difference $\hat{\sigma}^2_{G_{BLA}}(k) - \hat{\sigma}^2_{G_{BLA},n}(k)$ in (6-56) can become negative at frequencies with poor input signal-to-noise ratio.

The fast measurement procedure is illustrated in two exercises: first by an example where the input is spectrally pure (linear actuator and nonlinear system operating in open loop), and next by an example where the nonlinear system is captured in a feedback loop. The feedback loop jeopardizes the spectral purity of the input of the nonlinear system and, hence, complicates the interpretation of the output DFT spectrum.

Exercise 89.c (Fast method for noisy input/output measurements — open loop example)   Write a MATLAB® program for the fast measurement procedure. Apply this program to the noisy input/output observations $u(t)$, $y(t)$ of the setup in Figure 6-1 on page 184, where the actuator is equal to one, and the nonlinear plant is a discrete-time Wiener–Hammertein system (see Figure 5-13 on page 157) with linear dynamic blocks

$$G_1(z^{-1}) = \frac{1}{1 - 0.2z^{-1} + 0.9z^{-2}} \text{ and } G_2(z^{-1}) = \frac{1 + 0.5z^{-1}}{1 - 0.5z^{-1} + 0.9z^{-2}} \qquad (6\text{-}57)$$

and static nonlinear block

$$z(t) = x(t) + 0.1x^2(t) + 0.001x^3(t). \qquad (6\text{-}58)$$

The input $n_u(t)$ and output $n_y(t)$ errors are zero mean white noise sources with standard deviation 0.1 and 0.2, respectively. Choose $f_s = 4$ GHz and make the nonlinearity analysis (measurement scheme of Figure 6-2 on page 185 with $M = 1$ and $P = 6$) for odd and full bandpass random phase, random harmonic grid multisine excitations (6-44), with

- a linear (uniform) frequency distribution f_k in the band $[700\ \text{MHz}, 1100\ \text{MHz}]$,
- a frequency resolution of the excited harmonics of, respectively, 1 MHz (odd) and 0.5 MHz (full),
- one (odd) detection line randomly chosen in each group of three consecutive (odd) harmonics ($F_{group} = 3$),
- equal amplitudes A_k,
- an rms value equal to one,
- phases uniformly distributed in $[0, 2\pi)$ (see Exercise 89.a).

Plot the input and output DFT spectra from DC to Nyquist. Can you draw any conclusion from the output DFT spectrum? Justify your answer? Is it possible to classify the nonlinear distortions into even and odd contributions? Justify your answer? Plot the corrected output DFT spectrum. In which frequency band can it be shown? Justify your answer? Compare the uncertainty of the corrected output DFT spectrum with that of the original spectrum. Explain. Proof expression (6-56) assuming that (i) the BLA varies linearly between the interpolation

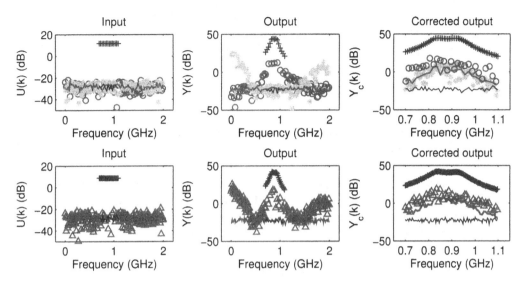

Figure 6-7   Response of a discrete-time Wiener–Hammertein system to a random phase multisine with random harmonic grid (one third detection lines): Input (left column) and output (middle column) DFT spectra of one signal period. Top row: Odd–random multisine. Bottom row: Full-random multisine. Black "+": Excited harmonics (top: odd only; bottom: odd and even). Dark gray "o": nonexcited odd harmonics. Light gray "*": Nonexcited even harmonics. Dark gray "Δ": Nonexcited harmonics. Solid lines: Standard deviation of the corresponding harmonics.

frequencies, (ii) the input signal-to-noise ratio is sufficiently large, and (iii) the signal energy at the input detection lines is small compared with that at the excited harmonics (*Hint*: first show that $\hat{Y}_c(l) \approx Y_S(l) + N_{\hat{Y}}(l) - G_{BLA}(j\omega_l)N_{\hat{U}}(l)$). Plot the estimated BLA (6-49), its noise variance (6-49), its total variance (6-55), and the variance of the stochastic nonlinear distortions (6-56).   □

Observations   Figure 6-7 shows the noisy input/output DFT spectra of one random phase multisine realization. Using the null hypothesis test (6-50), where $F_{0.95, 2, 2P-2} = 4.10$ for $P = 6$, it can be verified that the input is spectrally pure ((6-50) is satisfied at 95.0% of the input detection lines) while the output contains significant in-band and out-band nonlinear distortions ((6-50) is violated at almost all output detection lines).

Discussion   It shows that the plant behaves nonlinearly, and that it produces significant in-band and out-band nonlinear distortions. For the odd multisines the even and odd nonlinear contributions of PISPO systems are uniquely linked to the presence of signal energy at, respectively, the even and odd harmonics in the output spectrum (see Section 5.2.2).

Observations   From the top row of Figure 6-7 it can be seen that the in-band distortions ([700 MHz, 1100 MHz]) are odd (dark gray "o"), while the out-band distortions are even (light gray "*": [0 Hz, 400 MHz] and [1.4 GHz, 2 GHz]) or odd (dark gray "o": [300 MHz, 700 MHz] and [1.1 GHz, 1.5 GHz]).

Discussion   This is a direct consequence of the polynomial nonlinearity (6-58) and the fact that the 400 MHz bandwidth of the odd multisine is smaller than the 700 MHz of the smallest excited harmonic (see Exercise 77.b). This classification is in general impossible for the full multisine.

Observations   The corrected output DFT spectrum is shown in Figure 6-7. It can only be calculated in the excited frequency band because the correction (6-52) requires the best linear approximation. Compared with the original DFT spectrum, the uncertainty of the corrected output DFT spectrum at the detection lines is much larger (see Figure 6-7: in the right plots the dark gray solid line is much larger than the black solid line, while in the middle plots they are at exactly the same level). Note also that the even detection lines (light gray '*') in the corrected output DFT spectrum (upper right plot) are significantly larger than those in the original spectrum (upper middle plot).

Discussion The increased variability of the corrected output DFT spectrum is due to the noise on the input detection lines. The even nonlinear distortions are still zero after correction, because the noise standard deviations at the even detection lines are increased by the same amount (the light and dark gray solid lines coincide).

Equation (6-56) for estimating the stochastic nonlinear distortions is proven as follows. If the input signal amplitude at the nonexcited frequencies (detection lines) is small compared with that at the excited frequencies, then the input/output DFT spectra (6-48) at a detection line can be written as

$$
\begin{aligned}
\hat{U}(l) &= U_0(l) + N_{\hat{U}}(l), \\
\hat{Y}(l) &= G_{\mathrm{BLA}}(j\omega_l) U_0(l) + Y_{\mathrm{S}}(l) + N_{\hat{Y}}(l).
\end{aligned}
\tag{6-59}
$$

Assuming that $G_{\mathrm{BLA}}(j\omega)$ varies linearly between ω_k and ω_m, the estimated BLA $\hat{G}_{\mathrm{BLA}}(j\omega_l)$ at detection line l (6-51) is related to the true BLA $G_{\mathrm{BLA}}(j\omega_l)$ as

$$
\hat{G}_{\mathrm{BLA}}(j\omega_l) = G_{\mathrm{BLA}}(j\omega_l) + \Delta G_{\mathrm{BLA}},
\tag{6-60}
$$

where

$$
\Delta G_{\mathrm{BLA}} = \frac{(m-l)\Delta\hat{G}_{\mathrm{BLA}}(j\omega_k) + (l-k)\Delta\hat{G}_{\mathrm{BLA}}(j\omega_m)}{m-k}
\tag{6-61}
$$

with k, m the excited harmonics and

$$
\Delta\hat{G}_{\mathrm{BLA}}(j\omega_r) = G_{\mathrm{BLA}}(j\omega_r)\left(\frac{N_{\hat{Y}}(r)}{Y_0(r)} - \frac{N_{\hat{U}}(r)}{U_0(r)}\right) = \frac{N_{\hat{Y}}(r) - G_{\mathrm{BLA}}(j\omega_r)N_{\hat{U}}(r)}{U_0(r)}
\tag{6-62}
$$

for $r = k, m$ (see (6-24)). Using (6-59) and (6-60), the corrected output DFT spectrum $\hat{Y}_{\mathrm{c}}(l)$ (6-52) at detection line l becomes

$$
\hat{Y}_{\mathrm{c}}(l) = Y_{\mathrm{S}}(l) + N_{\hat{Y}}(l) - G_{\mathrm{BLA}}(j\omega_l)N_{\hat{U}}(l) + \Delta G_{\mathrm{BLA}}(U_0(l) + N_{\hat{U}}(l)).
\tag{6-63}
$$

Combining (6-61) and (6-62), we find an upper bound for $\Delta G_{\mathrm{BLA}}(U_0(l) + N_{\hat{U}}(l))$ in (6-63):

$$
\begin{aligned}
|\Delta G_{\mathrm{BLA}}(U_0(l) + N_{\hat{U}}(l))| \leq{}& \left|\frac{m-l}{m-k}\right| |N_{\hat{Y}}(k) - G_{\mathrm{BLA}}(j\omega_k)N_{\hat{U}}(k)| \left|\frac{U_0(l)}{U_0(k)} + \frac{N_{\hat{U}}(l)}{U_0(k)}\right| + \cdots + \\
& \left|\frac{l-k}{m-k}\right| |N_{\hat{Y}}(m) - G_{\mathrm{BLA}}(j\omega_m)N_{\hat{U}}(m)| \left|\frac{U_0(l)}{U_0(m)} + \frac{N_{\hat{U}}(l)}{U_0(m)}\right|.
\end{aligned}
\tag{6-64}
$$

Since the magnitudes of the noise $N_{\hat{U}}(l)$ and the input $U_0(l)$ at detection line l are much smaller than the magnitude of an excited input harmonic r, viz.

$$|U_0(l)/U_0(r)| \ll 1 \quad \text{and} \quad |N_{\hat{U}}(l)/U_0(r)| \ll 1 \quad \text{for} \quad r = k, m, \qquad (6\text{-}65)$$

and since $|(m-l)/(m-k)| < 1$ and $|(l-k)/(m-k)| < 1$, it follows from (6-64) and (6-65) that

$$|\Delta G_{\text{BLA}}(U_0(l) + N_{\hat{U}}(l))| \ll |N_{\hat{Y}}(l) - G_{\text{BLA}}(j\omega_l)N_{\hat{U}}(l)|$$

Hence,

$$\hat{Y}_c(l) \approx Y_S(l) + N_{\hat{Y}}(l) - G_{\text{BLA}}(j\omega_l)N_{\hat{U}}(l) \qquad (6\text{-}66)$$

and (6-55) is an estimate of the total variance of the best linear approximation.

Observations The left column of Figure 6-8 shows the estimated best linear approximation, its noise variance, its total variance, and the variance of the stochastic nonlinear distortions. Since $\hat{\sigma}^2_{G_{\text{BLA}},\,n}(k) \ll \hat{\sigma}^2_{G_{\text{BLA}}}(k)$ we have that $\text{var}(G_S(j\omega_k)) \approx \hat{\sigma}^2_{G_{\text{BLA}}}(k)$ for one multisine measurement and, hence, the light ($\text{var}(G_S(j\omega_k))$) and dark ($\hat{\sigma}^2_{G_{\text{BLA}}}(k)$) gray solid lines coincide. From the right column of Figure 6-8 it can be seen that the noise variance of the full-random multisine BLA is about 3 dB larger than that of the odd–random multisine BLA.

Discussion This is due to the fact that the signal power of the full-random multisine is distributed over about twice as many frequencies compared with the odd–random multisine, resulting in a decrease of the input and output signal-to-noise ratios of about 3 dB. However,

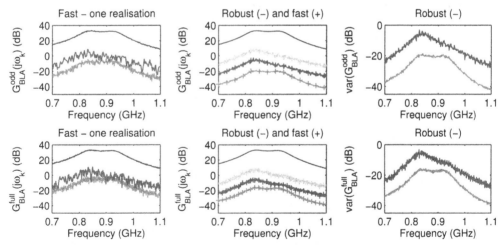

Figure 6-8   Measurement of the best linear approximation (BLA) of a discrete-time Wiener–Hammertein system using random phase multisines with random harmonic grid (one third detection lines). Left column: Fast method (one realization). Middle column: Comparison of the robust method (solid lines) and the fast method ("+") rms averaged over the realizations (both coincide). Right column: Zoom of the noise (medium gray) and total (dark gray) variances of the robust method. Top row: Odd multisines. Bottom row: Full multisines. Black: BLA. Dark gray: Total variance (sum noise and stochastic nonlinear distortions). Medium gray: Noise variance. Light gray: Variance of stochastic nonlinear distortions w.r.t. one multisine realization.

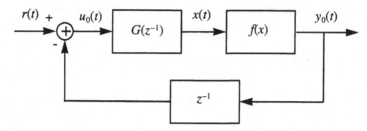

Figure 6-9   Discrete-time Wiener system (from $u_0(t)$ to $y_0(t)$) captured in a unity delay feedback.

the variances of the stochastic nonlinear distortions are the same for the full-random and odd–random multisine BLAs. The reader is referred to Section 6.2 for a detailed explanation.

Exercise 89.d (Fast method for noisy input/output measurements — closed loop example)   Apply the fast measurement procedure to noisy input/output observations $u(t)$, $y(t)$ of a Wiener system, consisting of the cascade of a first-order linear dynamic system $G(z^{-1})$

$$ G(z^{-1}) \;=\; \frac{100(1 - e^{-1/(\tau f_s)})}{1 - (1 - e^{-1/(\tau f_s)})z^{-1}} \tag{6-67} $$

($\tau = 0.5$ s) and a static nonlinearity $f(x) = \tanh(x)$, and captured in a unity delay feedback loop (see Figure 6-9). The noiseless input/output signals $u_0(t)$, $y_0(t)$ are disturbed by zero mean white measurement noise with standard deviation $1{\times}10^{-4}$ and $2{\times}10^{-4}$ respectively. Choose $f_s = 160$ Hz and make the nonlinearity analysis (measurement scheme of Figure 6-2 on page 185 with $M = 1$ and $P = 6$) for odd and full base band random phase, random harmonic grid multisine excitations (6-44), with

- a linear (uniform) frequency distribution f_k in the band [0.01 Hz, 2 Hz] ,
- a frequency resolution of the excited harmonics of, respectively, 0.01 Hz (odd) and 0.005 Hz (full),
- one (odd) detection line randomly chosen in each group of three consecutive (odd) harmonics ($F_{\text{group}} = 3$),
- equal amplitudes A_k ,
- an rms value equal to 0.2,
- phases uniformly distributed in $[0, 2\pi)$ (see Exercise 89.a).

Plot the input and output DFT spectra in the excited frequency band. Can you draw any conclusion from the output DFT spectrum? If not, why not? Plot the corrected DFT spectrum and compare the level of the nonlinear distortions with that of the original spectrum. Explain (*Hint*: replace the Wiener system in Figure 6-9 by the sum of the BLA and the stochastic nonlinear distortions as shown in Figure 5-25). Plot the estimated BLA (6-49), its noise variance (6-49), its total variance (6-55), and the variance of the stochastic nonlinear distortions (6-56). Calculate the FRF from the noiseless input $u_0(t)$ to the noiseless output $y_0(t)$ of the Wiener system at the nonexcited harmonics in the band [0.01 Hz, 16 Hz] and compare it to the transfer function $-1/z^{-1}$. What do you conclude? Explain (use the same hint as before).   □

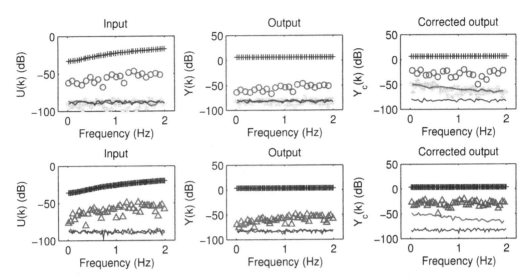

Figure 6-10   Response of a discrete-time Wiener system captured in a unity delay feedback system to a random phase multisine with random harmonic grid (one third detection lines): Input (left column) and output (middle column) DFT spectra of one signal period. Top row: Odd–random multisine. Bottom row: Full-random multisine. Black "+": Excited harmonics (top: odd only; bottom: odd and even). Dark gray "o": Nonexcited odd harmonics. Light gray "*": Nonexcited even harmonics. Dark gray "Δ": Nonexcited harmonics. Solid lines: Standard deviation of the corresponding harmonics.

Observations   Figure 6-10 (left and middle columns) shows the noisy input/output DFT spectra in the excited frequency band of one random phase multisine realization. As could be expected ($\tanh(x)$ is an odd function) the (odd) input and output detection lines are well above the noise level (solid lines), while the even detection lines of the odd–random multisine measurements are at the noise level ((6-50) is satisfied at about 95% of the even detection lines). Since the input detection lines contain significant signal energy, it is impossible to decide whether the nonlinear distortions at the output stem from the generator/actuator or the plant. Therefore, one should look at the corrected output DFT spectrum (right column). It can be seen that significant (odd) nonlinear distortions are present, showing the nonlinear behavior of the plant. Moreover, the nonlinear distortions in the corrected output (-30 dB to -20 dB) are much higher than in the original spectrum (-60 dB to -50 dB).

Discussion   To explain this phenomenon the Wiener system in Figure 6-9 is replaced by the sum of its BLA and the stochastic nonlinear distortions (see Figure 6-11). Note that the diagram of Figure 6-11 is exact, without any approximation (see Section 6.1.3). Using this diagram it can easily be verified that

$$Y_0(k) = \frac{G_{\text{BLA}}(j\omega_k)}{1 + z_k^{-1}G_{\text{BLA}}(j\omega_k)}R(k) + \frac{1}{1 + z_k^{-1}G_{\text{BLA}}(j\omega_k)}Y_\text{S}(k),$$

$$U_0(k) = \frac{1}{1 + z_k^{-1}G_{\text{BLA}}(j\omega_k)}R(k) - \frac{z_k^{-1}}{1 + z_k^{-1}G_{\text{BLA}}(j\omega_k)}Y_\text{S}(k),$$

(6-68)

where $z_k = \exp(j2\pi k/N)$. It shows that the contribution of the stochastic nonlinear distortions $Y_\text{S}(k)$ to the output $Y(k)$ is small if the open loop gain is large $|G_{\text{BLA}}(j\omega_k)| \gg 1$, which is the case in the band $[0, 2 \text{ Hz}]$ (see Figure 6-12). What we have shown here is a special case of a general property of a feedback loop with high open loop gain: it suppresses the influence

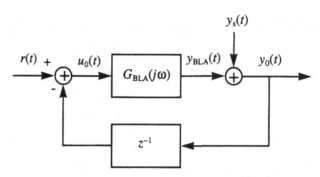

Figure 6-11   Discrete-time Wiener system of Figure 6-9 replaced by its best linear approximation (see Figure 5-25).

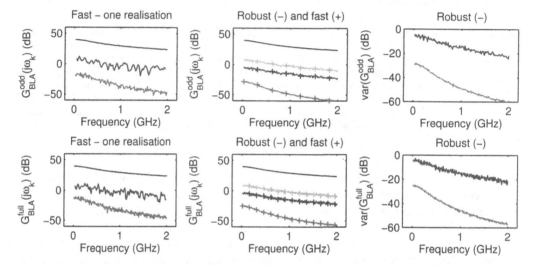

Figure 6-12   Measurement of the best linear approximation (BLA) of a discrete-time Wiener system captured in a unity delay feedback using random phase multisines with random harmonic grid (one third detection lines). Left column: Fast method (one realization). Middle column: Comparison of robust method (solid lines) and fast method ("+") rms averaged over the realizations. Right column: Zoom of the noise (medium gray) and total (dark gray) variances of the robust method. Top row: Odd–random multisines. Bottom row: Full-random multisines. Black: BLA. Dark gray: Total variance (sum noise and stochastic nonlinear distortions). Medium gray: Noise variance. Light gray: Variance stochastic nonlinear distortions w.r.t. one multisine realization.

of any disturbance (noise, time variation, nonlinearity) in the forward path (Goodwin *et al.*, 2001). From (6-68) it also follows that subtracting $G_{BLA}(j\omega_k)U_0(k)$ from the output $Y_0(k)$ recovers the exact stochastic nonlinear distortions $Y_S(k)$. Although this is true without any restriction on the signal-to-distortion ratio $|R(k)|/\mathrm{std}(Y_S(k))$ (see Section 6.1.3), one should keep in mind that the BLA should be estimated from the observed input/output data. In feedback, this nonparametric estimate is biased by the presence of the stochastic nonlinear distortions $Y_S(k)$ (see Chapter 3). This bias can be neglected if the signal-to-distortion ratio $|R(k)|/\mathrm{std}(Y_S(k))$ is sufficiently large, which is compatible with the assumption made for calculating the corrected output spectrum (6-52): The input signal-to-distortion ratio is at least 10 dB. Note that the bias on the BLA due to noisy inputs can be avoided via the indirect method for measuring an FRF (see Section 6.1.3).

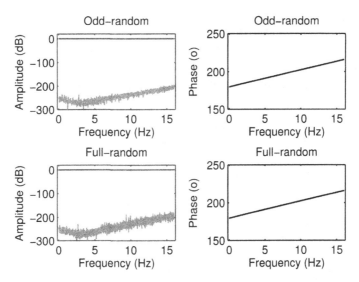

Figure 6-13   Discrete-time Wiener system captured in a unity delay feedback system and excited by a random phase multisine with random harmonic grid (one third detection lines). Black: Frequency response function (FRF) $Y_0(k)/U_0(k)$ calculated at the nonexcited in-band and out-band harmonics (detection lines). Left column: Amplitude FRF. Right column: Phase FRF. Gray: Magnitude of the complex difference between the FRF and $-1/z_k^{-1}$.

Observations   The left column of Figure 6-12 shows the estimated best linear approximation, its noise variance, its total variance, and the variance of the stochastic nonlinear distortions. Note that the results are similar as those in Figure 6-8.

Discussion   The same conclusions can be made as in Figure 6-8: (i) the variance of the stochastic nonlinear distortions (light gray) coincides with the total variance (dark gray), (ii) the noise variance (medium gray) of the full multisine BLA is about 3 dB larger than that of the odd multisine BLA (see Figure 6-12, right column), and (iii) the variance of the stochastic nonlinear distortions of full and odd multisine BLAs are the same (see Figure 6-12, right column).

Observations   Calculating the FRF from the noiseless input $u_0(t)$ to the noiseless output $y_0(t)$ of the Wiener system in Figure 6-9 at the nonexcited harmonics in the band [0.01 Hz, 16 Hz], gives the results shown in Figure 6-13. Comparing this FRF to $-1/z^{-1}$ shows that they are equal within the numerical precision of the MATLAB$^{\circledR}$ calculations.

Discussion   This can be explained using the diagram of Figure 6-11 and equation   (6-68). Indeed, at the detection lines, the reference signal is zero ($R(k) = 0$) while the stochastic nonlinear distortions are different from zero ($Y_S(k)$). Hence, at those lines the FRF, $Y_0(k)/U_0(k)$, where $Y_0(k)$ and $U_0(k)$ are defined in (6-68), reduces to

$$\frac{Y_0(k)}{U_0(k)} = -\frac{1}{z_k^{-1}},   \tag{6-69}$$

which is exactly minus one over the feedback transfer function z^{-1}. From the left column of Figure 6-13 it can be seen that the complex error (gray line) between the FRF $Y_0(k)/U_0(k)$ and $-1/z_k^{-1}$ increases above $5\ \text{Hz}$. This is due to the decreasing amplitude of the out-band output harmonics. Due to the poor signal-to-noise ratio at the detection lines, the ratio $Y(k)/U(k)$ is a (very) noisy estimate of (6-69).

C. Bias on the Estimated Levels of the Odd and Even Nonlinear Distortions

If the odd–random multisine contains signal energy at the even detection lines, then the estimated level of the odd or even nonlinear distortions in the corrected output spectrum can be biased. This is illustrated in the following exercise.

Exercise 89.e (Bias on the estimated odd and even distortion levels) Apply the fast method to the setup of Figure 6-1 on page 184, where the reference signal $r(t)$ is an odd–random multisine covering the band [4 Hz, 2 kHz] with a frequency resolution of 4 Hz, rms value equal to one, and one odd detection line randomly chosen in each group of two consecutive odd harmonics $F_{group} = 2$ (see Section 6.1.2.A for the design). At the sampling instances, the input/output signals are disturbed by normally distributed white noise with zero mean and standard deviation 1×10^{-4}. The actuator is a fourth-order analog Chebyshev filter with a passband ripple of 6 dB and a cutoff frequency of 2 kHz, and the nonlinear dynamic system is a Wiener–Hammerstein system (see Figure 5-13 on page 157) with linear dynamic blocks G_1, G_2 defined in (6-8), and with static nonlinear block

$$z(t) = x(t) + \alpha x^2(t) + \beta x^3(t) \tag{6-70}$$

Choose $f_s = 50\,\text{kHz}$, $\alpha = 0.01$, $\beta = 5 \times 10^{-3}$ and apply the fast method of Section 6.1.2.B with $P = 2$ for the following four odd–random multisine excitations with the same rms value

- Undistorted odd–random: no signal energy at the detection lines.
- Odd distorted odd–random: no signal energy at the even detection lines, and a distortion-to-signal ratio of -20 dB at the odd in-band detection lines.
- Even distorted odd–random: no signal energy at the odd detection lines, and a distortion-to-signal ratio of -20 dB at the even in-band detection lines.
- Fully distorted odd–random: a distortion-to-signal ratio of -20 dB at all in-band detection lines.

where the amplitudes of the distortion are constant and the phases are uniformly distributed in $[0, 2\pi)$. Repeat the procedure for $M = 1000$ realizations of the random phase multisines and calculate the rms value of (i) the estimated level of the odd $Y_S^{odd}(k)$ and even $Y_S^{even}(k)$ nonlinear distortions in the corrected output spectrum, and (ii) the estimated level of the stochastic nonlinear distortions $G_S(j\omega_k)$ on the best linear approximation (BLA). Compare the results of the three odd–random multisines. What do you conclude? Explain the results. Repeat the exercise for $\alpha = 0.01$ and $\beta = 5 \times 10^{-4}$. □

Observations Figure 6-14 compares the results of the undistorted multisine with those of the even distorted multisine. It can be seen that:

1. In the first example (Figure 6-14, top row) the even distorted multisine overestimates the level of the even nonlinear distortions in the band [500 Hz, 1200 Hz].
2. In the second example (Figure 6-14, bottom row) the even distorted multisine multisine overestimates the level of the odd nonlinear distortions in the bands [0 Hz, 500 Hz] and [1200 Hz, 2000 Hz].

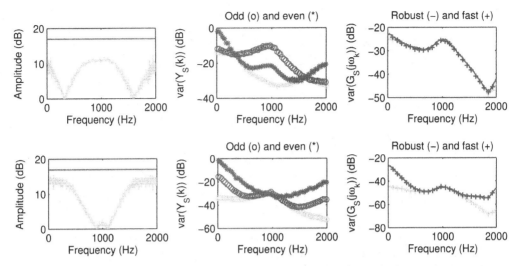

Figure 6-14   Fast method using odd multisines applied to the Wiener–Hammerstein system of Figure 5-13 on page 157 with G_1 and G_2 defined in (6-8), and with static nonlinearity $z(t) = x(t) + \alpha x^2(t) + \beta x^3(t)$. Left column: Comparison between $|db(std(Y_S^{odd})/std(Y_S^{even}))|$ (light gray) and the ratio of the effective input power spectral densities of the odd excited lines and the even detection lines (dark gray). Middle and right columns: Estimated level of the stochastic nonlinear distortions on, respectively, the output and the BLA for the undistorted (light gray) and distorted (dark gray) odd random multisines. Top row: $\alpha = 0.01$, $\beta = 5 \times 10^{-3}$. Bottom row: $\alpha = 0.01$, $\beta = 5 \times 10^{-4}$.

    3. In the two examples the even distorted multisine overestimates the level of the stochastic nonlinear distortions on the BLA in those frequency bands where the estimates of odd nonlinear distortions are overbiased.

In addition, it also turns out that (not shown in Figure 6-14):

    4. The results of the undistorted and odd distorted multisines are the same.

    5. The results of the even and fully distorted multisines coincide.

    Discussion All these observations can be explained as follows. Since $\alpha x^2(t)$ and $\beta x^3(t)$ combine, respectively, two and three frequencies of $x(t)$ (see Section 5.1.1.B), the dominant stochastic nonlinear contributions at the even detection lines $2m$ in $z(t)$ are of the type

$$
\begin{aligned}
\alpha x^2(t): &\quad \alpha X(2k_1 + 1)X(2k_2 + 1), &\quad 2k_1 + 2k_2 + 2 = 2m, \\
\beta x^3(t): &\quad \beta X(2k_3 + 1)X(2k_4 + 1)X(2l), &\quad 2k_3 + 2k_4 + 2l + 2 = 2m,
\end{aligned}
\tag{6-71}
$$

where $X(2k_i + 1)$, $i = 1, ..., 4$, are excited odd harmonics and $X(2l)$ is an even distortion line. If $|\beta X(2l)|$ is not much smaller than $|\alpha|$, then the third degree contribution in (6-71) will bias the estimated level of the even nonlinear distortions. As a rule of thumb this bias can be neglected if the ratio of the variances of the odd and even nonlinear distortions is much smaller than the ratio of the effective power spectral densities of the odd excited harmonics to the even distortion lines

$$\frac{\operatorname{var}(Y_{\S}^{\mathrm{odd}})}{\operatorname{var}(Y_{\S}^{\mathrm{even}})} \ll \frac{\dfrac{|U(2k+1)|^2}{(2f_{\mathrm{S}}/N)}\dfrac{F_{\mathrm{group}}-1}{F_{\mathrm{group}}}}{\dfrac{|U(2k)|^2}{(2f_{\mathrm{S}}/N)}} = \frac{|U(2k+1)|^2}{|U(2k)|^2}\frac{F_{\mathrm{group}}-1}{F_{\mathrm{group}}}, \tag{6-72}$$

where $U(2k+1)$ is an odd excited harmonic and $U(2k)$ an even distorted detection line.

Similarly, the dominant stochastic nonlinear contributions at the odd detection lines $2m+1$ in $z(t)$ are of the form

$$
\begin{aligned}
\alpha x^2(t)\colon &\quad \alpha X(2k_1+1)X(2l), &\quad 2k_1+2l+1 = 2m+1, \\
\beta x^3(t)\colon &\quad \beta X(2k_2+1)X(2k_3+1)X(2k_4+1), &\quad 2k_2+2k_3+2k_4+3 = 2m+1,
\end{aligned}
\tag{6-73}
$$

where $X(2k_i+1)$, $i = 1, \ldots, 4$, are excited odd harmonics and $X(2l)$ is an even distortion line. If $|\alpha|$ is not much smaller than $|\beta X(2k_2+1)|$, then the even contribution in (6-73) will bias the estimated level of the odd nonlinear distortions. As a rule of thumb this bias can be neglected if

$$\frac{\operatorname{var}(Y_{\S}^{\mathrm{even}})}{\operatorname{var}(Y_{\S}^{\mathrm{odd}})} \ll \frac{|U(2k+1)|^2}{|U(2k)|^2}\frac{F_{\mathrm{group}}-1}{F_{\mathrm{group}}}, \tag{6-74}$$

where $U(2k+1)$ is an odd excited harmonic and $U(2k)$ an even distorted detection line.

Observations The left column in Figure 6-14 plots the left (light gray)- and right (dark gray)-hand sides of inequalities (6-72) and (6-74). From the left and middle columns of Figure 6-14 it can be seen that the bias on estimated level of the even or odd nonlinear distortions is apparent in those frequency bands where, respectively, (6-72) or (6-74) is not satisfied. Comparing the middle and right columns of Figure 6-14 it can also be seen that the bias on $\operatorname{var}(G_{\mathrm{S}}(j\omega_k))$ is independent of the bias on $\operatorname{var}(Y_{\S}^{\mathrm{even}})$ and only depends on the bias on $\operatorname{var}(Y_{\S}^{\mathrm{odd}})$.

Discussion The latter is explained by the fact that for odd multisines, the variance of the stochastic nonlinear distortions $\operatorname{var}(G_{\mathrm{S}}(j\omega_k))$ on the BLA only depends on the stochastic nonlinear distortions at the odd harmonics. If the odd-multisine is not distorted at the even detection lines then, $\alpha x^2(t)$ and $\beta x^3(t)$ in (6-70) combine, respectively, two and three odd harmonics and, hence, influence only the even and odd harmonics respectively (see Section 5.1.1.B). It explains why the odd distorted multisine behaves as the undistorted multisine, and why the results of the even distorted multisine are similar to those of the fully distorted multisine.

6.1.3 The indirect method for measuring the best Linear approximation

In this section we introduce the indirect method for measuring the frequency response function (or best linear approximation) of a dynamic system. The approach requires the knowledge of the reference signal.

Consider a nonlinear system captured in a linear feedback loop (see Figure 6-15). Due to the feedback loop the input of the nonlinear system depends on the stochastic nonlinear distortions. Hence, the direct methods (5-20) and (5-36) for measuring the best linear approx-

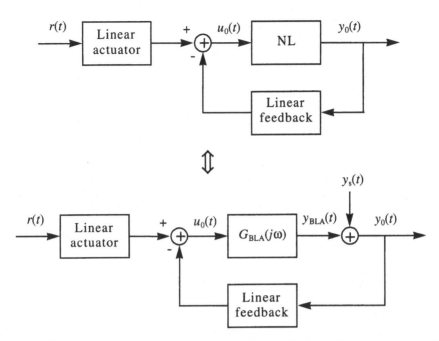

Figure 6-15 The output of a nonlinear system captured in a linear feedback loop (top diagram) can exactly be written as the sum of the output of the best linear approximation and the stochastic nonlinear distortions (bottom diagram). The stochastic nonlinear distortions $y_s(t)$ are uncorrelated with - but not independent of the reference signal $r(t)$.

imation (BLA) of a nonlinear system captured in a linear feedback loop will be biased. Therefore, the BLA of nonlinear systems operating in feedback is redefined as

$$G_{\text{BLA}}(j\omega) = \frac{S_{YR}(j\omega)}{S_{UR}(j\omega)}. \qquad (6\text{-}75)$$

Dividing the numerator and denominator of (6-75) by the auto-power spectrum $S_{RR}(j\omega)$ of the reference signal

$$G_{\text{BLA}}(j\omega) = \frac{S_{YR}(j\omega)/S_{RR}(j\omega)}{S_{UR}(j\omega)/S_{RR}(j\omega)} = \frac{G_{ry}(j\omega)}{G_{ru}(j\omega)} \qquad (6\text{-}76)$$

it follows that the BLA (5-50) of the nonlinear system is the ratio of the BLA from reference to output $G_{ry}(j\omega)$ by the BLA from reference to input $G_{ru}(j\omega)$. The so-called indirect method for measuring the frequency response function of a dynamic system (6-75) is closely related to the instrumental variables approach (Wellstead, 1981). For random phase multisines (6-75) simplifies to

$$G_{\text{BLA}}(j\omega) = \frac{\mathbb{E}\{Y(k)\bar{R}(k)\}}{\mathbb{E}\{U(k)\bar{R}(k)\}} = \frac{\mathbb{E}\{Y(k)e^{-j\angle R(k)}\}}{\mathbb{E}\{U(k)e^{-j\angle R(k)}\}}, \qquad (6\text{-}77)$$

where the second equality uses the fact that $|R(k)|$ is independent of the random phase realization.

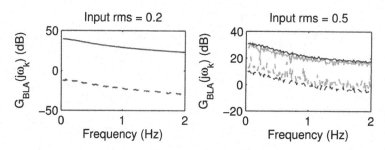

Figure 6-16   Measurement of the best linear approximation (BLA) of a discrete-time Wiener system operating in feedback (see Figure 6-9) for 2 different input rms values (left 0.2, and right 0.5). Gray lines: The fast method and the robust method without reference. Black lines: The robust method with known reference. Solid lines: Mean BLA. Dashed lines: Total variance (noise + NL distortions) of the mean BLA.

Exercise 90 (Indirect method for measuring the best linear approximation).
Repeat the simulation of Exercise 89.d for $M = 100$ independent random phase realizations of the multisine excitation. Calculate over these M realizations the mean and standard deviation of the BLA estimated via the fast method. Compare the results with those of the robust method with and without known reference signal (see Section 6.1.1). What do you conclude? Repeat the simulation with an input rms value of 0.5. Do the previous conclusions still hold? If not, why not? Show that the BLA measured via the robust method with known reference signal is an estimate of the BLA obtained via the indirect method (6-77). Is this also true for the fast method? What goes wrong in the fast method for small signal-to-distortion ratios? Show that the indirect method (6-75) reduces to the direct method (6-42) when the nonlinear system in Figure 6-15 operates in open loop (the feedback branch is removed). *Hint*: use the results of Exercise 86. Motivate the bottom block diagram of Figure 6-15.   □

Observations Figure 6-16 shows that the direct (fast and robust without reference) and indirect (robust with known reference) methods coincide for an input rms value of 0.2, while they differ significantly for an input rms value of 0.5. For the high input rms value the BLA of the direct method is underbiased and has a larger variability.

Discussion These observations can be explained as follows. Comparing (6-28) and (6-34) to (6-77) it can easily be seen that the BLA measured via the robust method with known reference signal (see Section 6.1.1.C) converges to (6-77) as the number of random phase realizations M tends to infinity. The fast method (see Section 6.1.2.B) and the robust method without reference (see Section 6.1.1.B) calculate (5-36) for respectively one and M random phase realization(s) and, hence, are estimates of the direct method (5-36). The resulting BLA will be unbiased if the input signal-to-total-distortion ratio is larger than 10 dB:

$$20\log_{10}\frac{|\hat{U}(k)|}{\sigma_{\hat{U}}(k)} > 10 \text{ dB} \tag{6-78}$$

with $\sigma_{\hat{U}}^2(k)$ the total variance (noise variance + variance nonlinear distortions). For the robust method with known reference signal (indirect method) this condition is relaxed to

$$20\log_{10}\frac{\sqrt{M}|\hat{U}_R(k)|}{\sigma_{\hat{U}_R}(k)} > 10 \text{ dB} \tag{6-79}$$

with $\sigma^2_{\hat{U}_R}(k)$ the total variance (motivation: use (6-32) and (6-33)). Equation (6-79) can be used for determining the minimum number of realizations M needed to get an unbiased estimate of the BLA.

If the nonlinear system in Figure 6-15 operates in open loop (the feedback branch is removed), then the BLA $G_{\text{BLA}}^{\text{indirect}}(j\omega)$ measured by the indirect method (6-75) equals the BLA $G_{\text{BLA}}^{\text{direct}}(j\omega)$ of the direct method (6-42). Indeed, using definition (6-75) and the property that the BLA of the cascade of a linear and a nonlinear system equals the FRF of the linear system multiplied by the BLA of the nonlinear system (see Exercise 86), we find

$$G_{\text{BLA}}^{\text{indirect}}(j\omega) = G_{ry}(j\omega)G_{ru}^{-1}(j\omega) = G_{\text{BLA}}^{\text{direct}}(j\omega)G_{\text{act}}(j\omega)G_{\text{act}}^{-1}(j\omega) = G_{\text{BLA}}^{\text{direct}}(j\omega),$$

where $G_{\text{act}}(j\omega)$ is the frequency response function of the actuator.

The bottom block diagram of Figure 6-15 is motivated as follows. The indirect method (6-75) calculates the ratio of the BLA from reference to output to the BLA from reference to input (see (6-76)). Hence, the input/output DFT spectra are related to the reference signal as

$$\begin{aligned} Y(k) &= G_{ry}(j\omega_k)R(k) + \tilde{Y}_s(k), \\ U(k) &= G_{ru}(j\omega_k)R(k) + \tilde{U}_s(k), \end{aligned} \tag{6-80}$$

where the stochastic nonlinear distortions $\tilde{U}_s(k)$ and $\tilde{Y}_s(k)$ are uncorrelated with $R(k)$. Multiplying the second equation by $G_{\text{BLA}}(j\omega_k)$ and subtracting the result from the first equation gives

$$Y(k) - G_{\text{BLA}}(j\omega_k)U(k) = \tilde{Y}_s(k) - G_{\text{BLA}}(j\omega_k)\tilde{U}_s(k) \equiv Y_s(k), \tag{6-81}$$

where $Y_s(k)$ is uncorrelated with $R(k)$ because $\tilde{U}_s(k)$ and $\tilde{Y}_s(k)$ are uncorrelated with $R(k)$. This shows the validity of the block diagram of Figure 6-15.

6.1.4 Comparison of the fast and robust methods

In this section we compare and discuss the pros and cons of the robust (Section 6.1.1.B) and fast (Section 6.1.2) methods for estimating the best linear approximation, its noise variance, and the variance of the stochastic nonlinear distortions, starting from noisy input/output measurements without reference signal. We also verify the conditions under which both approaches lead to the same results.

Exercise 91 (Comparison robust and fast methods)   Calculate in Exercise 89.e the variance of the stochastic nonlinear distortions on the estimated BLA also via the robust method, and compare the results with those of the fast method. What do you conclude? Repeat Exercises 89.c and 89.d for $M = 20$ different random phase realizations of full-random and odd–random multisine excitations (measurement scheme of Figure 6-2 on page 185 with $M = 20$ and $P = 6$). Note that the same choice of the random harmonic grid and the same rms value should be used for all realizations. Calculate for each realization the BLA, its noise variance, its total variance, and the variance of the stochastic nonlinear distortions. Compare the mean value over the M realizations (arithmetic mean for the BLA, and rms averaging for the variances) with the results of the robust method for noisy input/output signals without reference signal (see Section 6.1.1.B). What do you conclude? Discuss the pros and cons of the fast and the robust methods. Repeat Exercises 89.c and 89.d with $M = 20$ and $P = 6$ for the following two classes of odd multisine excitations:

- Odd random harmonic grid: one odd detection line randomly chosen in each group of two consecutive odd harmonics ($F_{group} = 2$).

- Odd–rodd harmonic grid: every second odd harmonic is a detection line ($4k - 1$ with $k = 1, 2, \dots$).

Compare the mean BLA and its uncertainty bounds with the results of the robust method. What do you observe? Explain.   □

Observations From the middle and right columns of Figure 6-14 on page 212 it can be seen that, although the estimated level of the odd nonlinear distortions is overbiased in some frequency bands (middle column), the fast method still coincides with the robust method applied to the distorted multisine (right column). Note also that the difference between $\text{var}(G_S(j\omega_k))$ of the undistorted and distorted multisines is only visible in those frequency bands where $\text{var}(Y_S^{odd}(k))$ is biased.

Observations Figure 6-8 on page 206 and Figure 6-12 on page 209 show the simulation results for the full and odd random harmonic grid multisines ($F_{group} = 3$: one (odd) detection line randomly chosen in each group of three consecutive (odd) harmonics). It can be seen that the fast estimates of the BLA and its uncertainty bounds, averaged over the M realizations, coincide with the robust estimates. Note also that the variability of the fast estimates using one realization is much larger than that of the robust estimates.

The *fast method* has the following advantages: (i) representation of the in-band and out-band distortions; (ii) classification of the nonlinearities in even and odd contributions (odd multisine only); and (iii) smaller measurement time ($P \geq 7$ periods are needed). Its disadvantages are: (i) larger uncertainty of the variance estimates; (ii) the interpolation of the BLA and the stochastic nonlinear distortion over the frequency which requires a sufficiently high frequency resolution; (iii) the input signal-to-nonlinear distortion ratio should be sufficiently high; and (iv) the level of the even or odd nonlinear distortions are overbiased if the even input detection lines contain too much distortion signal (inequalities (6-72) and (6-74) are not satisfied).

The main advantages of the *robust method* are: (i) no interpolation is involved and hence there are no constraints on the frequency resolution; (ii) no constraints on the input signal-to-distortion ratio; (iii) works for any type of random phase multisine; and (iv) smaller uncertainty of the variance estimates. Its disadvantages are: (i) longer measurement time ($P \geq 2$ periods and $M \geq 7$ realizations are needed); (ii) no classification of the nonlinear distortions; and (iii) no estimate of the out-band distortions.

It is of course possible to combine the advantages of both methods by using the robust method with an odd random harmonic grid multisine as we did in Exercise 91. However, if the measurement time is an issue, then the fast method is the prime choice.

Observations Figure 6-17 illustrates the importance of the random position of the detection lines. For the Wiener–Hammerstein system operating in open loop (top row) it can be seen that the odd–rodd harmonic grid multisine predicts a linear in-band behavior (top right): the total variance (dark gray "+") coincides with the noise variance (medium gray "+"); while

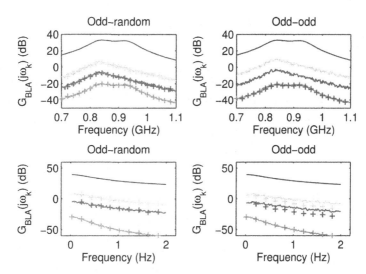

Figure 6-17   Measurement of the best linear approximation (BLA) and its uncertainty bounds using random phase multisines with (i) odd random harmonic grid (left column), and (ii) odd–rodd harmonic grid (right column). Comparison between the robust method (solid lines) and the fast method ("+") rms averaged over the realizations. Top row: Wiener–Hammerstein system of Exercise 89.c. Bottom row: Wiener system of Exercise 89.d. Black: BLA. Dark gray: Total variance (sum of noise and stochastic nonlinear distortions). Medium gray: Noise variance. Light gray: Variance of stochastic nonlinear distortions of one multisine realization.

the odd random harmonic grid multisine predicts the correct distortion level (top left). For the Wiener system operating in feedback (bottom row) the odd–rodd harmonic grid multisine seriously underestimates the level of the stochastic nonlinear distortions (bottom right: the dark and light gray "+" are below the dark and light gray solid lines); which is not the case for the odd random harmonic grid multisine (bottom left).

Discussion   This can be explained as follows. Due to the periodicity of the detection lines in the odd–rodd multisine, more nonlinear combinations of frequencies fall at the excited harmonics and less at the detection lines. In the extreme case of a bandpass odd–rodd multisine whose bandwidth is smaller than the smallest excited harmonic, it can easily be verified that all the odd excited frequency combinations all fall at the excited odd harmonics and none at the odd detection lines. Take, for example, a third degree polynomial nonlinearity, and three in-band excited odd–rodd harmonics $4k_1 + 1$, $4k_2 + 1$, and $4k_3 + 1$. Any third degree frequency combination resulting in an in-band frequency

$$(4k_p + 1) - (4k_q + 1) + (4k_r + 1) = 4(k_p - k_q + k_r) + 1 \quad \text{with} \quad p, q, r \in \{1, 2, 3\} \quad (6\text{-}82)$$

always falls at an excited odd–rodd harmonic. It can be concluded that the randomness of the detection lines is an essential condition for predicting the correct level of the stochastic nonlinear distortions via the fast method. The reader is referred to Schoukens *et al.* (2009a) for an in depth theoretical analysis.

6.1.5 Calculation of confidence bounds on the measured best linear approximation

Using the robust (Section 6.1.1) and fast (Section 6.1.2) measurement procedures results in the estimates of the best linear approximation,

$$\hat{G}_{\text{BLA}}(j\omega_k) = G_{\text{BLA}}(j\omega_k) + N_G(k) + G_S(j\omega_k) \tag{6-83}$$

its noise variance $\text{var}(N_G(k))$, its total variance $\text{var}(\hat{G}_{\text{BLA}}(j\omega_k))$, and the variance of the stochastic nonlinear distortions $\text{var}(G_S(j\omega_k))$. The exercise in the sequel to this section addresses how to construct uncertainty bounds for $\hat{G}_{\text{BLA}}(j\omega_k)$ with a given confidence level from these estimates. The basic assumption made to construct the uncertainty bounds are that $N_G(k)$ and $G_S(j\omega_k)$ are circular complex normally distributed. The latter is asymptotically true for the number of excited frequencies approaching infinity (Exercise 87.b), while the former is true if the input signal-to-noise ratio (SNR) before division of the input/output DFT spectra is at least 20 dB (Pintelon *et al.*, 2003). For frequency response function (FRF) data, input/output (I/O) data without reference signal, and input/output data with reference signal, the conditions on the input signal-to-noise ratio $\text{SNR}_U(k) = |U_0(k)| / \text{std}(N_U(k))$ of one signal period are, respectively,

FRF data:	$\text{SNR}_U(k) \geq 20 \text{ dB}$,
I/O data without ref.:	$\text{SNR}_U(k) \geq 20 \text{ dB} - 10\log_{10}(P)$,
I/O data with ref.:	$\text{SNR}_U(k) \geq 20 \text{ dB} - 10\log_{10}(MP)$.

$$\tag{6-84}$$

These three cases cover the three robust methods as well as the fast method (input/output data without reference).

Exercise 92 (Confidence intervals for the BLA) Assuming that $\hat{G}_{\text{BLA}}(j\omega_k)$ is circular complex normally distributed with mean value $G_{\text{BLA}}(j\omega_k)$ and known variance $\sigma^2_{\hat{G}_{\text{BLA}}}(k)$, construct the most compact $100 \times p \%$ confidence region (*Hint*: $|\hat{G}_{\text{BLA}} - G_{\text{BLA}}|^2 / \sigma^2_{\hat{G}_{\text{BLA}}}$ is chi-squared distributed with two degrees of freedom.) In practice, only a sample estimate $\hat{\sigma}^2_{\hat{G}_{\text{BLA}}}$ of $\sigma^2_{\hat{G}_{\text{BLA}}}$ is available and, hence, the confidence region should be constructed using $\hat{G}_{\text{BLA}}(j\omega_k)$ and $\hat{\sigma}^2_{\hat{G}_{\text{BLA}}}$. This is done in three steps. First, find the distribution of the sample variances $\hat{\sigma}^2_{\hat{G}_{\text{BLA}},n}(k)$, $\hat{\sigma}^2_{\hat{G}_{\text{BLA}}}$, and $\text{var}(G_S)$. Next, find the distribution of $|\hat{G}_{\text{BLA}} - G_{\text{BLA}}|^2 / \hat{\sigma}^2_{\hat{G}_{\text{BLA}}}$ (*Hint*: the ratio of two independent chi-squared distributed random variables is F-distributed.) Finally, calculate the confidence level of a circular confidence region. ☐

For normally distributed random variables $x \in \mathbb{R}^n$ with mean value μ and covariance matrix C_x, the quadratic form $(x - \mu)^T C_x^{-1}(x - \mu)$ is chi-squared (χ^2) distributed with n degrees of freedom (Stuart and Ord, 1987). Hence, a $100 \times p \%$ ($0 < p < 1$) ellipsoidal confidence region can be constructed as

$$\text{Prob}((x - \mu)^T C_x^{-1}(x - \mu) \leq \chi^2_{p,n}) = p, \tag{6-85}$$

where $\chi^2_{p,n}$ is the $100 \times p \%$ percentile of a χ^2_n-distributed random variable ($\chi^2_{p,n}$ can be calculated via the MATLAB® function `chi2inv(p,n)`). Among all possible $100 \times p \%$ confidence regions, (6-85) is for Gaussian random variables, the one with the smallest volume (Kendall and Stuart, 1979). For circular complex normally distributed noise z, the ellipsoids (6-85) reduce to circles because the real and imaginary parts of z are uncorrelated and have equal variance ($C_x = (\sigma^2_z/2)I_2$ with $\sigma^2_z = \text{var}(z)$ and $x = [\text{Re}(z), \text{Im}(z)]^T$).

Since $\hat{G}_{\text{BLA}}(j\omega_k)$ is circular complex normally distributed, the ratio

$$\left| \hat{G}_{\text{BLA}}(j\omega_k) - G_{\text{BLA}}(j\omega_k) \right|^2 / \sigma_{G_{\text{BLA}}}^2(k) \tag{6-86}$$

is chi-squared distributed with two degrees of freedom (Stuart and Ord, 1987). Hence, the most compact $100 \times p\%$ confidence region for $\hat{G}_{\text{BLA}}(j\omega_k)$ is a circle with centre $\hat{G}_{\text{BLA}}(j\omega_k)$ and radius $\sigma_{G_{\text{BLA}}}(k)\sqrt{-\log(1-p)}$,

$$\text{Prob}\left(\left| G - \hat{G}_{\text{BLA}} \right| \leq \sigma_{G_{\text{BLA}}} \sqrt{-\log(1-p)} \right) = p, \tag{6-87}$$

where $-2\log(1-p)$ is the $100 \times p\%$ percentile of an χ_2^2-distributed random variable ($-2\log(1-p)$ can also be calculated via the MATLAB® function `chi2inv(p,2)`).

Now, consider the case where the true variance of the BLA is replaced by the sample variance $\hat{\sigma}_{G_{\text{BLA}}}^2$ based on R independent observations. Since $N_G(k)$ and $G_S(j\omega_k)$ in (6-83) are circular complex normally distributed, the sample variance $\hat{\sigma}_{G_{\text{BLA}}}^2$ is chi-squared distributed with $2R - 2$ degrees of freedom. Hence,

$$\left| \hat{G}_{\text{BLA}}(j\omega_k) - G_{\text{BLA}}(j\omega_k) \right|^2 / \hat{\sigma}_{G_{\text{BLA}}}^2(k) \tag{6-88}$$

is the ratio of two independent chi-squared distributed random variables with 2 and $2R - 2$ degrees of freedom respectively (the sample mean and sample variance of Gaussian random variables are independent). Hence, (6-88) is $F_{2,\,2R-2}$-distributed, and a $100 \times p\%$ confidence region for $\hat{G}_{\text{BLA}}(j\omega_k)$ can be constructed as a circle with center $\hat{G}_{\text{BLA}}(j\omega_k)$ and radius $\hat{\sigma}_{G_{\text{BLA}}}(k)\sqrt{F_{p,2,2R-2}}$,

$$\text{Prob}\left(\left| G - \hat{G}_{\text{BLA}} \right| \leq \hat{\sigma}_{G_{\text{BLA}}} \sqrt{F_{p,2,2R-2}} \right) = p, \tag{6-89}$$

where $F_{p,2,2R-2}$ is the $100 \times p\%$ percentile of an $F_{2,2R-2}$-distributed random variable ($F_{p,2,2R-2}$ can be calculated via the MATLAB® function `finv(p,2,2*R-2)`). If R is sufficiently large ($R \geq 20$), then $F_{p,2,2R-2} \approx -\log(1-p)$, and (6-89) reduces to (6-87). In the sequel of this section we discuss the use of (6-89) for the robust and fast methods (see also Table 6-1).

For the robust method (6-89) is valid with $R = MP$ for the noise variance $\hat{\sigma}_{G_{\text{BLA}}}^2 = \hat{\sigma}_{G_{\text{BLA, n}}}^2$ (equations (6-4), (6-16), or (6-34)), and $R = M$ for the total variance $\hat{\sigma}_{G_{\text{BLA}}}^2 = \hat{\sigma}_{G_{\text{BLA}}}^2(k)$ (equations (6-3), (6-14), or (6-34)) and the variance of the stochastic nonlinear distortions $\hat{\sigma}_{G_{\text{BLA}}}^2 = \text{var}(G_S(j\omega_k))$ (equations (6-7), (6-22), or (6-39)).

For the fast method (6-89) is valid with $R = P$ for the noise variance $\hat{\sigma}_{G_{\text{BLA}}}^2 = \hat{\sigma}_{G_{\text{BLA, n}}}^2$ (6-49), and $2 < R \leq 3$ for the total variance $\hat{\sigma}_{G_{\text{BLA}}}^2 = \hat{\sigma}_{G_{\text{BLA}}}^2$ (6-55) and the variance of the stochastic nonlinear distortions $\hat{\sigma}_{G_{\text{BLA}}}^2 = \text{var}(G_S)$ (6-56). The reason for the latter is that $\hat{\sigma}_{G_{\text{BLA}}}^2$ (6-55) is obtained via linear interpolation (6-54) of the power of the nearest output detection lines. Indeed, if $m - k = k - l$ in (6-54), then $2R - 2 = 4$; and if $m - k \gg k - l$ or $m - k \ll k - l$, then $2R - 2 = 2$. To reduce the radius of the circle (6-89) for $\hat{\sigma}_{G_{\text{BLA}}}^2 = \hat{\sigma}_{G_{\text{BLA}}}^2, \text{var}(G_S)$, it is recommended to average (6-55) and (6-56) over K (for example, two or three) neighboring frequencies. In (6-89) $2R - 2$ should then be replaced by $K(2R - 2)$.

6.1.6 Deviation from the true underlying linear system

Assuming that the input/output signals are observed without errors, the best linear approxi-

TABLE 6-1 Use of the confidence region of the BLA (6-88).

	Noise var $\hat{\sigma}^2_{G_{BLA}} = \hat{\sigma}^2_{G_{BLA,n}}$	Total var or var NL distortions $\hat{\sigma}^2_{G_{BLA}} = \hat{\sigma}^2_{G_{BLA}}$ or $\hat{\sigma}^2_{G_{BLA}} = \text{var}(G_S)$
FAST	$R = P$	$R \in (2, 3]$
ROBUST	$R = MP$	$R = M$

mation (BLA) of a PISPO system measured using a random phase multisine can be written as

$$\hat{G}_{BLA}(j\omega_k) = G_{BLA}(j\omega_k) + G_S(j\omega_k) \tag{6-90}$$

with $G_{BLA}(j\omega_k)$ the BLA, and $G_S(j\omega_k)$ the stochastic nonlinear distortions (see Section 5.2.3). Defining the underlying linear system $G_0(j\omega)$ as the limit value of the BLA when the rms value of the input tends to zero, then, the BLA can be split in two contributions:

$$G_{BLA}(j\omega_k) = G_0(j\omega_k) + G_B(j\omega_k) \tag{6-91}$$

with $G_B(j\omega_k)$ the nonlinear bias. Since for random phase multisines $G_B(j\omega_k)$ depends exclusively on the odd nonlinear distortions, one can wonder as to whether its order of magnitude can be predicted via the odd stochastic nonlinear distortions. This is explored in the following exercise.

Exercise 93 (Prediction of the bias contribution in the BLA) Consider the discrete-time Wiener–Hammerstein system of Exercise 89.c. Show that the underlying linear system $G_0(j\omega)$ equals $G_1(e^{-j\omega T_s})G_2(e^{-j\omega T_s})$. Calculate the best linear approximation $G_{BLA}(j\omega)$, the variance of the odd stochastic nonlinear distortions, and the nonlinear bias contribution $G_B(j\omega) = G_{BLA}(j\omega) - G_0(j\omega)$ for the following two classes of bandpass random phase multisines (6-44):

- Odd multisines: all odd in-band harmonics are excited.
- Odd random harmonic grid multisines: one odd detection line randomly chosen in each group of two consecutive odd in-band harmonics ($F_{group} = 2$).

Choose in (6-44) $f_s = 4\,\text{GHz}$, a linear (uniform) frequency distribution f_k in the band $[700\,\text{MHz}, 1.1\,\text{GHz}]$, a frequency resolution of the excited harmonics of $1\,\text{MHz}$, equal amplitudes A_k, rms value equal to one, and phases uniformly distributed in $[0, 2\pi)$. Apply the robust method for frequency response functions with $P = 1$ (no noise is added to the input/output signals) and $M = 400$ (see Section 6.1.1.A). Why can the robust method be used for estimating the standard deviation of the in-band odd stochastic nonlinear distortions? Plot the BLA, its standard deviation, and the nonlinear bias contribution. Compare the results of the two classes of excitation signals. What do you conclude? □

Since the bandwidth of the multisine excitation ($400\,\text{MHz}$) is smaller than the lowest excited frequency ($700\,\text{MHz}$), the even frequency combinations $f_k \pm f_l$ all fall outside the band $[700\,\text{MHz}, 1.1\,\text{GHz}]$. Hence, the in-band nonlinear distortions only stem from the odd part of the static nonlinearity, independent of the harmonic content of the excitation (odd, even, or odd and even). It explains why the robust method measures solely the odd stochastic nonlinear contributions in this example. However, in general this is not true, and the standard

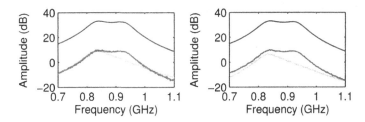

Figure 6-18 Best linear approximation (black) together with the standard deviation of the stochastic nonlinear distortions (light gray) and the bias contribution (dark gray). Left: Odd multisine. Right: Odd random harmonic grid multisine with $F_{group} = 2$.

deviation of the odd stochastic nonlinear distortions is then measured via the fast method — possibly averaged over different realizations — using odd random harmonic grid multisines.

Observations The simulation results are shown in Figure 6-18. It can be seen that $G_{BLA}(j\omega)$ and $G_B(j\omega)$ are the same for both classes of multisines, which is in agreement with the findings of Section 5.2. The variances of the stochastic nonlinear distortions, however, differ by 3 dB (see Section 6.2 for a detailed explanation). From Figure 6-18 it also follows that $std(G_S(j\omega))$ predicts $G_B(j\omega)$ in some frequency bands quite well, but in other bands it can be 6 dB (odd) or 9 dB (odd random harmonic grid) off, which is in agreement with the results of Schoukens *et al.* (2002).

Discussion Extensive simulations have shown that the ratio $std(G_S(j\omega))/|G_B(j\omega)|$ can be as large as ± 20 dB . Therefore, it can be concluded that $std(G_S(j\omega))$ is only suitable to predict the order of magnitude (in the broad sense) of the bias contribution $G_B(j\omega)$. The reader is referred to Schoukens *et al.* (2010b) and Schoukens and Pintelon (2010) for a detailed analysis.

Note that the true underlying linear system $G_0(j\omega)$ in (6-91), which is defined as the limit value of the BLA when the rms value of the input tends to zero, does not always exist. This is illustrated in the following exercise.

Exercise 94 (True underlying linear system) Consider the following static nonlinear system:

$$y(u) = \text{sgn}(u) = \begin{cases} -1, & u < 0, \\ 0, & u = 0, \\ 1, & u > 0, \end{cases} \qquad (6\text{-}92)$$

excited by zero mean white Gaussian noise with variance σ_u^2 . Calculate analytically the BLA $y(t) = g_{BLA}u(t)$ via (5-18), and study the behavior of g_{BLA} as a function of σ_u. Explain the result. □

Minimizing (5-18) w.r.t. g_{BLA} gives $g_{BLA} = \mathbb{E}\{|u|\}/\mathbb{E}\{u^2\} = \sqrt{2/\pi}\sigma_u^{-1}$. For $\sigma_u \to 0$, the BLA g_{BLA} increases to infinity and, hence, the true underlying linear system g_0 does not exist. This is due to the infinite slope of the sign function (6-92) at the origin. For $\sigma_u \to \infty$ we find $g_{BLA} = 0$ and, therefore, $g_0 = 0$. This is due to the zero slope of the sign function (6-92) at $\pm\infty$.

6.1.7 What have we learned in Section 6.1?

■ Measuring the best linear approximation (BLA) and its uncertainty bounds (noise and stochastic nonlinear distortions) of a PISPO (period in, same period out) system for the class of Gaussian inputs with a given power spectrum (rms value and coloring), can be done in two different ways using random phase multisine excitations. The first approach (= *robust method*) is based on $P \geq 2$ successive periods of the steady-state response to $M \geq 7$ different realizations of a random phase multisine (see Section 6.1.1.). The second approach (= *fast method*) uses $P \geq 7$ periods of the steady-state response to one odd or full random phase, random harmonic grid multisine (see Section 6.1.2).

■ If the input signal-to-noise ratio and the number of excited harmonics are sufficiently large, then $100 \times p \%$ circular confidence bounds can easily be constructed for the measured BLA (Exercise 92).

■ The *robust method* starts either from frequency response (FRF) data (Exercise 88.a), input/output data without reference signal (Exercise 88.b), or input/output data with reference signal (Exercise 88.c). The information about the stochastic nonlinear distortions is obtained through averaging over different random phase realizations of the multisine excitation. The plant behaves nonlinearly if the total variance of the BLA is significantly larger than its noise variance. If the reference signal is available, then the spectral purity of the input can be verified by comparing the total input variance with the noise input variance (Exercise 88.c).

■ The *fast method* starts from input/output data without reference signal (Exercises 89.c and 89.d). The information about the stochastic nonlinear distortions is obtained via the detection lines (nonexcited harmonics) in the output DFT spectrum. If these detection lines are randomly chosen among the (odd) excited harmonics (Exercises 89.a and 89.b) then, the distortion level predicted by the fast method coincides with that of the robust method (Exercise 91). Compared with the full random harmonic grid multisines, the odd ones have the advantage to allow for a classification of the nonlinear distortions in even and odd contributions (Exercise 89.c). The plant behaves nonlinearly if the detection lines at the corrected output DFT spectrum contain significant signal energy. This is verified via the null hypothesis test (6-50). The same test is used to verify the spectral purity of the input signal. If the distortion at the even input detection lines is too large, then the estimated level of the even or odd nonlinear distortions in the corrected output spectrum is overbiased (Exercise 89.e). The predicted total variance of the best linear approximation, however, is still correct. To avoid a bias in the estimated level of the odd or even nonlinear distortions, the spectral purity of the even input detection lines should be imposed. This is possible via an iterative procedure based on measurements with a spectrum analyzer (Rabijns *et al.*, 2004).

■ The pros and cons of the robust and fast methods are summarized in Table 6-2. The advantages of both methods can be combined by using the robust procedure with random harmonic grid multisines (Exercise 91). The only drawback is the increased measurement time.

■ The BLA of a nonlinear system operating in feedback is defined as the ratio of the BLA from reference to output by the BLA from reference to input.

■ If the measured NL system is captured in a linear feedback loop, then the frequency response function at the detection lines (nonexcited harmonics) is exactly equal to minus one over the feedback transfer function (Exercise 89.d).

TABLE 6-2 Pros and cons of the robust and fast measurement methods

	Robust method with known reference	Fast method
Pros	- no constraints on the frequency resolution - no constraints on the input signal-to-distortion ratio - any type of random phase multisine - smaller uncertainty estimates	- smaller measurement time - classification of the nonlinear distortions in even and odd contributions - estimation in-band and out-band distortions - easy verification of the input spectral purity
Cons	- larger measurement time - no classification of the nonlinear distortions in even and odd contributions - no estimation out-band distortions - the reference signal is needed	- larger uncertainty estimates - sufficiently high input signal-to-distortion ratio - sufficiently high frequency resolution - estimates odd or even nonlinear distortions are overbiased if the distortion at even input detection lines is too large

- The *fast method* and the *robust method without reference* are direct methods for measuring the BLA. These methods are biased if either the input signal-to-noise-ratio or the input signal-to-distortion ratio are smaller than 10 dB (Exercise 90). The bias introduced by the input noise can always be removed by measuring enough periods P. The bias introduced by the input stochastic nonlinear distortions can only be removed by averaging over sufficient different random phase realizations M.

- The *robust method with known reference* is an indirect method for measuring the BLA, where the reference signal is used as instrumental variable. This method is biased if either the input signal-to-noise ratio or the input signal-to-distortion ratio are smaller than 10 dB. The bias introduced by the input noise or input stochastic nonlinear distortions can always be removed by measuring enough periods P or random phase realizations M (Exercise 90).

- The order of magnitude of the bias contribution $G_B(j\omega_k)$ on the BLA can be predicted from the standard deviation of the stochastic nonlinear contribution $G_S(j\omega_k)$

$$\gamma^{-1}\text{std}(G_S(j\omega_k)) \leq |G_B(j\omega_k)| \leq \gamma\text{std}(G_S(j\omega_k))$$ (6-93)

As a rule of thumb, one can use $\gamma = 10$, however, this bound should be used with great care (Exercise 93).

6.2 MEASURING THE NONLINEAR DISTORTIONS

In Section 5.2.2 (see Exercise 84.c) we learned that the best linear approximation (BLA) does not depend on the harmonic content of the random phase multisine, as long as the effective power per frequency band is the same, and the number of excited frequencies is sufficiently large. However, the variance of the stochastic nonlinear distortions $G_S(j\omega_k)$ strongly depends on the harmonic content. Indeed, from Figure 5-16 on page 163 it can be seen that var($G_S(j\omega_k)$) of the full multisine BLA differs considerably from that of the odd multisine BLA (top left plot), which in turn differs from that of the odd–random (harmonic grid) multisine BLA (bottom left plot). In this section, we first give a detailed explanation of this strange behavior; next, show how to predict the var($G_S(j\omega_k)$) of the full, odd, and full-random (harmonic grid) multisine BLAs via the fast measurement method using odd random harmonic grid multisines; and finally we discuss the pros and cons of full-random (harmonic grid) and odd–random multisines.

In the following exercise we analyze, in detail, the influence of the harmonic content of the random phase multisine on the variance of the stochastic nonlinear distortions. As a result, a procedure is constructed for predicting $\text{var}(G_S(j\omega_k))$ of the full, odd, and full-random multisine measurements starting from the odd–random multisine measurements.

Exercise 95 (Prediction of the nonlinear distortions using random harmonic grid multisines) Consider the generalized Wiener–Hammerstein system of Figure 5-13 on page 157, with linear dynamic blocks $G_1(z^{-1})$ and $G_2(z^{-1})$ defined in (5-33), NFIR block

$$z(t) = x(t) + (\text{atan}(x(t))x^2(t-1) + 10|x(t-2)|)/2000 \qquad (6\text{-}94)$$

and where the input/output signals are observed without errors. Using the robust method for noisy FRF measurements with $M = 1000$ and $P = 1$ (see Section 6.1.1.A), calculate the best linear approximation and its standard deviation of this PISPO system for the following classes of random phase multisines (6-44), where $\omega_0 = 2\pi f_s/N$, $N = 1024$, and $A_k = \alpha|L(z_k^{-1})|$ with $L(z^{-1})$ defined in (5-34) and α chosen such that $u_{\text{rms}} = 1.1653$,

- Full: all harmonics k are excited, $k = 1, 2, ..., N/2 - 1$.
- Odd: all odd harmonics $2k - 1$ are excited, $k = 1, 2, ..., N/4 - 1$.
- Full-random: full random harmonic grid with $F_{\text{group}} = 3$ (one-third detection lines).
- Odd–random: odd random harmonic grid with $F_{\text{group}} = 3$ (one-third odd detection lines).

Plot the ratio of the standard deviations of the full and full-random BLAs (*Hint*: use the MATLAB® function `interp1` to get the values at the same frequencies). Do the same for the odd and odd–random BLAs. What do you conclude? Explain (*Hint*: Gaussian input signals $u(t)$ with the same power spectral density V^2/Hz, generate stochastic nonlinear distortions $y_s(t)$ with the same power spectral density V^2/Hz). Plot the difference between the variances of the full and odd BLAs. What does it represent? Explain. Using the fast method for input/output measurements (see Section 6.1.2.B), calculate the variance of the output stochastic nonlinear distortions for the full-random and odd–random multisines. Average the results over the M realizations. What kind of averaging should be used? Explain. Add the variances of the even and odd output distortions of the odd–random multisine and compare the result with the variance of the output distortions of the full-random multisine (*Hint*: use the MATLAB® function `interp1` to get the values at the same frequencies). What do you observe? Explain. What is the ratio of signal amplitudes of the full-random and odd–random multisines at the excited harmonics? Explain. What do you conclude concerning the signal-to-distortion ratio? Using all previous results, predict the variance of the robust odd and full multisine measurements from the fast odd–random multisine measurement. Compare the values to those obtained by the robust method.   □

The two key observations to understand the simulation results of Exercise 84.c and Exercise 95 are

1. Gaussian input signals $u(t)$ with the same power spectral density V^2/Hz generate stochastic nonlinear distortions $y_s(t)$ with exactly the same power spectral density V^2/Hz.

2. Full, odd, full-random, and odd–random multisines with the same rms value, the same coloring of the amplitude spectrum, and a sufficiently large number of frequencies have approximately the same effective power spectral density.

While observation 1 is trivial, observation 2 remains to be proven. To justify observation 2 we calculate the input signal power in a frequency band whose width Δf depends on the harmonic content of the random phase multisine

$$
\Delta f = \begin{cases}
\dfrac{f_s}{N}, & \text{full multisine,} \\[2ex]
2f_s/N, & \text{odd multisine,} \\[2ex]
F_{\text{group}}(f_s/N), & \text{full-random multisine,} \\[2ex]
F_{\text{group}}(2f_s/N), & \text{odd-random multisine.}
\end{cases}
\tag{6-95}
$$

with N/f_s the signal period. The effective power spectral density $\text{PSD}_{\text{eff}}(k)$ is then the ratio of the signal power in the given frequency band divided by its width Δf (6-95)

$$
\text{PSD}_{\text{eff}}(k) = \begin{cases}
\dfrac{|U_{\text{full}}(k)|^2}{f_s/N}, & \text{full multisine,} \\[3ex]
\dfrac{|U_{\text{odd}}(k)|^2}{2f_s/N}, & \text{odd multisine,} \\[3ex]
\dfrac{(F_{\text{group}} - 1)|U_{\text{full-random}}(k)|^2}{F_{\text{group}}(f_s/N)}, & \text{full-random multisine,} \\[3ex]
\dfrac{(F_{\text{group}} - 1)|U_{\text{odd-random}}(k)|^2}{F_{\text{group}}(2f_s/N)}, & \text{odd-random multisine.}
\end{cases}
\tag{6-96}
$$

For notational simplicity, the signal power in (6-96) is located in the middle of the frequency band, and it is assumed that the power of the excited harmonics of the full-random and odd–random multisines is constant within each group of F_{group} consecutive (odd) harmonics. If the latter is not true, then $|U_{\text{full-random}}(k)|^2$ and $|U_{\text{odd-random}}(k)|^2$ in (6-96) should be interpreted as the mean power of the $(F_{\text{group}} - 1)$ excited harmonics. Since the full, odd, full-random, and odd–random multisines have, by assumption, the same rms value (= same power), and the same coloring of the amplitude spectrum; and since their number of excited frequencies is given by

$$
\text{no. of excited harmonics} = \begin{cases}
F, & \text{full multisine,} \\[2ex]
F/2, & \text{odd multisine,} \\[2ex]
\dfrac{F_{\text{group}} - 1}{F_{\text{group}}}F, & \text{full-random multisine,} \\[2ex]
\dfrac{F_{\text{group}} - 1}{F_{\text{group}}}(F/2), & \text{odd-random multisine.}
\end{cases}
\tag{6-97}
$$

where F is sufficiently large; the amplitude squared of their DFT spectra are approximately related as

$$|U_{\text{full}}(k)|^2 = \frac{|U_{\text{odd}}(k)|^2}{2} = \frac{F_{\text{group}}-1}{F_{\text{group}}}|U_{\text{full-random}}(k)|^2 = \frac{F_{\text{group}}-1}{F_{\text{group}}}\frac{|U_{\text{odd-random}}(k)|^2}{2}. \qquad (6\text{-}98)$$

This relationship is exact if the signal power is continuously distributed over infinite frequencies. Combining (6-96) and (6-98) reveals that the four classes of multisines have approximately the same power spectral density.

The direct consequence of observations 1 and 2 is that the effective power spectral density of the stochastic nonlinear distortions $y_s(t)$ is approximately the same for the four classes of random phase multisines. Taking into account that (i) the even Y_S^{even} and odd Y_S^{odd} stochastic nonlinear distortions are uncorrelated (Pintelon and Schoukens, 2001), and (ii) for full and full-random multisines Y_S^{even} and Y_S^{odd} are distributed over all harmonics, while for odd and odd–random multisines Y_S^{even} and Y_S^{odd} only appear at, respectively, the even and odd harmonics (see Exercise 80), it immediately follows that the variances of the stochastic nonlinear distortions $Y_S(k)$ are approximately related as

$$
\begin{aligned}
\text{var}(Y_{\text{S, full}}(k)) &= \text{var}(Y_{\text{S, full-random}}(k)) \\
\text{var}(Y_{\text{S, odd}}^{\text{even}}(k)) &= \text{var}(Y_{\text{S, odd-random}}^{\text{even}}(k)) \\
\text{var}(Y_{\text{S, odd}}^{\text{odd}}(k)) &= \text{var}(Y_{\text{S, odd-random}}^{\text{odd}}(k)) \\
\text{var}(Y_{\text{S, full}}(k)) &= (\text{var}(Y_{\text{S, odd}}^{\text{even}}(k)) + \text{var}(Y_{\text{S, odd}}^{\text{odd}}(k)))/2
\end{aligned}
\qquad (6\text{-}99)
$$

This relationship is exact if the signal power, and hence also the stochastic nonlinear distortion power, is continuously distributed over infinitely many frequencies.

Combining (6-98) and (6-99) shows that the variance of the stochastic nonlinear distortions $\text{var}(G_S(j\omega_k)) = \text{var}(Y_S(k))/|U(k)|^2$ of the BLA measurements are related as

$$
\begin{aligned}
\text{var}(G_{\text{S, full}}(j\omega_k)) &= \frac{F_{\text{group}}}{F_{\text{group}}-1}\text{var}(G_{\text{S, full-random}}(j\omega_k)) \\[2mm]
\text{var}(G_{\text{S, odd}}^{\text{even}}(j\omega_k)) &= \frac{F_{\text{group}}}{F_{\text{group}}-1}\text{var}(G_{\text{S, odd-random}}^{\text{even}}(j\omega_k)) \\[2mm]
\text{var}(G_{\text{S, odd}}^{\text{odd}}(j\omega_k)) &= \frac{F_{\text{group}}}{F_{\text{group}}-1}\text{var}(G_{\text{S, odd-random}}^{\text{odd}}(j\omega_k)) \\[2mm]
\text{var}(G_{\text{S, full}}(j\omega_k)) &= \text{var}(G_{\text{S, odd}}^{\text{even}}(j\omega_k)) + \text{var}(G_{\text{S, odd}}^{\text{odd}}(j\omega_k))
\end{aligned}
\qquad (6\text{-}100)
$$

where G_S^{even} and G_S^{odd} stand for the respective even and odd contributions. From (6-100) it can be seen that the variance of the full multisine BLA can be predicted as

$$\text{var}(G_{\text{S, full}}(j\omega_k)) = \frac{F_{\text{group}}}{F_{\text{group}}-1}(\text{var}(G_{\text{S, odd-random}}^{\text{even}}(j\omega_k)) + \text{var}(G_{\text{S, odd-random}}^{\text{odd}}(j\omega_k))) \qquad (6\text{-}101)$$

where $\text{var}(G_{\text{S, odd-random}}^{\text{even}}(j\omega_k))$ and $\text{var}(G_{\text{S, odd-random}}^{\text{odd}}(j\omega_k))$ are estimated from input/output measurements with one realization of an odd random harmonic grid multisine (use (6-54) to (6-56) at the respective even and odd detection lines).

Observations Formula (6-100) explains the simulations results shown in Figure 5-16 (Exercise 84.c) and Figure 6-19 (Exercise 95). Indeed, from the bottom left plot of Figure 5-16, it can be seen that the ratio between the variances of the odd and odd–random

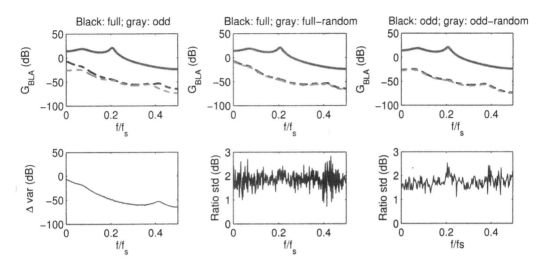

Figure 6-19   Best linear approximation (solid lines) and standard deviation of the stochastic nonlinear distortions (dashed lines) of a generalized Wiener–Hammertein system, calculated via the robust method for the following classes of random phase multisines: full, odd, full random harmonic grid, and odd random harmonic grid (top row). Bottom row: Difference (left), and ratio (middle and right) of the variances of the BLAs in the top row.

multisine BLAs is about 3 dB, while the same ratio is about 1.8 dB in Figure 6-19 (bottom right plot). Since $F_{group} = 2$ in Exercise 84.c and $F_{group} = 3$ in Exercise 95, the observed ratios correspond to the factor $F_{group}/(F_{group} - 1)$ in (6-100). Note that the same conclusion holds for the ratio of the variances of the full and full-random BLAs in Figure 6-19 (bottom middle plot). The last formula of equation (6-100) also shows that the difference between the variances of the full and odd multisine BLAs (top left plot of Figures 5-16 and 6-19) is equal to $var(G_{S, odd}^{even}(j\omega_k))$, the variance of the even nonlinear contributions. This is the quantity drawn in the bottom left plot of Figure 6-19.

Observations   Equations (6-98) and (6-99) explain the simulation results of Figure 6-20. Indeed, the amplitude ratio at the excited harmonics of the odd–random and full-random multisines is about 3 dB (compare the bottom right plot to the first formula of equation (6-98)), and the same is true for the ratio of the total variance (even + odd) of the stochastic nonlinear distortions (compare the bottom right plot to the last formula of equation (6-99)). It shows that the odd–random multisine measurements can easily be used to predict the distortion level of a full-random multisine experiment.

Observations   Finally, in Figure 6-21, the variances of the full and odd multisine BLAs are predicted via (6-100) and (6-101) using the fast measurement method with odd–random multisines (see Section 6.1.2.B). It can be seen that the predictions coincide with the actual values obtained via the robust measurement procedure with full and odd multisines.

Note that in order to ease the comparison with the results of the robust method, the magnitude of the nonlinear distortion levels in Figures 6-20 and 6-21 obtained via the fast method with full-random and odd–random multisines have been smoothed via rms averaging over the M realizations. The reason for rms averaging is that the magnitude squared (= power) of a detection line is an estimate of the variance of the stochastic nonlinear distortion.

In Exercises 89.c, 89.d, and 95 full-random and odd–random multisines have been used for measuring the best linear approximation, its noise variance, its total variance (noise + nonlinear distortions), and the variance of the stochastic nonlinear distortions. In the follow-

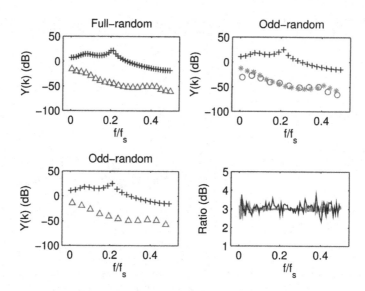

Figure 6-20 Response (top row) of a generalised Wiener–Hammertein system to random phase multisines with, respectively, full (top left) and odd (top right) random harmonic grid (one third detection lines): Excited harmonics of one realization ("+"); and nonexcited harmonics ("Δ", even "*", and odd "o") rms averaged over the realizations. Bottom left: Total variance "Δ" (sum of the even "*" and the odd "o" variances of the top right plot) predicted by the odd random phase response. Bottom right: Ratio of the excited harmonics (black) and the total variances "Δ" (gray) of the odd by the full random phase responses.

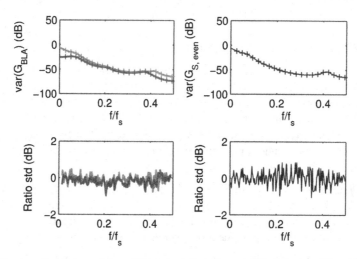

Figure 6-21 Top row: Comparison of the odd (dark gray line), the even (black line), and the total (light gray) variances of the BLAs obtained by the robust method using full and odd multisines, with the variance of the BLAs predicted by the fast method using odd random harmonic grid multisines ("+"). Bottom row: Ratio of the actual (robust method) and the predicted (fast method) standard deviations in the top row.

ing exercise the advantages and disadvantages of both classes of multisines are analyzed. It is also verified as to which signal class leads to the smallest uncertainty on the estimates.

Exercise 96 (Pros and cons full-random and odd–random multisines)   Compare in Figures 6-7 and 6-10 the results of the full-random and odd–random multisines. What do you observe? Taking into account the conclusions of Exercises 89.c, 89.d, and 95, give some pros and cons of both classes of multisines. In Figures 6-8 and 6-12 the signal-to-noise ratio of the full-random multisine BLA (medium gray lines) is 3 dB smaller than that of the odd–random multisine BLA, while the signal-to-distortion ratios of both BLAs (light gray lines) are exactly the same. Explain. Do the same conclusions hold for the classes of full and odd multisines? Which of the two BLA measurements - full-random or odd–random - contains most information? Explain. Can this result be generalized to any PISPO system? If not, show under which condition(s) the full-random multisine BLA measurement contains more information than the odd–random multisine BLA measurement. Is this result also valid for full and odd multisine BLA measurements?   □

The major advantage of the odd–random multisine is its ability to distinguish between even and odd nonlinear distortions (see Figure 6-7) and, hence, to predict the variance of the stochastic nonlinear distortions of the full and odd multisine BLA measurements (see Figure 6-21). Its main disadvantage is the loss in frequency resolution by a factor of two compared with the full-random multisine (see Figures 6-7, 6-10, and 6-20).

Since the signal period of both classes of multisines is the same, the input/output noise (co-)variances (6-48) are exactly the same. Taking into account that the excited harmonics of the odd–random multisines contain twice as much power (see equation (6-98)), it follows that the noise variance (6-49) of the odd–random multisine BLA measurement is a factor 2 (3 dB) smaller than that of the full-random multisine BLA measurement. Applying a similar reasoning to the classes of the full and odd multisines gives

$$\text{var}(N_{G,\text{full}}(k)) = \frac{F_{\text{group}}}{F_{\text{group}} - 1}\text{var}(N_{G,\text{full-random}}(k)),$$

$$\text{var}(N_{G,\text{odd}}(k)) = \frac{F_{\text{group}}}{F_{\text{group}} - 1}\text{var}(N_{G,\text{odd-random}}(k)),$$

$$\text{var}(N_{G,\text{full}}(k)) = 2\text{var}(N_{G,\text{odd}}(k)),$$

$$(6\text{-}102)$$

where $N_G(k)$ is the noise (6-20) on the BLA measurement.

In Exercise 89.c, there are no even in-band distortions (see Figure 6-7), while in Exercise 89.d the PISPO system contains no even nonlinear terms. Hence, only odd nonlinear distortions contribute to the variance of the stochastic nonlinear distortions of the full-random multisine BLA measurement. Taking into account that the odd–random multisine BLA measurement is only influenced by the odd stochastic nonlinear distortions, it follows that the variance of the stochastic nonlinear distortions of the full-random multisine BLA equals that of the odd–random multisine BLA

$$\text{var}(G_{S,\text{full-random}}(j\omega_k)) = \text{var}(G_{S,\text{odd-random}}^{\text{odd}}(j\omega_k)) \qquad (6\text{-}103)$$

(see equation (6-100)). This shows that the signal-to-distortion ratio is the same for both classes of multisines. Combining (6-100) and (6-103) proves that the same is true for the classes of full and odd multisines.

What about the information content of the BLA measurements in Exercises 89.c and 89.d? Suppose that the frequency resolution is large enough such that the full-random BLA measurement can be averaged over two neighboring frequencies without introducing significant bias errors. The variance of the resulting BLA $\hat{G}^{\text{ave}}_{\text{BLA, full-random}}$ is then reduced by a factor two. Hence, the noise variance of $\hat{G}^{\text{ave}}_{\text{BLA, full-random}}$ equals that of the odd–random multisine BLA $\hat{G}_{\text{BLA, odd-random}}$, while the variance of the stochastic nonlinear distortions is half that of the odd–random multisine BLA

$$\text{var}(G^{\text{ave}}_{\text{S, full-random}}(j\omega_k)) = \text{var}(G^{\text{odd}}_{\text{S, odd-random}}(j\omega_k))/2 \qquad (6\text{-}104)$$

(see equation (6-103)). This shows that the total variance of the averaged full-random multisine BLA is smaller than that of the odd-multisine BLA and, therefore, the full-random multisine BLA measurements in Exercises 89.c and 89.d contain more information than the odd–random ones. Note that a parametric estimator of the BLA measurements implicitly performs the averaging over the frequencies without introducing bias errors (see Chapter 5 and 7). The effect of the presence of even nonlinear distortions is discussed in the sequel of this section.

If the full-random multisine measurement is disturbed by even nonlinear distortions, then, depending on the level of these distortions, the full-random multisine BLA measurement is more or less informative than that of the odd–random one. However, if the even nonlinear distortions are smaller than or equal to the odd nonlinear distortions

$$\text{var}(G^{\text{even}}_{\text{S}}(j\omega_k)) \leq \text{var}(G^{\text{odd}}_{\text{S}}(j\omega_k)), \qquad (6\text{-}105)$$

then, using (6-100), we obtain

$$\text{var}(G_{\text{S, full-random}}(j\omega_k)) \leq 2\,\text{var}(G^{\text{odd}}_{\text{S, odd-random}}(j\omega_k)). \qquad (6\text{-}106)$$

and, hence, following the same reasoning as before,

$$\text{var}(G^{\text{ave}}_{\text{S, full-random}}(j\omega_k)) = \frac{\text{var}(G_{\text{S, full-random}}(j\omega_k))}{2} \leq \text{var}(G^{\text{odd}}_{\text{S, odd-random}}(j\omega_k)) \qquad (6\text{-}107)$$

Since the noise variances are equal, (6-107) shows that the total variance of the averaged full-random BLA measurement is smaller than or equal to that of the odd–random BLA measurement. It can be concluded that the full-random BLA measurements are more informative than the odd–random ones if (6-105) is satisfied. The same conclusion holds for the classes of full and odd multisines, because the full, full-random, odd, and odd–random BLA variances are related as

$$\frac{\text{var}(G_{\text{S, full}}(j\omega_k))}{\text{var}(G_{\text{S, full-random}}(j\omega_k))} = \frac{\text{var}(G^{\text{odd}}_{\text{S, odd}}(j\omega_k))}{\text{var}(G^{\text{odd}}_{\text{S, odd-random}}(j\omega_k))} = \frac{\text{var}(G^{\text{even}}_{\text{S, odd}}(j\omega_k))}{\text{var}(G^{\text{even}}_{\text{S, odd-random}}(j\omega_k))}$$
$$\frac{\text{var}(N_{G, \text{full}}(k))}{\text{var}(N_{G, \text{full-random}}(k))} = \frac{\text{var}(N_{G, \text{odd}}(k))}{\text{var}(N_{G, \text{odd-random}}(k))} \qquad (6\text{-}108)$$

(see equations (6-100) and (6-102)).

6.2.1 What have we learned in Section 6.2?

■ Using the steady-state response of a PISPO system to one realization of a random phase, *odd–random* (harmonic grid) multisine, one can estimate the noise variance and the variance of the stochastic nonlinear distortions of the odd–random, odd, full-random, and full multisine BLA measurements. First, the noise variance $\text{var}(N_{G,\text{ odd-random}}(k))$ and the variance of the even and odd stochastic nonlinear distortions $\text{var}(G_{S,\text{ odd-random}}^{\text{even}}(j\omega_k))$ and $\text{var}(G_{S,\text{ odd-random}}^{\text{odd}}(j\omega_k))$ of the odd–random multisine BLA measurement are estimated via, respectively, (6-49), and (6-54) to (6-56) applied to the respective even and odd detection lines. Next, the noise variances and the variances of the stochastic nonlinear distortions of the other multisine BLA measurements are obtained via, respectively, (6-102) and (6-100).

■ Using the steady-state response of a PISPO system to one realization of a random phase, *full-random* (harmonic grid) multisine excitation, one can estimate the noise variance and the variance of the stochastic nonlinear distortions of the full-random, and full multisine BLA measurements. First, the noise variance $\text{var}(N_{G,\text{ full-random}}(k))$ and the variance of the stochastic nonlinear distortions $\text{var}(G_{S,\text{ full-random}}(j\omega_k))$ of the full-random multisine BLA measurement are estimated via, respectively, (6-49), and (6-54) to (6-56) applied to all the detection lines. Next, the noise variance and the variance of the stochastic nonlinear distortions of the full multisine BLA measurement are obtained via, respectively, (6-102) and (6-100).

■ The extrapolation ability of the odd–random multisine experiment to the odd, full-random, and full multisine experiments is a consequence of the fact that these four classes of multisines have the same effective power spectral density. The same holds for the extrapolation of the full-random to the full multisine experiment.

■ If the variance of the even stochastic nonlinear distortions is smaller than that of the odd stochastic nonlinear distortions, then full(-random) multisine experiments result in parametric transfer function estimates with a smaller uncertainty than that of odd(-random) multisine experiments. This condition is typically satisfied for band-pass multisine excitations whose signal bandwidth is smaller than the lowest excited frequency. If the condition is not satisfied and/or a classification of the nonlinear distortions in even and odd contributions is needed, then odd(-random) multisines should be used.

■ The pros and cons of the odd–random and full-random are summarized in Table 6-3.

TABLE 6-3 Pros and cons of the odd–random and full-random multisines. The properties in italic are also valid for the odd and full multisines.

	Odd–random multisine	Full-random multisine
Pros	- classification in even and odd nonlinear contributions - extrapolation to odd, full-random, and full multisine experiments	- *larger frequency resolution* - *smaller uncertainty transfer function model estimate if the even distortions are smaller than the odd distortions*
Cons	- *reduced frequency resolution* - *larger uncertainty transfer function model estimate if the even distortions are smaller than the odd distortions*	- no classification of the nonlinear contributions - extrapolation limited to full multisine experiments

Together with Table 6-2 it allows for a motivated choice of the excitation signal.

6.3 GUIDELINES

The best linear approximation (BLA) of a PISPO system depends on the power spectrum (coloring and rms value) and the multivariate probability density function (pdf) of the excitation. Hence, the operational perturbations of the system should belong to the class of signals (power spectrum, pdf) used for identifying the system. In general, the BLA of a static nonlinear system is dynamic.

If the operational perturbations are normally distributed, then the identification experiment can be performed using the class of random phase multisine excitations with the same effective power spectral density (coloring and rms value) and a sufficiently large number of excited frequencies.

- If the measurement time is not the limiting factor, then the robust method for input/output measurements with reference signal is recommended because it allows for the poorest input signal-to-noise ratio. Advantages of the robust method: no assumptions concerning the input signal-to-distortion ratio, nor the frequency resolution.

- If the measurement time is a critical issue and/or a classification of the nonlinear distortions in even and odd contributions is required, then the fast method using odd–random (harmonic grid) multisines should be applied. Advantage of the fast method: it allows us to predict the noise and distortion levels of the odd, full-random, and odd–random multisine BLA measurements.

- In some applications it is known that the even distortions are smaller than the odd ones. This is often the case for bandpass excitations where the bandwidth is smaller than the lowest excited frequency. Full (-random) multisine measurements result in parametric transfer function estimates with a smaller uncertainty than that of the odd (-random) multisine experiments.

The Gaussian BLA of a static nonlinear system is static.

The predictive power of the best linear approximation is limited by the stochastic nonlinear distortions. This is the difference between the actual output of the nonlinear system and the output predicted by the BLA. If the level of the stochastic nonlinear distortions is too high for the intended application, then the linear framework is insufficient and a nonlinear model should be identified.

6.4 PROJECTS

Project 1 (Use of Full and Odd Multisines for Parametric Estimation)   Consider the static nonlinear system $y(t) = u^3(t)$. Calculate the best linear approximation and its variance for the classes of odd and full random phase multisines (6-44) with $N = 1024$, amplitudes $A_k = 1$, phases ϕ_k uniformly distributed in $[0, 2\pi)$, and rms value equal to one. Use the robust measurement procedure of Section 6.1.1.A with $P = 1$ and $M = 10^4$. Compare the signal-to-distortion ratios of the BLA measurements. What do you conclude? Explain the result. Using the input–output DFT spectra at the excited harmonics of one multisine realization, estimate the static gain $\hat{g}_{BLA}$ of the BLA by minimizing

$$\sum_{k=1}^{F} |Y(k) - g_{BLA} U(k)|^2 \tag{6-109}$$

w.r.t. g_{BLA}, where F represents the number of excited harmonics. Note that g_{BLA} is a parametric model of the BLA $G_{BLA}(j\omega_k)$ (5-36). Calculate the minimizer of (6-109) for each of the $M = 10^4$ multisine realizations. Compare the sample means and sample variances over the M realizations of the odd and full multisine estimates. What is the ratio of the sample means, and the ratio of the sample variances? Explain the results. Repeat the calculations for the static nonlinearity $y(t) = \alpha u^2(t) + u^3(t)$, where α is chosen such that the odd and even stochastic nonlinear distortions have the same power (*Hint*: show that $\alpha = 1.9711$.) What is the ratio of the sample means, and the ratio of the sample variances of estimates? Explain the results. What is the general conclusion?   □

Project 2 (Prediction Ability of Odd Random Phase Multisines)   Consider   the nonlinear system shown in Figure 5-13 on page 157 where

$$G_1(s) = 1, \; z(t) = x(t) + \alpha x^2(t) + \beta x^3(t) \text{ and } G_2(s) = 1/(1 + s/(2\pi f_0)) \qquad (6\text{-}110)$$

with $\alpha = 3\times10^{-3}$, $\beta = 1\times10^{-3}$, and $f_0 = 300\,\text{Hz}$ (= Hammerstein system). Design an odd random phase multisine (6-44) with rms value equal to one, equal harmonic amplitudes A_k, and about a hundred logarithmically distributed frequencies f_k in the band $[0\,\text{Hz}, 10\,\text{kHz}]$ with frequency ratio of 1.05 (= logtone). Make an appropriate choice for the sampling frequency f_s and the number of time domain samples N in one period. Using the robust measurement procedure of Section 6.1.1.A with $P = 1$ (the data is noiseless) and $M = 100$, calculate the stochastic nonlinear distortions $Y_S(k)$ and $G_S(j\omega_k)$ at the output and best linear approximation respectively. Design, further, an odd random phase, random harmonic grid ($F_{group} = 2$) multisine with uniform frequency distribution (= lintone) that has the same effective power spectral density (*Hint*: the total power of both signals should be the same in each band $[f_k, f_{k+1})$ of the logtone). Show that the overall behavior of the amplitudes of the lintone is given by $A_k \div 1/\sqrt{f_k}$. Using the fast measurement procedure of Section 6.1.2.B with $P = 1$ (the data is noiseless), predict the stochastic nonlinear distortions $Y_S(k)$ and $G_S(j\omega_k)$ of the logtone measurement (*Hint*: Y_S and G_S of the logtone measurement only depend on the odd stochastic nonlinear distortions). Average the predictions over $M = 100$ random phase realizations and compare with the logtone results. What do you conclude? Replace the odd logtone by a full logtone and repeat the exercise (*Hint*: Y_S and G_S of the logtone measurement depend on the odd and even stochastic nonlinear distortions). Add measurement noise to the input and output signals and repeat the exercise with $P > 1$. What do you conclude as far as the signal-to-noise ratios of the logtone and lintone measurements are concerned? Which signal is to be preferred? Motivate your choice.   □

Project 3 (Nonlinear Model Selection via BLA Measurements)   Consider the following nonlinear block structures:

- Wiener system: see Exercise 88.a, equations (6-8) and (6-9), where $G_2(s) = 1$.
- Hammerstein system: see Project 2, equation (6-110).
- Wiener–Hammerstein system: see Exercise 88.a, equations (6-8) and (6-9).
- Generalized Wiener–Hammerstein system: see Exercise 84.a, equation (5-33).
- Nonlinear feedback system: see Exercise 79.a, equation (5-12), where the right-hand-side is replaced by $u(t)$.

Using the robust measurement procedure of Section 6.1.1.A, calculate the best linear approximation (BLA) and its variance of the five nonlinear block structures for the class of odd random phase multisines. For each nonlinear system, perform the calculations for different settings of the input power spectrum (rms value and coloring)

■ Vary the rms value of the input while maintaining the coloring (shape) of the power spectrum.

■ Vary the shape of the power spectrum while keeping the rms value of the input constant.

Compare the amplitude and phase of the BLAs over the different rms values and shapes of the input power spectrum. What do you observe? Does the phase change or not? Does the shape of the amplitude remain the same or not? Summarize your observations in a table. Which of the nonlinear block structures can be distinguished via the BLA measurements with different rms values and shapes of the input power spectrum? Compare your results to Lauwers *et al.* (2008). □

Project 4 (Best Linear Approximation of a Multivariable Nonlinear PISPO System) Consider a multivariable nonlinear PISPO system with n_u inputs and n_y outputs. The system operates in open loop and is excited by Gaussian noise $u(t)$ with auto-power spectrum $S_{UU}(j\omega) \in \mathbb{C}^{n_u \times n_u}$. The best linear approximation is then defined as

$$G_{\text{BLA}}(j\omega) = S_{YU}(j\omega)S_{UU}^{-1}(j\omega) \tag{6-111}$$

with $S_{YU}(j\omega) \in \mathbb{C}^{n_y \times n_u}$ the cross-power spectrum, and $G_{\text{BLA}}(j\omega)$ the $n_y \times n_u$ frequency response matrix. In practice the auto- and cross-power spectra are calculated from a record of $M \times N$ input/output samples as

$$\hat{S}_{YU}(j\omega_{k+1/2}) = \frac{1}{M}\sum_{m=1}^{M} Y_{\text{diff}}^{[m]}(k)(U_{\text{diff}}^{[m]}(k))^H \tag{6-112}$$

with $X_{\text{diff}}(k) = X(k+1) - X(k)$, and $X^{[m]}(k)$ the DFT spectrum of the mth subrecord. This results in the following BLA estimate

$$\hat{G}_{\text{BLA}}(j\omega_{k+1/2}) = \hat{S}_{YU}(j\omega_{k+1/2})\hat{S}_{UU}^{-1}(j\omega_{k+1/2}). \tag{6-113}$$

Assuming that the input is known exactly, the covariance matrix of the estimated BLA (6-113) is obtained as

$$\text{Cov}(\text{vec}(\hat{G}_{\text{BLA}}(j\omega_{k+1/2}))) = \frac{2}{M}(\hat{S}_{UU}^{-1}(j\omega_{k+1/2}) \otimes \hat{C}_V(k)),$$

$$\hat{C}_V(k) = \frac{M}{2(M-n_u)}(\hat{S}_{YY}(j\omega_{k+1/2}) - \hat{S}_{YU}(j\omega_{k+1/2})\hat{S}_{UU}^{-1}(j\omega_{k+1/2})\hat{S}_{YU}^H(j\omega_{k+1/2})), \tag{6-114}$$

where $\text{vec}(x)$ stacks the columns of the matrix x on top of each other (MATLAB® instruction x(:)), and with $C = A \otimes B$ the Kronecker product of two matrices (MATLAB® instruction C = kron(A,B)): C is a matrix with block entry $A_{[i,j]}B$. The reader is referred to Brewer (1978) for an exhaustive overview of the properties of the Kronecker product.

■ Show that $\hat{C}_V(k)$ in (6-114) is an estimate of the covariance matrix of the output disturbances (noise and/or nonlinear distortions). *Hint*: for each block of N samples we have after differencing the input/output DFT spectra

$$Y_{\text{diff}}^{[m]}(k) \approx G_{\text{BLA}}(j\omega_k)U_{\text{diff}}^{[m]}(k) + Y_{s\,\text{diff}}^{[m]}(k) + N_{Y\text{diff}}^{[m]}(k).$$

with $Y_s^{[m]}(k)$ and $N_Y^{[m]}(k)$ respectively the output stochastic nonlinear distortions and the output noise of the mth block. Next, show that $\text{Cov}(X_{\text{diff}}(k)) \approx 2\text{Cov}(X(k))$ for stochastic processes $X(k)$ with a smooth power spectrum. Finaly, show that the factor $M/(M - n_u)$ in (6-114) removes the bias in the expected value of $\hat{C}_V(k)$.

■ Proof the formula for the covariance matrix of $\text{vec}(\hat{G}_{\text{BLA}}(j\omega_{k+1/2}))$. *Hint*: use the following properties of the Kronecker product:

$$\text{vec}(ABC) = (C^T \otimes A)\text{vec}(B),\ (A \otimes B)(C \otimes D) = (AC) \otimes (BD),\ \text{and}$$
$$(v \otimes A)B = (v \otimes A)(1 \otimes B) = v \otimes (AB),$$

where A, B, C, and D are matrices of appropriate size, and where v is a vector.

From one single experiment with periodic inputs

$$Y(k) = G_{\text{BLA}}(j\omega_k)U(k), \qquad\qquad (6\text{-}115)$$

one can in general not estimate the BLA. Indeed, (6-115) contains n_y equations with $n_y n_u$ unknowns. Therefore, n_u experiments must be performed with linearly independent sets of input signals giving

$$\left[Y^{[1]}(k)\ Y^{[2]}(k)\ ...\ Y^{[n_u]}(k)\right] = G_{\text{BLA}}(j\omega_k)\left[U^{[1]}(k)\ U^{[2]}(k)\ ...\ U^{[n_u]}(k)\right] \qquad (6\text{-}116)$$

with $U^{[m]}(k)$ the $n_u \times 1$ input DFT spectrum of the mth experiment. The BLA is then estimated as

$$\hat{G}_{\text{BLA}}(j\omega_k) = \mathbf{Y}(k)\mathbf{U}^{-1}(k) \qquad\qquad (6\text{-}117)$$

with $\mathbf{X}(k) = \left[X^{[1]}(k)\ X^{[2]}(k)\ ...\ X^{[n_u]}(k)\right]$ and $X = Y, U$. The input signals of the n_u experiments are constructed as

$$\mathbf{U}(k) = \Lambda_k \begin{bmatrix} U_1(k) & 0 & ... & 0 \\ 0 & U_2(k) & ... & 0 \\ ... & ... & ... & ... \\ 0 & 0 & ... & U_{n_u}(k) \end{bmatrix} T \begin{bmatrix} e^{j\phi_1(k)} & 0 & ... & 0 \\ 0 & e^{j\phi_2(k)} & ... & 0 \\ ... & ... & ... & ... \\ 0 & 0 & ... & e^{j\phi_{n_u}(k)} \end{bmatrix}, \qquad (6\text{-}118)$$

where each column of $\mathbf{U}(k)$ defines the input DFT spectra of 1 experiment. $U_r(k)$, $r = 1, 2, ..., n_u$, are random phase multisines with rms values equal to 1 and constant amplitude spectra; $T \in \mathbb{C}^{n_u \times n_u}$ is an arbitrary orthogonal matrix ($T^H = T^{-1}$); $\phi_m(k)$, $m = 1, 2, ..., n_u$, are random (over the frequency k and the experiments m) phases such

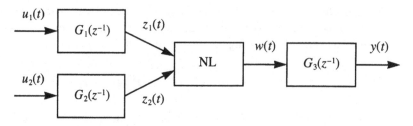

Figure 6-22 Nonlinear two-input, one-output discrete-time system: $G_1(z^{-1})$, $G_2(z^{-1})$, and $G_3(z^{-1})$ are single-input, single output LTI systems, and NL is a static7 two-input, one-output static nonlinear system.

that $E\{e^{j\phi_r(k)}\} = 0$; and the matrix $\Lambda_k \in \mathbb{C}^{n_u \times n_u}$ defines the coloring of the amplitude spectra of the multisines and the correlation among the inputs as a function of the frequency. Equation (6-118) defines the so-called full random orthogonal multisines.

■ The output DFT spectrum of the mth experiment can be written as

$$Y^{[m]}(k) = G_{\text{BLA}}(j\omega_k)U^{[m]}(k) + Y_s^{[m]}(k)$$

Show that for the choice (6-118), the $n_y \times 1$ vector of the stochastic nonlinear distortions $Y_s^{[m]}(k)$ is uncorrelated over the experiments m, $m = 1, 2, ..., n_u$ (*Hint*: using the fact that the phases $\phi_m(k)$ are randomly chosen over the experiments m and the frequency k, show that two different columns of $\mathbf{U}(k)$ are mutually uncorrelated and uncorrelated over the frequency).

■ If the matrix $\Lambda_k \in \mathbb{C}^{n_u \times n_u}$ in (6-118) is chosen such that

$$\Lambda_k\Lambda_k^H = \frac{f_s}{N}S_{UU}(j\omega_k),$$

then the BLA obtained via (6-117) equals the BLA (6-111). Proof this statement (*Hint*: show that the covariance matrix of each column of $\mathbf{U}(k)$ equals $\Lambda_k\Lambda_k^H$, and use the fact that two different columns of $\mathbf{U}(k)$ are uncorrelated).

Consider the two-input, one-output system of Figure 6-22, with linear dynamic blocks

$$G_1(z^{-1}) = \frac{1}{1 - 0.5z^{-1} + 0.9z^{-2}}, G_2(z^{-1}) = \frac{z^{-1} + 0.5z^{-2}}{1 - 1.5z^{-1} + 0.7z^{-2}}, \text{ and}$$

$$G_3(z^{-1}) = \frac{z^{-1} - 0.2z^{-2}}{1 - 0.8z^{-1} + 0.1z^{-2}}$$

and static nonlinear block

$$w(t) = z_1(t) + z_2(t) + 0.01z_1(t)z_2(t) + 0.02z_1^2(t)z_2(t) - 0.02z_1(t)z_2^2(t).$$

■ Excite the system in Figure 6-22 with independent (over time and inputs) and identically distributed Gaussian noise with zero mean and variance equal to one, and calculate the response for $M \times N$ data points with $M = 4000$ and $N = 1024$. Estimate the BLA and its variance via (6-113) and (6-114).

■ Construct full random orthogonal multisines (6-118), with $n_u = 2$, $\Lambda_k = I_{n_u}$ for $k = 1, 2, ..., N/2 - 1$, and

$$ T = \frac{1}{\sqrt{2}} \begin{bmatrix} 1 & 1 \\ 1 & -1 \end{bmatrix} $$

Show that T is an orthogonal matrix ($T^H = T^{-1}$). Calculate the steady-state response of the two-input, one-output system in Figure 6-22 to the full random orthogonal multisines for one period of N samples and $M/2 = 512$ different random phase realizations. Calculate the mean value of the BLA (6-117) and its variance over these 512 different random phase realizations.

Plot the BLAs and their variances obtained via the Gaussian white noise inputs and the full random orthogonal multisines. What do you conclude? Are the variances of both BLAs the same? If so, explain why (*Hint*: for the full random orthogonal multisines the stochastic nonlinear distortions are averaged over the 2 experiments and the $M/2$ random phase realizations). Is the measurement time of both Gaussian noise and full random orthogonal multisine experiments the same? Generalize your conclusions for a multivariable nonlinear system with n_u inputs: How many random phase realizations of the full random orthogonal multisines are needed in order to have the same covariance of the stochastic nonlinear distortions of the BLA estimate as for the Gaussian excitations?   □

7

Identification of Parametric Models in the Presence of Nonlinear Distortions

What you will learn: In this chapter we bring all the methods and ideas that were explained in the Chapters 4, 5, and 6 together. We identify the best linear approximation of a nonlinear system. First we will use random excitations, next we focus on the use of periodic excitation signals. In the latter case it is possible to tell much more about the presence and the level of the nonlinear distortions. This allows the user to decide about the risks and problems when using a linear model in a nonlinear setting. The reader will learn more about the following topics:

- Identifying the best linear approximation starting from experiments with random excitations using time domain methods with parametric noise models (Exercise 97), and frequency domain methods with nonparametric noise models (Section 7.4).

- Study of the impact of nonlinear distortions on the uncertainty interval (Exercise 98).

- Identifying the best linear approximation starting from experiments with periodic excitations, using only one realization (Exercise 99), or multiple realizations (Exercise 100).

7.1 INTRODUCTION

In the previous chapters we learned how to identify a model for a linear dynamic system under idealized conditions: estimation a linear model for a linear system. However, in real life applications, the user faces nonlinear distortions because most systems behave nonlinearly. The best choice to describe such a system would be to built a nonlinear model, but that is often too expensive or time consuming. For that reason, linear system identification is also often used to model (weakly) nonlinear systems. This is a sound solution as long as the user knows what is the impact of the violation of the linearity assumption on the quality of the model. Also the use of these models is affected and limited by the presence of nonlinear distortions. The exercises in this chapter address some of these questions, so that the reader will be better equipped and trained to deal with real life situations.

In each system identification process, the user has to address three major questions: (1) How to design the experiment? (2) What model complexity should be used (number of poles and zeros)? (3) What noise weighting should be used when matching model and data?

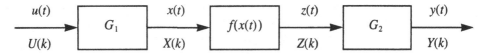

Figure 7-1   Wiener–Hammerstein (WH) system: Cascade of a first linear dynamic block G_1, a static nonlinear system $z(t) = f(x(t))$, and a second linear dynamic block G_2.

The answer to these questions is strongly influenced by the presence of nonlinear distortions. The exercises in this chapter will learn the reader how to operate under these conditions.

7.2 IDENTIFICATION OF THE BEST LINEAR APPROXIMATION USING RANDOM EXCITATIONS

In this exercise we identify a parametric plant model for the best linear approximation G_{BLA} of a nonlinear system. As was shown in Chapter 6, G_{BLA} strongly depends on the applied excitation signal. In this exercise we use filtered Gaussian distributed noise. The user can shape the power spectrum using a generator filter, but the distribution is fixed. The data of this experiment will be processed two times: a first time in the time domain using a Box Jenkins model that makes use of a parametric noise model; a second time in the frequency domain using a nonparametric noise model that is estimated first in the nonparametric preprocessing step. Both results will be compared to each other, a linear validation test will be made, and the observed standard deviation of the model will be compared to the theoretical expected value. As a test system we will use a Wiener–Hammerstein system (see Figure 7-1). For such a system it can be shown that for Gaussian excitations $G_{\text{BLA}}(j\omega) = \alpha G_1(j\omega)G_2(j\omega)$.

Exercise 97 (Parametric estimation of the best linear approximation G_{BLA})   Estimate the best linear approximation of a Wiener–Hammerstein system using a filtered Gaussian random noise excitation.

(i) Process a single data set, the output is disturbed by additive white noise

■ Generate a Wiener–Hammerstein system (see Figure 7-1) with:
  G_1: `[b10,a10]` = `butter(2,2*0.25);`
  G_2: `[b20,a20]` = `cheby1(2,10,2*0.25);`
  $f(x) = x + x^2 + x^3$

■ Generate the signals, using filtered Gaussian noise as excitation signal:
  G_{gen}: `[bGen,aGen]` = `butter(3,2*f0.45);`
  `u0 = filter(bGen,aGen,randn(N+NTrans,1));`
  `y0 = WienerHammerstein(u0);`
  Add white disturbing Gaussian noise $N0$, $\sigma^2 = 0.05^2$ to the output.
  Choose $N = 5000$ and $N_{\text{Trans}} = 500$. Delete at the end of the generation of the signals the first N_{Trans} data points to eliminate the transient effects of the simulation.

■ Estimate the BLA using the time domain identification toolbox:
  `bj([y u0],[OrderG+1 OrderNoise OrderNoise OrderG 0],...`
  `'maxiter',IterMax,'tol',1e-6,'lim',0);`
  with `OrderG = 4` and `OrderNoise = 5`. Put `IterMax = 100`.

■ Estimate the BLA using the frequency domain identification toolbox. Use the arbitrary excitation option in the setting of the toolbox. A nonparametric noise model is automatically generated by the toolbox during the preprocessing of the data.

Plot the FRF of the estimated transfer functions. Compare the result with the theoretic value $G_{BLA}(j\omega) = \alpha G_1(j\omega)G_2(j\omega)$.

(ii) Process 100 data sets, the output is disturbed by additive white noise.

■ Repeat this NRep=100 times. Use in that case also the command line implementations for the frequency domain identification, making use of the following routines:

```
U0 = fft(u0(:))/sqrt(N); Y = fft(y(:))/sqrt(N);
U0(N/2+1:end) = []; Y(N/2+1:end) = [];
Data.U = U0(:).'; Data.Y = Y(:).';
Data.Freq = [0:N/2]'/N;
Method.moment = 6;
[CY, Ym, TY, Gnp] = LocalPolyAnal(Data,Method);
YVar=squeeze(CY.m_nt);
Ynt=squeeze(Ym.m_nt);
fs= 1; f = [0:N/2-1]'/N*fs;
Fdat = fiddata(Ynt,U0,f,YVar,0);
nb = OrderG; na = OrderG;
MFD = elis(Fdat,'z',nb,na,struct('fs',1));
```

■ Verify visually that none of the individual simulation results ended in a local minimum. To do so plot the difference between the estimated and the median FRF and eliminate the outliers.

Plot the mean value and the observed standard deviation of the amplitude of the estimated transfer functions and compare it to the theoretical value obtained with the Box–Jenkins method (see also Exercise 69 in Chapter 4).

(iii) Validation: Process a single data set, no disturbing noise is added to the output

Run a single realization of the simulation and put the disturbing noise equal to zero so that only the nonlinear distortions are present. Make a validation test using
`e = resid(Mbj,[y u0]);`

Observations: We study successively (i) the presence of systematic errors, (ii) the standard deviation, and (iii) the linear validation test.

Systematic errors? The mean value G_{BLAm} of the transfer function over the 100 repeated simulations is shown in Figure 7-2. The results for the time and frequency domain coincide completely on this figure. Also the ratio G_{BLAm}/G_{BLAth} is shown. As expected it is a

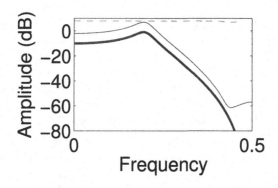

Figure 7-2 Estimated best linear approximation. Bold black: G_{BLA0}. Thin black line: Estimated $\hat{G}_{BLA}$ with Box–Jenkins (parametric noise model) and with frequency domain (nonparametric noise model). Broken black $\hat{G}_{...}/G$

constant. This proves also that the estimated BLA coincides with G_{BLAth} (verify also that the phases are equal).

Statistical properties: Calculate the mean value and standard deviation. In Figure 7-3

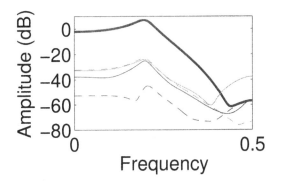

Figure 7-3   Study of the theoretic and simulated standard deviation of $\hat{G}_{BLA}$. Bold black: G_{BLA0}. Thin black line: Theoretic standard deviation of Box–Jenkins (parametric noise model). Gray line: Simulated standard deviation with Box–Jenkins. Broken gray line: Simulated stdandard deviation. with frequency domain (nonparametric noise model). Broken black line: Error

the mean value, the theoretic and observed standard deviation are shown. Also the difference between the averaged time and frequency domain results is added to the plot. From the plot it can be seen that the theoretical predicted standard deviation is about 3 to 8 dB too small with respect to the actual standard deviation obtained from the simulations.

The difference between the averaged time and frequency domain result is 20 dB smaller than the standard deviation, hence there is no statistical significant difference (even by taking into account that the standard deviation of the averaged values is scaled down by 20 dB with respect to that of a single realization).

Validation test: In this test the disturbing noise was put equal to zero, the only errors are due to the nonlinear distortions. The validation results under these conditions are shown in Figure 7-4. The autocorrelation of the residuals is shown in the top figure. As can be seen this is almost zero within the uncertainty bounds. This is an indication that the coloring of the nonlinear distortions is well captured by the BJ noise model, the residuals are white. This is confirmed by the fft analysis of the residuals as shown in Figure 7-5. Also the cross-correlation between the input and the residuals is shown in Figure 7-4. Here it can be concluded that it is not significantly different from zero. This lead to the final conclusion that the linear model passes the classical validation test. The presence of the nonlinear distortions is not detected.

Discussion From this exercise three lessons can be learned on the behavior of linear identification methods in the presence of nonlinear distortions using random excitations. The first lesson is that the Box–Jenkins model is flexible enough to include that part of the nonlinear distortions that was not captured by the linear model in the noise model. For the user, these errors seem to behave as another additive noise source. This is also why the validation test fails to detect the presence of the nonlinear distortions: during the analysis of the residuals no flag is turned up to warn the user (lesson 2). This creates a potentially dangerous situation because the user will be confident in the linear model. From the analysis of the variability of the identified models, it turns out that the theoretical calculated standard deviation under estimates the actual standard deviation. For a cubic system, the provided uncertainty bounds

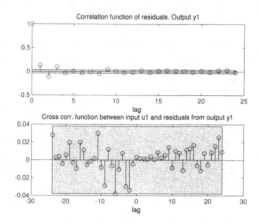

Figure 7-4   Validation of the Box–Jenkins model (plots taken from the Mathworks System Identification toolbox). Upper figure: Autocorrelation of the prediction errors. Lower figure: Cross-correlation between input and the prediction errors. The gray boxes indicate the 99% confidence regions.

can be a factor 3 too small. For nonlinearities of a higher degree, the under estimation can be even worse. The user should account for these too optimistic bounds during the rest of the design (lesson 3).

7.3 GENERATION OF UNCERTAINTY BOUNDS?

In the previous exercise, we observed that the theoretic uncertainty bounds are too small with respect to the true variance of the estimated models. In this exercise we study this phenomenon in more detail on a very simple system: $y(t) = u_0(t)^n$ without disturbing noise.

Exercise 98   Variance of the parametric estimate of the BLA of $y = u_0^n$.

Part 1: a cubing system $y = u_0^3$

■ Generate a random noise excitation $u_0 \sim N(0, 1)$ with length N, and calculate $y_0 = u_0^3$. No disturbing noise is added to the output.

■ Estimate the a_{BLA} in least-square-sense: $y_{BLA} = a_{BLA}u_0$:

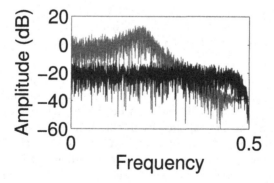

Figure 7-5   Study of the prediction errors. Black lines: FFT of the prediction errors. Gray line: FFT of the output.

$$\hat{a}_{\text{BLA}} = \frac{\sum_{k=1}^{N} y_0(k) u_0(k)}{\sum_{k=1}^{N} u_0(k)^2}. \tag{7-1}$$

■ Calculate the error signal $e(t) = y_0(t) - \hat{a}_{\text{BLA}} u_0(t)$ and the theoretical variance:

$$\sigma_{\text{th}}^2 = \frac{\sigma_e^2}{\sum_{k=1}^{N} u_0(k)^2} \approx \frac{1}{N} \frac{\sum_{k=1}^{N} e(t)^2}{\sum_{k=1}^{N} u_0(k)^2}. \tag{7-2}$$

■ Repeat this 100 times and estimate the actual variance of $\hat{a}_{\text{BLA}}$ from these results.

■ Plot the estimated variance and the theoretical variance σ_{th}^2 for each of the 100 realizations of the input.

Part 2: $y = u_0^n$

■ Repeat the previous simulations for $y = u_0^n$, $n = 3, 5, 7, 9$, and $N = 1000$. Estimate from 100 realizations the actual variance of $\hat{a}_{\text{BLA}}$, and an averaged estimate of σ_{th}^2.

■ Plot the results.

Observations (see Figures 7-6 and 7-7) From Figure 7-6, it can be seen that the calculated standard deviation varies a lot from one realization to the other, especially for short data records. This is due to the random input u_0. For a growing length of the data record, the result becomes more stable. From the figure it is also seen that the actual standard deviation is about a factor 2.5 larger than the theoretical one. From Figure 7-7 it seen that the under estimation of the variance grows with the degree of the nonlinearity.

Discussion As mentioned at the end of the previous exercise, the uncertainty bounds that are calculated on the parametric BLA estimates under estimate the actual standard devia-

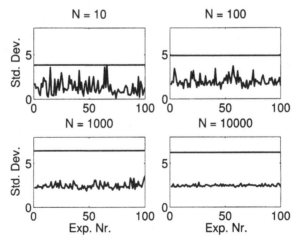

Figure 7-6   Study of the observed and theoretical standard deviation of the best linear approximation of $y = u^3$. The normalized observed standard deviation ($\sigma_{\text{th}} \sqrt{N}$) is plotted for 100 realizations of an experiment with varying record length $N = 10, 100, 1000$, and $10{,}000$, and compared to the actual values estimated from the 100 realizations (horizontal line).

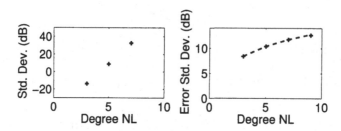

Figure 7-7   Study of the experimental and theoretical standard deviation of the best linear approximation of $y = u^n$ as a function of the nonlinear degree n. Left: Simulated standard deviation; Right: Ratio of the simulated and theoretical standard deviation.

tion. When the nonlinear distortions are the dominating error source, it can be shown that this underestimation is $\sigma_{\hat{\sigma}}^2 / \sigma_{th}^2 = 2n + 1$, with n the degree of the nonlinearity. This factor goes to 1 when the disturbing noise dominates. The user should keep this under estimation in mind whenever it is not sure that there are no nonlinear distortions present. This can be verified using a nonparametric distortion analysis using the methods of Section 6.2. The under estimation of the variance is NOT present in a nonparametric analysis. The reader is referred to Schoukens *et al.* (2010a, 2010b).

7.4 IDENTIFICATION OF THE BEST LINEAR APPROXIMATION USING PERIODIC EXCITATIONS

In Chapter 6 it was explained that it is possible, using well designed multisines, to measure the FRF, and to detect simultaneously the presence of nonlinear distortions. Two possibilities, the fast and the robust method, were discussed. In the fast method, a well designed multisine is used to measure the FRF at the excited frequencies. At the same time the nonlinear distortions are detected, qualified, and quantified a the detection lines (a set of well-chosen nonexcited frequencies). In the robust method, multiple realizations of a random phase multisine are used to measure separately the nonlinear distortions and the disturbing noise level. In both cases we can use this information to estimate G_{BLA}, using a frequency weighting that includes also the nonlinear distortion levels. This is different compared to the frequency domain identification methods that were illustrated in Section 4.4. In that case the frequency weighting was set solely by the variance of the disturbing noise. In Section 7.2, the parametric noise model captured both the noise and nonlinear distortions, but the user didn't get any warning for the presence of the nonlinear distortions. That is the major difference with the approach presented in this section, where explicit information of the nonlinear distortions is passed to the user. In the next exercises we will illustrate how to use the fast and the robust method to estimate a parametric model for G_{BLA}. In opposite to the other exercises in this book we will give here a guided tour, using the GUI of the frequency domain identification toolbox. We advice the reader to follow this tour on her/his own computer, and to experience directly the impact of the choices that are made by varying some of the settings of the parameters that we advice.

7.4.1 Single realization of a well designed random phase multisines

Exercise 99 (Estimate a parametric model for the best linear approximation G_{BLA} using the Fast Method)   In this exercise we illustrate the fast method using the frequency domain system identification toolbox FDident. First the special multisine excitation is designed using the GUI of FDident. Next the nonlinear simulation is done, and the simulation results are captured in the GUI using the "Read Time Domain Data" window. These data are further processed in the "Variances; Nonlin. Anal.; Averaging" window before a parametric transfer function estimate is made using the "Computer Aided Model Scan" window..

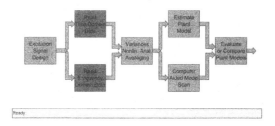

Figure 7-8   Basic window of the GUI of the FDident toolbox.

To make this exercise, the following options should be switched on in the general window: The signal type should be switched to "periodic only," the model type to "nonlinear errors," and the user level should be set to "advanced."

A. Generate a random phase multisine with a random harmonic grid for the fast nonlinear analysis

In this step we generate a well-designed odd multisine as explained in Section 6.1. A number of randomly selected odd frequencies is also not excited. The parametric model will be estimated using the excited frequencies, while the nonexcited frequencies are used to detect and analyze the nonlinear distortions. Because we use an odd multisine, the even nonlinear distortions will not disturb the measurements and the estimated linear model (see Section 5.1). The signal design parameters are set by opening the "Excitation Signal Design" window (see Figure 7-9). Make the following settings: Number of samples is 1000; clock frequency is 1000 Hz; fmin = 1 and fmax = 499 Hz. Set the number of generated periods to "Reps" = 8, and only 1 realization is made by setting "Exps"=1. Push next "Adjust freqs." to create the random harmonic grid, followed by "Design." The resulting spectrum can be visualized with "view spect." Once the design is finished, the window should be closed and the arrow data exported to the MATLAB® work space by clicking on the highlighted arrow (see Figure 7-10).

B. Simulate the nonlinear system

Extract the excitation signal from the time domain object (u0 = name.input) and make the following simulation on a Wiener–Hammerstein system $y0 = G2(f(G1(u_0)))$:

- [b10,a10] = butter(OrderG1,2*fMax1); % G1
- [b20,a20] = cheby1(OrderG2,10,2*fMax1); % G2
- StdNoise = 0.01;   % std. dev. of the disturbing noise
- NTrans = 1000; % to eliminate the simulation transients

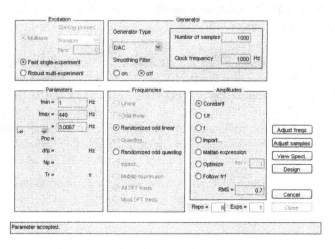

Figure 7-9   Step 2: Setting of the excitation parameters.

- ■ p0 = filter(b10,a10,u0); % first linear system G1
- ■ q0 = p0+p0.^2+p0.^3; % static nonlinearity function
- ■ y0 = filter(b20,a20,q0); % 2nd linear system G2
- ■ y = y0+StdNoise*randn(size(y0)); % add disturbing noise
- ■ u0(1:NTrans) = []; y(1:NTrans) = []; %eliminate transients
- ■ tdData = tiddata(y(:),u0(:),1); % create object
tdData.periodlength = 1000; % set the period length

After the elimination of the simulation transients, 7 periods remain available. These data are loaded in the GUI using the "Read Time Domain Data" window. Next the data are preprocessed: the 7 periods are split in individual subrecords, the FFT is calculated, and the frequency band of interest is selected (see Figure 7-11)

The data are now ready to be processed in the "Variances; Nonlin. Anal.; Averaging" window (see Figure 7-12). It is in this window that the user can choose to extract the nonlinear frequency weighting information from the non excited frequencies by setting the thumb nail "Interp. Nonlin. Var. Anal.". In practical experiments, it often happens that the detection lines are not perfectly equal to zero. A first order correction is automatically applied to compensate for these errors (see Section 6.1.2.B). The result of this nonparametric preprocessing step is shown in Figure 7-13. The measured FRF, the impact of the nonlinear distortions and the disturbing noise is plotted. It is clear that in this example the nonlinear distortions are the dominating errors that are far more important than the disturbing noise distortions. These re-

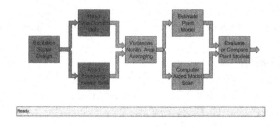

Figure 7-10   Step 3: The excitation signal is ready, export the result to the MATLAB® work space.

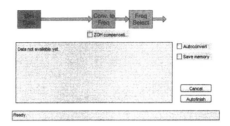

Figure 7-11   Step 4: Preprocessing of the time domain data.

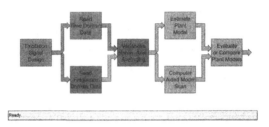

Figure 7-12   Step 5: The time domain data are transformed to the frequency domain.

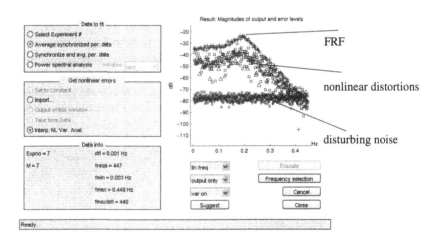

Figure 7-13   Step 6: Nonparametric estimate of the FRF, the nonlinear distortions, and the disturbing noise level.

sults are used to estimate a parametric model for G_{BLA}, using a frequency weighting that accounts for the nonlinear distortions and the disturbing noise.

C. Parametric estimate of G_{BLA}

In the next step, the preprocessed data can be used to identify the best linear approximation starting from the averaged input and output discrete Fourier transform data, using the nonparametric distortion analysis results as a frequency weighting. The nonlinear distortions are considered as an additive noise source to the output, the input is assumed to be exactly

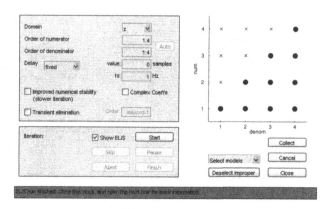

Figure 7-14   Step 7: Identify all proper models up to degree 4.

known. Make a scan over all proper model structures for the plant model up to order 4. This can be done using the "Computer Aided Model Scan" window (see Figure 7-14). Study the identification results, select the best model using the AIC-criterion; analyze the residuals.

Observations (see Figures 7-15, 7-16, and 7-17). From the first figure, it is seen that the best model is of order $n_a = 4$ and $n_b = 4$. The theoretical value of the cost function is close to the theoretical expected value of the cost function. The errors between the model and the measurements are well described by the 50% and the 95% error bounds. In the next figure, the whiteness of the normalized residues is verified. Also here a reasonable good agreement is found. In the last figure the poles and zeros of $\hat{G}_{BLA}$ are shown.

Discussion Identification in the presence of nonlinear distortions The models that are found using this approach are very similar to those obtained using a random excitation with the same flat power spectrum as was done in Section 7.2 using the Box–Jenkins model. The major difference is that the user gets much more information about the level of the nonlinear distortions that is present, using the periodic excitation approach. This is very valid information because this nonlinear level provides a natural bound on the reliability of the model. It does not make much sense to push a linear design beyond the nonlinear distortions level.

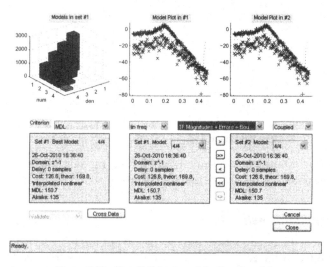

Figure 7-15   Step 8: Selection of the "best" model.

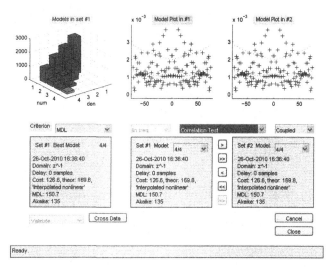

Figure 7-16   Step 9: Study of the residuals.

Changing the nature or the level of the excitation signal will create variations in the model that are in the same order of magnitude as what is provided by this bound. We encourage the reader to process the same data also using the linear setting, extracting the variance information of the disturbing noise from the repeated periods only (the detection lines are not used to extract information about the nonlinear distortions level).

From Figure 7-13, it can be seen that the disturbing noise is white in the frequency band of interest at a level that is 30 to 40 dB below the nonlinear distortion level. In Figure 7-18, the result of FDident using this setting is shown. The cost function was 1000 times larger than the theoretical expected value, although the fit is quite good. This corresponds to the nonlinear distortions that are about 30 dB above the noise floor. It can be seen that the residuals are also much larger than the error bounds that are obtained on the basis of the noise variance information. This points very strongly to the presence of nonlinear distortions, which

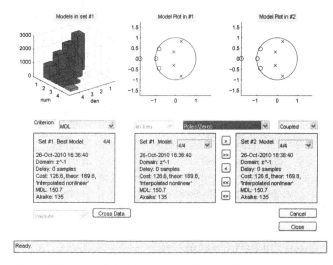

Figure 7-17   Step 10: Poles and zeros of the identified parametric model.

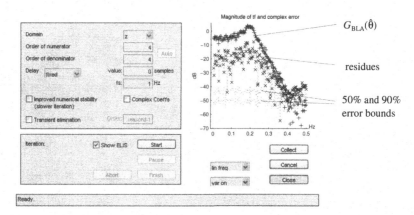

Figure 7-18   Identification result if only the disturbing noise variance is used as frequency weighting.

are known to be present from the previous analysis. In practice, the user can use this as an indication to switch to a nonlinear analysis of the data.

7.4.2 Multiple realizations of the random phase multisine

In this section we use the robust method to characterize the noise and the nonlinear distortion level. The reader is referred to Section 6.1.1 for a detailed discussion. The starting point is to generate M realizations of a random phase multisine, and for each realization we measure P periods in steady state. From the repeated periods, the disturbing noise variance is estimated. Next the variance of the nonlinear noise distortions plus the variance of the disturbing noise is estimated from the variations over the realizations. In each step a proper scaling of the variances should be made to account for the averaging effects. In order to get an idea of the absolute distortion levels, it is necessary to scale the estimated variances back to a single realization with only one period measured. However, for the estimation of the parametric G_{BLA} model, the variance on the averaged data should be used as a frequency weighting because it are these data that are used as raw measurement input to the identification scheme.

Exercise 100 (Estimating a parametric model for the best linear approximation G_{BLA} using the robust method)

A. Generate a set of random phase multisines

In the first step we generate $M = 7$ realizations of a random phase multisine. This can be done either with the routines written before, or using the "Excitation Signal Design" of the GUI (see Figure 7-9). Select the option "Robust multi-experiment" and use the following parameters for the signal: number of points in a period $N = 1000$, clock frequency $f_s = 1000\,\text{Hz}$, $f_{min} = 1\,\text{Hz}$, $f_{max} = 449\,\text{Hz}$, number of repetitions $M = 7$, and number of periods $P = 3$ (one period will be used to eliminate the transients in the simulation). Push next "Design" and export the data from the highlighted arrow to the workspace.

B. Simulate the nonlinear system

Apply the simulation to each of these $M = 7$ realizations of the random phase multisine, using the settings of Exercise 99 and delete each time the first period to eliminate the transient

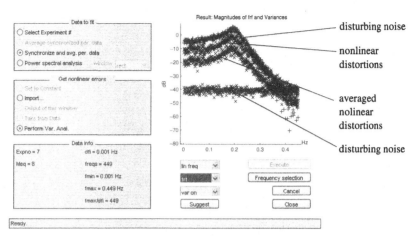

Figure 7-19   Robust method: results of the noise and distortion analysis.

effects of the simulation. Load the data in the GUI using the "Read Time Domain Data" window and apply the preprocessing (see Figure 7-11). In the next step the distortion and noise analysis using the "Variances, Nonlin. Anal., Averaging" window (see Figure7-13). The result of this analysis is shown in Figure 7-19 The FRF is shown together with the standard deviation of the disturbing noise and the nonlinear distortion. The latter are the dominating errors in this example. Two nonlinear distortion curves are plotted. The first shows the actual level of the distortions as they are present at the output of the system. However, to estimate the model, the data that are averaged over the 7 realizations. Although averaging does not eliminate the presence of the nonlinear distortions, their impact on the variance of the estimated FRF of G_{BLA} is reduced and it is the latter variance that should be used during the identification since the averaged data are used in the identification step.

C. Parametric Estimate G_{BLA}
The preprocessed data are used to identify the parametric model for G_{BLA}. Again a scan over the model order can be made. The results for the model of order $n_a = 4$ and $n_b = 4$ is shown in Figure 7-20. The analysis of the residues is shown in Figure 7-21.

Observations (see Figures 7-20 and 7-21) From these figures it can be seen that the results are again very similar to those that were obtained before. Also in this case, the cost function is very close to its expected value. The correlation test of the residues does not indicate the existence of remaining errors, the residues are white within the uncertainty bounds.

Discussion Compared with the previous Exercise 99, the reliability of the disturbance levels in this test are more reliable because the nonlinearity levels are obtained from $M = 7$ realizations of the random phase input, instead of extrapolating the results from a small number of detection lines as was done before. The reader should realize that also in this case the confidence intervals are under estimating the true variability of the parametric estimates, the conclusions obtained in Exercise 98 remain also valid here.

7.5 ADVISES AND CONCLUSIONS

In this chapter we applied the linear identification approach in the presence of nonlinear distortions.

For Gaussian noise or random phase multisine excitations, the estimate converges to the parametric best linear approximation $G_{BLA}(\theta)$ without needing to change the use of the algorithms. Depending upon the choice of the excitation signal and the selected preprocess-

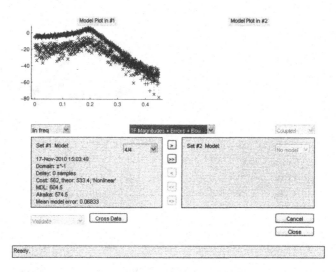

Figure 7-20   Identification result using the weighting obtained from the robust distortion analysis method.

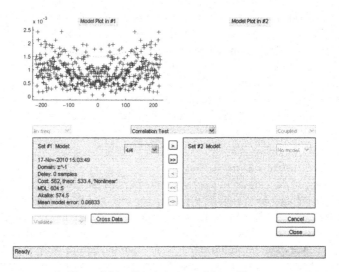

Figure 7-21   Correlation test of the residues.

ing procedure, the user gets whether or not a warning for the presence of nonlinear distortions.

For random excitations, the (non)parametric noise variance estimate that is retrieved is $\sigma^2_{\text{noise}} + \sigma^2_{\text{NLdistortion}}$. The linear model that is estimated, using this noise weighting, passes all validation tests without indicating the presence of the nonlinear distortions.

For periodic excitations, the result depends upon the followed procedure, but in all cases it is possible to warn the user for the presence of nonlinear distortions if the methods are properly applied. Below we discuss briefly the three possible situations:
(i) If the variance is estimated over successively measured periods of a single realization of the random phase multisine, only the disturbing noise variance σ^2_{noise} is retrieved. The presence of the nonlinear distortions is visible in the value of the cost function that will be (much) larger than the theoretical value, while the whiteness test is still ok.

(ii) If the variance is estimated over multiple realizations of a random phase multisine, again the sum $\sigma_{noise}^2 + \sigma_{NLdistortion}^2$ is retrieved and the presence of the nonlinear distortions is no longer reflected in the value of the cost function. However, if for each realization at least two periods are measured, it is still possible to separate both variance contributions. This gives the user already in the preprocessing step access to the full analysis of the nonlinear distortions and the noise.

(iii) If a random grid multisine is used, it is possible to detect the presence of the nonlinear distortions at the "detection lines," while the variance of the noise can still be measured from the successively measured periods. In that case, it is also possible to generate a frequency weighting that accounts for the nonlinear distortions, but this method is less robust than the method based upon multiple realizations of the excitation signal.

From the exercises it also became clear that the theoretic uncertainty bounds that are calculated from the linear identification theory are underestimating the variability of the model that is induced by the nonlinear distortions. The user should account for that when using these bounds in later designs.

REFERENCES

Akaike, H. (1974). A new look at the statistical model identification. *IEEE Trans. Autom. Contr.*, vol. 19, pp. 716–723.

Antoni, J., and J. Schoukens (2007). A comprehensive study of the bias and variance of frequency-response-function measurements: optimal window selection and overlapping strategies. *Automatica*, vol. 43, no. 10, pp. 1723-1736.

Antoni, J., and J. Schoukens (2009). Optimal settings for measuring frequency response functions with weighted overlapped segment averaging. *IEEE Trans. Instrum. Meas.*, vol. 58, no. 9, pp. 3276-3287.

Bendat, J. S., and A. G. Piersol (1980). *Engineering Applications of Correlations and Spectral Analysis*. Wiley, New York.

Billingsley, P. (1995). *Probability and Measure*. Wiley, New York.

Bombois, X., G. Scorletti, M. Gevers, P.M.J. Van den Hof, and R. Hildebrand (2006). Least costly identification experiment for control. *Automatica*, vol. 42, no. 10, pp. 1651-1662.

Brewer, J. W. (1978). Kronecker products and matrix calculus in system theory, *IEEE Transactions on Circuits and Systems*, vol. 25, no. 9, pp. 772-781.

Brigham, E. O. (1974). *The Fast Fourier Transform*. Prentice-Hall, Englewood Cliffs, NJ.

Brillinger, D. R. (1981). *Time Series: Data Analysis and Theory*. McGraw-Hill, New York.

Bultheel, A., M., Van Barel, Y. Rolain, and R. Pintelon (2005). Numerically robust transfer function modeling from noisy frequency response data, *IEEE Trans. Autom. Contr.*, vol. 50, no. 11, pp. 1835-1839.

Chow, Y. S., and H. Teicher (1988). *Probability Theory: Independence, Interchangeability, Martingales* (2nd ed.). Springer-Verlag, New York.

Dobrowiecki, T. P., and J. Schoukens (2002). Cascading Wiener–Hammerstein systems, *Proceedings IEEE Intrumentation and Measurement Technology Conference - IMTC2002*, Anchorage, AK, USA, May 21-23, pp. 881-886.

Dobrowiecki, T. P., J. Schoukens, and P. Guillaume (2006). Optimized excitation signals for MIMO frequency response function measurements. *IEEE Trans. Instrum. Meas.*, vol. 55, no. 6, pp. 2072-2079.

Enqvist, M. (2005). *Linear Models of Nonlinear Systems*. Ph.D. dissertation, Dept. of Electrical Engineering, Linköping University, Sweden.

Enqvist, M., and L. Ljung (2005). Linear approximations of nonlinear FIR systems for separable input processes. *Automatica*, vol. 41, no. 3, pp. 459-473.

Eykhoff, P. (1974). *System Identification, Parameter and State Estimation*. Wiley, New York.

Fedorov, V. V. (1972). *Theory of Optimal Experiments*. Academic Press, New York.

Gevers, M., A.S. Bazanella, X. Bombois, and L. Miskovic (2009). Identification and the information matrix: How to get just sufficiently rich? (2009). *IEEE Trans. Autom. Contr.*, vol. 54, no. 12, pp. 2828-2840.

Godfrey, K. R. (1969). The theory of the correlation method of dynamic analysis and its application to industrial processes and nuclear power plant. *Measurement and Control*, vol. 2, pp. T65–T72.

Godfrey, K. R. (1980). Correlation methods. *Automatica*, vol. 16, pp. 527–534.

Godfrey, K. R., editor (1993). *Perturbation Signals for System Identification*. Prentice-Hall, London.

Godfrey, K. R., Tan, A. H., Barker, and B. Chong (2005). A survey of readily accessible perturbation signals for system identification in the frequency domain. *Control Engineering Practice*, vol. 13, pp. 1391-1402.

Golub, G. H., and C. F. Van Loan (1996). *Matrix Computations* (3rd ed.). John Hopkins University Press, Baltimore.

Goodwin, G. C., and R. L. Payne (1977). *Dynamic System Identification. Experimental Design and Data Analysis.* Academic Press, New York.

Goodwin, G. C., S. F. Graebe, and M. E. Salgado (2001). *Control System Design.* Prentice-Hall, Upper Saddle River (NJ).

Guillaume, P., R. Pintelon, and J. Schoukens (1992). Non-parametric frequency response function estimators based on nonlinear averaging techniques. *IEEE Trans. Instrum. Meas.*, vol. 41, no. 6, pp. 739–746.

Guillaume, P., R. Pintelon, and J. Schoukens (1997). Accurate estimation of multivariable frequency response functions. *Proceedings 13th IFAC World Congress*, pp. 423-428.

Guillaume, P., J. Schoukens, R. Pintelon, and I. Kollár (1991). Crest-factor minimization using nonlinear chebyshev approximation methods. *IEEE Trans. Instrum. Meas.*, vol. 40, no. 6, pp. 982–989.

Harris, F. (1978). On the use of windows for harmonic analysis with discrete Fourier transform. *Proceedings of the IEEE,* vol. 66, pp. 51-83.

Heath, W. P. (2001). Bias of indirect non-parametric transfer function estimates for plants in closed loop. *Automatica,* vol. 37, no. 10, pp. 1529-1540.

Henrici, P. (1974). *Applied and Computational Complex Analysis,* Vol. 1. Wiley, New York.

Hjalmarsson, H. (2009). System identification of complex and structured systems. *European Journal of Control*, vol. 15, no. 3-4, pp. 275-310.

Johansson, R. (1993). *System Modeling and Identification.* Prentice Hall, Englewood Cliffs, NJ.

Kendall, M., and A. Stuart (1979). *Inference and Relationship,* Vol. 2 of *The Advanced Theory of Statistics* (4th ed.). Charles Griffin, London.

Kollár, I. (1994). *Frequency-Domain System Identification Toolbox for Use with Matlab.* The Mathworks, Natick, MA.

Kreider, D. L., R. G. Kuller, D. R. Ostberg, and F. W. Perkins (1966). *An Introduction to Linear Analysis.* Addison-Wesley, Reading, MA.

Lauwers, L., J. Schoukens, R. Pintelon, and M. Enqvist (2008). A nonlinear block structure identification procedure using the best linear approximation, *IEEE Trans. Instrum. Meas.*, vol. 57, no. 10, pp. 2257-2264.

Ljung, L. (1999). *System Identification: Theory for the User* (2nd ed.). Prentice-Hall, Upper Saddle River, NJ.

Lukacs, E. (1975). *Stochastic Convergence.* Academic Press, New York.

Oppenheim, A. V., A. S Willsky, and S. H. Nawab (1997). *Signals and Systems.* Prentice-Hall, London.

Papoulis, A. (1981). *Probability, Random Variables, and Stochastic Processes.* McGraw-Hill, New York.

Pintelon, R., and I. Kollár (2005). On the frequency scaling in continuous-time modeling, *IEEE Trans. Instrum. Meas.,* vol. 54, no. 1, pp. 318-321.

Pintelon, R., and J. Schoukens (2001). *System Identification: A Frequency Domain Approach.* IEEE Press, New York.

Pintelon, R. and J. Schoukens (2002). Measurement and modeling of linear systems in the presence of non-linear distortions. *Mechanical Systems and Signal Processing,* vol. 16, no. 5, pp. 785-801.

Pintelon, R., Y. Rolain and W. Van Moer (2003). Probability density function for frequency response function measurements using periodic signals. *IEEE Trans. Instrum. Meas.,* vol. 52, no. 1, pp. 61-68.

Pintelon, R., J. Schoukens, and G. Vandersteen (1997). Frequency domain system identification using arbitrary signals. *IEEE Trans. Autom. Contr.,* vol. 42, no. 12, pp. 1717–1720.

Pintelon, R., J. Schoukens, G. Vandersteen, and K. Barbé (2010a). Estimation of nonparametric noise and FRF models for multivariable systems—Part I: Theory. *Mechanical Systems and Signal Processing,* vol. 24, no. 3, pp. 573-595.

Pintelon, R., J. Schoukens, G. Vandersteen, and K. Barbé (2010b). Estimation of nonparametric noise and FRF models for multivariable systems—Part II: Extensions, applications. *Mechanical Systems and Signal Processing,* vol. 24, no. 3, pp. 596-616.

Rabijns, D., W. Van Moer, and G. Vandersteen (2004). Spectrally pure excitation signals: Only a dream? *IEEE Trans. Instrum. Meas.,* vol. 53, no. 5, pp. 1433-1440.

Rabiner, L. R., and B. Gold (1975). *Theory and Application of Digital Signal Processing.* Prentice-Hall, New York.

Schetzen, M. (1980). *The Volterra and Wiener Theories of Nonlinear Systems.* Wiley, New York.

Schoukens, J., K. Barbé, L. Vanbeylen, and R. Pintelon (2010a). Nonlinear induced variance of the frequency response function measurements, *IEEE Trans. Instrum. Meas.,* vol. 59, no. 9, pp. 2468-2474.

Schoukens, J., and T. Dobrowiecki (1998). Design of broadband excitation signals with a user imposed power spectrum and amplitude distribution. *Proceedings IEEE Instrumentation and Measurement Technology Conference,* pp. 1002-1005.

Schoukens, J., T. Dobrowiecki, and R. Pintelon (1998). Identification of linear systems in the presence of nonlinear distortions. A frequency domain approach. *IEEE Trans. Autom. Contr.,* vol. 43, no. 2, pp. 176–190.

Schoukens, J., T. Dobrowiecki, Y. Rolain, and R. Pintelon (2010b). Upper bounding variations of best linear approximations of nonlinear systems in power sweep measurements, *IEEE Trans. Instrum. Meas.,* vol. 59, no. 5, pp. 1141-1148.

Schoukens, J., J. Lataire, R. Pintelon, G. Vandersteen, and T. Dobrowiecki (2009a). Robustness issues of the equivalent linear representation of a nonlinear system, *IEEE Trans. Instrum. Meas.,* vol. 58, no. 5, pp. 1737-1745.

Schoukens J., R. Pintelon, and T. Dobrowiecki (2002). Linear modeling in the presence of nonlinear distortions, *IEEE Trans. Instrum. Meas.,* vol. 51, no. 4, pp. 786-792.

Schoukens, J., R. Pintelon, T. Dobrowiecki, and Y. Rolain (2005). Identification of linear systems with nonlinear distortions, *Automatica,* vol. 41, no. 2, pp. 491-504.

Schoukens, J., and R. Pintelon (2010). Study of the variance of parametric estimates of the best linear approximation of nonlinear systems, *IEEE Trans. Instrum. Meas.,* vol. 59, no. 12, pp. 3159-3167.

Schoukens, J., R. Pintelon, and Y. Rolain (2000). Broadband versus stepped sine FRF measurements. *IEEE Trans. Instrum. Meas.,* vol. 49, no. 2, pp. 275–278.

Schoukens, J., R. Pintelon, G. Vandersteen, and P. Guillaume (1997). Frequency-domain system identification using non-parametric noise models estimated from a small number of data sets. *Automatica,* vol. 33, no. 6, pp. 1073–1086.

Schoukens, J., Y. Rolain, and R. Pintelon (2006). Analysis of windowing/leakage effects in frequency response function measurements. *Automatica,* vol. 24, no. 1, pp. 27-38.

Schoukens, J., G. Vandersteen, K. Barbé, and R. Pintelon (2009b). Nonparametric preprocessing in system identifcation: a powerful tool. *European Journal of Control,* vol. 15, no. 3-4, pp. 260-274.

Schroeder, M. R. (1970). Synthesis of low peak factor signals and binary sequences with low autocorrelation. *IEEE Trans. Inform. Theory,* vol. IT-16, pp. 85–89.

Selby, S. M. (1973). *Standard Mathematical Tables.* The Chemical Rubber Co., Cleveland.

Söderström, T. (2007). Errors-in-variables identification in system identification. *Automatica,* vol. 43, no. 6, pp. 939-958.

Söderström, T., and P. Stoica (1989). *System Identification.* Prentice-Hall, Englewood Cliffs, NJ, p. 256.

Sorenson, H. W. (1980). *Parameter Estimation: Principles and Problems.* Marcel Dekker, New York.

Stuart, A., and J. K. Ord (1987). *Advanced Theory of Statistics. vol. 1: Distribution Theory.* Charles Griffin, London (UK).

Tan, A.H., and K. Godfrey (2009). A guide to the design and selection of perturbation signals. *Proceedings 48th IEEE Conference on Decision and Control,* pp. 464-469.

van den Bos, A. (2007). *Parameter Estimation for Scientists and Engineers*. Wiley-Interscience, Hoboken, NJ.

Van den Hof, P. M. J., and R. J. P. Schrama (1995). Identification and control—Closed loop issues, *Automatica,* vol. 31, no. 12, pp. 1751–1770.

Wellstead, P. E. (1977). Reference signals for closed-loop identification. *Int. J. Control,* vol. 26, no. 6, pp. 945-962.

Wellstead, P. E. (1981). Non-parametric methods of system identification. *Automatica,* vol. 17, no. 1, pp. 55-69.

Zarrop, M. B. (1979). Optimal experiment design for dynamic system identification. *Series of Lecture Notes in Control and Information Sciences,* vol. 21. Springer-Verlag, Berlin.

Zygmund, A. (1979). *Trigoniometric Series*. Cambridge University Press, Cambridge, UK.

SUBJECT INDEX

REFERENCE INDEX